PHYTOCHEMICALS FROM MEDICINAL PLANTS

Scope, Applications, and Potential Health Claims

Innovations in Plant Science for Better Health: From Soil to Fork

PHYTOCHEMICALS FROM MEDICINAL PLANTS

Scope, Applications, and Potential Health Claims

Edited by

Hafiz Ansar Rasul Suleria
Megh R. Goyal
Masood Sadiq Butt

Apple Academic Press Inc.
4164 Lakeshore Road
Burlington ON L7L 1A4
Canada

Apple Academic Press Inc.
1265 Goldenrod Circle NE
Palm Bay, Florida 32905
USA

© 2020 by Apple Academic Press, Inc.

First issued in paperback 2021

Exclusive worldwide distribution by CRC Press, a member of Taylor & Francis Group
No claim to original U.S. Government works

ISBN 13: 978-1-77463-470-7 (pbk)
ISBN 13: 978-1-77188-795-3 (hbk)

Library and Archives Canada Cataloguing in Publication

Title: Phytochemicals from medicinal plants : scope, applications, and potential health claims / edited by Hafiz Ansar Rasul Suleria, Megh R. Goyal, Masood Sadiq Butt.

Names: Suleria, Hafiz, editor. | Goyal, Megh Raj, editor. | Butt, Masood Sadiq, editor.

Series: Innovations in plant science for better health.

Description: Series statement: Innovations in plant science for better health: from soil to fork | Includes bibliographical references and index.

Identifiers: Canadiana (print) 20190163321 | Canadiana (ebook) 2019016333X | ISBN 9781771887953 (hardcover) | ISBN 9780429203220 (ebook)

Subjects: LCSH: Phytochemicals. | LCSH: Medicinal plants.

Classification: LCC QK861 .P65 2020 | DDC 572/.2—dc23

Library of Congress Cataloging-in-Publication Data

Names: Suleria, Hafiz, editor. | Goyal, Megh Raj, editor. | Butt, Masood Sadiq, editor.

Title: Phytochemicals from medicinal plants : scope, applications, and potential health claims / edited by Hafiz Ansar Rasul Suleria, Megh R. Goyal, Masood Sadiq Butt.

Other titles: Innovations in plant science for better health.

Description: Oakville, ON ; Palm Bay, Florida : Apple Academic Press, [2020] | Series: Innovations in plant science for better health : from soil to fork | Includes bibliographical references and index. | Summary: "Phytochemicals from Medicinal Plants: Scope, Applications, and Potential Health Claims explores the importance of medicinal plants and their potential benefits for human health. This book looks at bioactive compounds from medicinal plants, the health benefits of bioactive compounds, and the applications of plant-based products in the food and pharmaceutical industries. The first section discusses available sources of bioactive compounds from medicinal plants, biochemistry, structural composition, potential biological activities, and how bioactive molecules are isolated from medicinal plants. The authors examine the applications of bioactive molecules from a health perspective, looking at the pharmacological aspects of medicinal plants, the phytochemical and biological activities of different natural products, and ethnobotany/and medicinal properties, and also present a novel dietary approach for disease management. The book goes on to examine how the plant-based products are used in various sectors of the food and pharmaceutical industries. This book volume helps to shed light on the vast potential of functional foods for human health from different technological aspects. This compendium will be useful for researchers, scientists, academia, and students as well as industry professionals within the food, nutraceutical, and herbal industries"-- Provided by publisher.

Identifiers: LCCN 2019034484 (print) | LCCN 2019034485 (ebook) | ISBN 9781771887953 (hardcover) | ISBN 9780429203220 (ebook)

Subjects: MESH: Phytochemicals | Plants, Medicinal

Classification: LCC RS164 (print) | LCC RS164 (ebook) | NLM QV 766 | DDC 615.3/21--dc23

LC record available at https://lccn.loc.gov/2019034484

LC ebook record available at https://lccn.loc.gov/2019034485

Apple Academic Press also publishes its books in a variety of electronic formats. Some content that appears in print may not be available in electronic format. For information about Apple Academic Press products, visit our website at **www.appleacademicpress. com** and the CRC Press website at **www.crcpress.com**

OTHER BOOKS ON PLANT SCIENCE FOR BETTER HEALTH FROM APPLE ACADEMIC PRESS, INC.

Book Series: *Innovations in Plant Science for Better Health: From Soil to Fork*
Editor-in-Chief: Hafiz Ansar Rasul Suleria, PhD

ABOUT THE EDITORS

Hafiz Ansar Rasul Suleria, PhD

*Alfred Deakin Research Fellow,
Deakin University, Melbourne, Australia;
Honorary Fellow, Diamantina Institute
Faculty of Medicine, The University of
Queensland, Australia*

Hafiz Anasr Rasul Suleria, PhD, is currently working as the Alfred Deakin Research Fellow at Deakin University, Melbourne, Australia. He is also an Honorary Fellow at the Diamantina Institute, Faculty of Medicine, The University of Queensland, Australia.

Recently he worked as a postdoc research fellow in the Department of Food, Nutrition, Dietetic and Health at Kansas State University, USA.

Previously, he has been awarded an International Postgraduate Research Scholarship (IPRS) and an Australian Postgraduate Award (APA) for his PhD research at the University of Queens School of Medicine, the Translational Research Institute (TRI), in collaboration with the Commonwealth and Scientific and Industrial Research Organization (CSIRO, Australia).

Before joining the University of Queens, he worked as a lecturer in the Department of Food Sciences, Government College University Faisalabad, Pakistan. He also worked as a research associate in the PAK-US Joint Project funded by the Higher Education Commission, Pakistan, and the Department of State, USA, with the collaboration of the University of Massachusetts, USA, and National Institute of Food Science and Technology, University of Agriculture Faisalabad, Pakistan.

He has a significant research focus on food nutrition, particularly in the screening of bioactive molecules—isolation, purification, and characterization using various cutting-edge techniques from different plant, marine, and animal sources; and in vitro, in vivo bioactivities; cell culture; and animal modeling. He has also done a reasonable amount of work on functional foods and nutraceuticals, food and function, and alternative medicine.

Dr. Suleria has published more than 50 peer-reviewed scientific papers in different reputed/impacted journals. He is also in collaboration with more than ten universities where he is working as a co-supervisor/special member for PhD and postgraduate students and is also involved in joint publications, projects, and grants. He is Editor-in-Chief for the book series Innovations in Plant Science for Better Health: From Soil to Fork, published by AAP.

Readers may contact him at: hafiz.suleria@uqconnect.edu.au.

ABOUT THE SENIOR EDITOR-IN-CHIEF

Megh R. Goyal, PhD

Retired Professor in Agricultural and Biomedical Engineering, University of Puerto Rico, Mayaguez Campus; Senior Acquisitions Editor, Biomedical Engineering and Agricultural Science, Apple Academic Press, Inc.

Megh R. Goyal, PhD, PE, is a Retired Professor in Agricultural and Biomedical Engineering from the General Engineering Department in the College of Engineering at the University of Puerto Rico–Mayaguez Campus; and Senior Acquisitions Editor and Senior Technical Editor-in-Chief in Agriculture and Biomedical Engineering for Apple Academic Press, Inc. He has worked as a Soil Conservation Inspector and as a Research Assistant at Haryana Agricultural University and Ohio State University.

During his professional career of 50 years, Dr. Goyal has received many prestigious awards and honors. He was the first agricultural engineer to receive the professional license in Agricultural Engineering in 1986 from the College of Engineers and Surveyors of Puerto Rico. In 2005, he was proclaimed as "Father of Irrigation Engineering in Puerto Rico for the Twentieth Century" by the American Society of Agricultural and Biological Engineers (ASABE), Puerto Rico Section, for his pioneering work on micro irrigation, evapotranspiration, agroclimatology, and soil and water engineering. The Water Technology Centre of Tamil Nadu Agricultural University in Coimbatore, India, recognized Dr. Goyal as one of the experts "who rendered meritorious service for the development of micro irrigation sector in India" by bestowing the Award of Outstanding Contribution in Micro Irrigation. This award was presented to Dr. Goyal during the inaugural session of the National Congress on "New Challenges and Advances in Sustainable Micro Irrigation" held at Tamil Nadu Agricultural University. Dr. Goyal received the Netafim Award for Advancements in Microirrigation: 2018 from the American Society of Agricultural Engineers at the ASABE International Meeting in August 2018.

A prolific author and editor, he has written more than 200 journal articles and textbooks and has edited over 70 books. He is the editor of three book series published by Apple Academic Press: Innovations in Agricultural & Biological Engineering, Innovations and Challenges in Micro Irrigation, and Research Advances in Sustainable Micro Irrigation. He is also instrumental in the development of the new book series Innovations in Plant Science for Better Health: From Soil to Fork.

Dr. Goyal received his BSc degree in engineering from Punjab Agricultural University, Ludhiana, India; his MSc and PhD degrees from Ohio State University, Columbus; and his Master of Divinity degree from Puerto Rico Evangelical Seminary, Hato Rey, Puerto Rico, USA.

Masood Sadiq Butt, PhD

Dean of the Faculty of Food, Nutrition & Home Sciences, University of Agriculture, Faisalabad, Pakistan

Masood Sadiq Butt, PhD, is serving as Dean of the Faculty of Food, Nutrition & Home Sciences at the University of Agriculture in Faisalabad, Pakistan. Previously, he served as Director General of the National Institute of Food Science and Technology and Director of Graduate Studies at the University of Agriculture, Faisalabad. He was also a visiting Professor at the Department of Food Science, University of Massachusetts, Amherst, USA.

Dr. Butt is credited for the establishment of the Functional and Nutraceutical Food Research Section at the National Institute of Food Science and Technology under a PAK-US joint project in collaboration with the University of Massachusetts, Amherst. Earlier he was a member of the Scientific Panel of Punjab Food Authority. Additionally, he is a professional member of various national/international associations/societies of food science and nutrition. He has received the Research Productivity Award from the Pakistan Council for Science and Technology several times and has also received the coveted prize of top-cited research paper from the University of Agriculture, Faisalabad. He is also a recipient of the Star Laureate Award from South Asia Publications. He has participated in many national and international conferences/workshops and has made contributions toward the development of curriculum for degree programs leading to degrees in Food Science and Technology and Human Nutrition Dietetics. To date, he has mentored 29 PhD and 110 MSc students in addition to publishing more than 200 research publications in journals.

Professor Masood Sadiq Butt holds a PhD degree in Food Technology (UAF, Pakistan) and postdoctorate from North Carolina State University, Raleigh, USA. He is currently working in the domain of food science and nutrition with special reference to functional and nutraceutical foods.

Readers may contact him at: drmsbutt@yahoo.com.

CONTENTS

CONTRIBUTORS

Munawar Abbas, MS
Research Associate, Institute of Home and Food Sciences, Government College University, Faisalabad, Pakistan. E-mail: foodian2007@gmail.com

Godwin Ojochogu Adejo, PhD
Associate Professor and Head, Department of Biochemistry and Molecular Biology/Deputy Director of Academic Planning and Quality Assurance, Federal University, PMB 5001 Dutsin-ma, Katsina State, Nigeria. E-mail: dradejogod@gmail.com

Javee Anand
PhD Research Scholar, Centre for Advanced Studies in Botany, University of Madras, Guindy Campus, Chennai 600025, Tamil Nadu, India. E-mail: anandjavee@gmail.com

Muhammad Sajid Arshad, PhD
Assistant Professor, Institute of Home and Food Sciences, Government College University, Faisalabad, Pakistan. E-mail: sajid_ft@yahoo.com

Vengatesh Babu
PhD Research Scholar, Centre for Advanced Studies in Botany, University of Madras, Guindy Campus, Chennai 600025, Tamil Nadu, India. E-mail: vsquare.5656@gmail.com

Masood Sadiq Butt, PhD
Professor, National Institute of Food Science and Technology, Faculty of Food, Nutrition and Home Sciences, University of Agriculture, Faisalabad, Pakistan.
E-mail: drmsbutt@yahoo.com

Vivek K. Chaturvedi, MSc
PhD Research Scholar, Centre of Biotechnology, University of Allahabad, Allahabad 211002, UP, India. E-mail: vivekchaturvedi2013@gmail.com

Sushil K. Dubey, MSc
PhD Research Scholar, Centre of Biotechnology, University of Allahabad, Allahabad 211002, UP, India. E-mail: sushildubey40@gmail.com

Jyotsana Dwivedi, MPharm
Senior Research Fellow, Pharmacognosy and Ethnopharmacology Division, CSIR—National Botanical Research Institute, Lucknow 226001, UP, India. E-mail: dwivedijyotsana2010@gmail.com

Sunday Ene-Ojo Atawodi
FAS, Professor, Department of Biochemistry, Faculty of Science, Ahmadu Bello University, Zaria, Kaduna State, Nigeria. E-mail: atawodi.se@gmail.com

Megh R. Goyal, PhD, PE
Senior Editor-in-Chief, Apple Academic Press Inc., PO Box 86, Rincon 00677-0086, Puerto Rico; E-mail: goyalmegh@gmail.com

Sadia Hassan
PhD Scholar, Institute of Home and Food Sciences, Government College University, Faisalabad, Pakistan. E-mail: sadiahassan88@gmail.com

Abdullah Ijaz Hussain, PhD
Director, Hi-Tech Lab, Government College University Faisalabad, Pakistan.
E-mail: abdullahijaz@gcuf.edu.pk

Muhammad Bilal Hussain
PhD Scholar, Institute of Home and Food Sciences, Government College University, Faisalabad, Pakistan. E-mail: itsmee1919@gmail.com

Syed Makhdoom Hussain, PhD
Department of Zoology, Government College University Faisalabad, Pakistan.
E-mail: drmakhdoom90@gmail.com

Ali Imran, PhD
Assistant Professor, Institute of Home and Food Sciences, Government College University, Faisalabad, Pakistan. E-mail: dr.aliimran@gcuf.edu.pk

Sana Khalid, MS
Research Associate, Department of Pharmaceutical Sciences, Government College University, Faisalabad, Pakistan. E-mail: sanakhalid436@gmail.com

Urooj Khan, MS
Institute of Home and Food Sciences, Government College University, Faisalabad, Pakistan.
E-mail: uroojkhan162@gmail.com

Bhanu Kumar, MSc
Senior Research Fellow, Pharmacognosy and Ethnopharmacology Division,
CSIR—National Botanical Research Institute, Lucknow 226001, UP, India.
E-mail: bhanu.kumar85@gmail.com

M. S. Latha
Professor, School of Biosciences, Mahatma Gandhi University, Kottayam 686560, Kerala, India.
E-mail: mslathasbs@yahoo.com

S. Nagaraj, PhD
Assistant Professor, Centre for Advanced Studies in Botany, University of Madras, Guindy Campus, Chennai 600025, Tamil Nadu, India. E-mail: nagalilly@gmail.com

Nida Nazar
PhD Scholar, Natural Product and Synthetic Chemistry Lab., Department of Chemistry,
Government College University Faisalabad, Pakistan. E-mail: nidanazar786@gmail.com

Poonam Singh Nigam, PhD
School of Biomedical Sciences, Biomedical Science Research Institute, University of Ulster, Coleraine, UK. E-mail: p.singh@ulster.ac.uk

Oluwafemi Adeleke Ojo, PhD
Department of Biochemistry, Afe Babalola University, Ado-Ekiti, PMB 5454, Nigeria.
E-mail: oluwafemiadeleke08@gmail.com

Kingsley Okoyomoh
Lecturer 2, Federal Polytechnic, Nasarawa, Nasarawa State, Nigeria.
E-mail: kingsleyokoyomuh0211@gmail.com

M. K. Preetha
Research Scholar, Pharmacognosy Laboratory, School of Biosciences, Mahatma Gandhi University, Kottayam 686560, Kerala, India. E-mail: preethakmattathil111@gmail.com

R. N. Raji
Research Scholar, Pharmacognosy Laboratory, School of Biosciences, Mahatma Gandhi University, Kottayam 686560, Kerala, India. E-mail: mitra.rrn@gmail.com

Karuppan Ramamoorthy
PhD Research Scholar, Centre for Advanced Studies in Botany, University of Madras, Guindy Campus, Chennai 600 025, Tamil Nadu, India. E-mail: ramamoortthy@gmail.com

A. K. S. Rawat, PhD
Head and Senior Principal Scientist, Pharmacognosy and Ethnopharmacology Division, CSIR—National Botanical Research Institute, Lucknow 226001, UP, India.
E-mail: rawataks@gmail.com

Malairaj Sathuvan, PhD
Post-Doctoral Fellow, National Centre for Meat Quality and Safety Control (MOST) and National Centre of Meat Processing (MOA), College of Food Science and Technology, Nanjing Agricultural University, No. 1, Weigang, Nanjing 210095, Jiangsu Province, China.
E-mail: sathuvansjc@gmail.com

Umair Shabbir, MS
Research Associate, National Institute of Food Science and Technology, University of Agriculture, Faisalabad, Pakistan. E-mail: umair336@gmail.com

Mohammad Ali Shariati, PhD
Senior Associate Researcher, Laboratory of Biocontrol and Antimicrobial Resistance, Orel State University named after I.S. Turgenev, Orel 302026, Russia.
E-mail: Shariatymohammadali@gmail.com

Monika Sharma, MPharm, PhD
DST-INSPIRE Fellow, Pharmacognosy and Ethnopharmacology Division, CSIR-National Botanical Research Institute, Lucknow 226001, UP, India.
E-mail: monika.nbri@gmail.com

Bramhanand Singh, PhD
Senior Scientist, Pharmacognosy and Ethnopharmacology Division, CSIR—National Botanical Research Institute, Lucknow 226001, UP, India.
E-mail: bn.singh@nbri.res.in

M. P. Singh, PhD
Professor of Centre of Biotechnology, Coordinator of Centre of Bioinformatics, University of Allahabad, Allahabad 211002, UP, India. E-mail: mpsingh.16@gmail.com

K. G. Sreekala
PhD Research Scholar, Centre for Advanced Studies in Botany, University of Madras, Guindy Campus, Chennai 600025, Tamil Nadu, India. E-mail: scorpionic.srushti@gmail.com

Hafiz Ansar Rasul Suleria, PhD
Alfred Deakin Postdoctoral Research Fellow, Centre for Chemistry and Biotechnology, School of Life and Environmental Sciences, Faculty of Science, Engineering and Built Environment, Deakin University, 75 Pigdons Road, Geelong, VIC 3216, Australia. E-mail: hafiz.suleria@uqconnect.edu.au

D. Suma
Research Scholar, Pharmacognosy Laboratory, School of Biosciences, Mahatma Gandhi University, Kottayam 686560, Kerala, India. E-mail: sumadwarai@yahoo.com

N. Tabassum, MSc
PhD Research Scholar, Centre of Biotechnology, University of Allahabad, Allahabad 211002, UP, India. E-mail: nazishtabassumau@gmail.com

A. Vysakh
Research Scholar, Pharmacognosy Laboratory, School of Biosciences, Mahatma Gandhi University, Kottayam 686560, Kerala, India. E-mail: vysakh15@gmail.com

Marwa Waheed, PhD
Scholar, Institute of Home and Food Sciences, Government College University, Faisalabad, Pakistan. E-mail: wmarwa31@yahoo.com

ABBREVIATIONS

3,4-DHS	3,4-dihydroxy-*trans*-stilbene
4,40-DHS	4-hydroxy-*trans*-stilbene
ABA	abscisic acid
ABTS	2,2-azinobis-3-ethylbinzothiazoline-6-sulfonic acid
ACE	angiotensin-converting enzyme
ACEi	angiotensin-converting enzyme inhibitor
ACF	aberrant crypt foci
AChE	acetylcholinesterase
ACR	acrylamide
ADP	adenosine diphosphosphate
AGE	aged garlic extract
AGPs	antibiotic growth promoters
AIDS	acquired immune deficiency syndrome
ALA	alpha lipoic acid
ALT	alanine aminotransferase
AMR	antimicrobial resistance
APC	allophycocyanin
AST	aspartate aminotransferase
ATP	adenosine triphosphate
ATPase	adenosine triphosphatase
AYUSH	Ayurvedic Pharmacopoeia of India
BAD	Bcl-2-associated death promoter
BCBD	β-carotene bleaching test
BCL-2	B-cell lymphoma 2
BDNF	brain-derived neurotrophic factor
bGH	bovine growth hormones
BHA	butylated hydroxyl anisole
BHA	beta-hydroxy acid
BHT	butylated hydroxytoluene
BOD	biodegradable
B-PE	B-phycoerythrin
CAM	complementary and alternative medicine
CaMKII	Ca^{2+}/calmodulin-dependent protein kinase II

cAMP	cyclic adenosine monophosphate
CaMV	cauliflower mosaic virus
CAT	catalase
CCl_4	carbon tetrachloride
CD4	cluster of differentiation-4
CHD	coronary heart diseases
CI	confidence interval
CNS	central nervous system
CoA	coenzyme A
COX	cyclooxygenase
CP	cyclophosphamide
CSE	conventional solvent extraction
CTB	cholera toxin B
CVD	cardiovascular disease
DADS	diallyl disulfide
DAS	diallyl sulfide
DATS	diallyl trisulfide
DEAE	diethylaminoethyl
DHA	docosahexaenoic acid
DMH	1,2-dimethylhydrazine
DMPD	*N,N*-dimethyl *P* phenylenediamine
DPPH	1,1-diphenyl-2-picrylhydrazyl
DSSC	dye-sensitized solar cell
DW	dry weight
EAAE	enzyme-assisted aqueous extraction
EACC	Ehrlich ascites carcinoma cell
EACP	enzyme-assisted cold pressing
EAE	enzyme-assisted extraction
ECM	extracellular matrix
EDR	endothelium-dependent relaxation
EE	ether extract
EFSA	European Food Safety Authority
ELISA	enzyme-linked immune sorbent assay
eNOS	endothelial nitric oxide synthase
EPA	eicosapentaenoic acid
EPA	Environmental Protection Agency
EPCG	(−)-epigallocatechin-3-gallate
ER	endoplasmic reticulum

ERK	extracellular signal-regulated kinases
ESVAC	European Surveillance of Veterinary Antimicrobial Consumption
EU	European Union
EVs	edible vaccines
FAS	fatty acid synthase enzyme
FCAT	Freund's complete adjuvant test
FDA	Food and Drug Administration
FRAP	ferric reducing antioxidant power
FTIR	Fourier transform infrared spectroscopy
FYGL	Fudan–Yueyang *G. lucidum*
GABA	gamma-aminobutyric acid
GADD	Growth arrest and DNA damage
GAP	good agriculture practices
GC–MS	gas chromatography–mass spectrometry
GCP	good collection practices
GEP	good ethical practices
GIT	gastrointestinal tract
GK	glucokinase
GLUT	glucose transporter
GM	genetically modified
GPP	good procurement practices
GPx	glutathione peroxidase
GR	glutathione reductase
GRAS	generally recognized as safe
GRP	glucose-regulated protein
GRx	glutathion reductase
GSH	glutathione
GSH-Px	glutathione peroxidase
GSP	good storage practices
H_2O_2	hydrogen peroxide
HBsAg	hepatitis B surface antigen
HBV	hepatitis B virus
HCC	hepatocellular carcinoma
HCT	human colorectal carcinoma cell line
HCT116	human colon cancer cell
HDFa	human dermal fibroblasts adult
HDL	high-density lipoproteins

HepG2	human hepatocarcinoma cell
HFD	high fructose diet
hGH	human growth hormones
HHP	high hydrostatic pressure
HHPE	high hydrostatic pressure extraction
HIF-1α	hypoxia-inducible factor-1α
HIV	human immunodeficiency virus
HLA	human leukocyte antigen
HMGCR	3-hydroxy-3-methylglutaryl-coenzyme A reductase
HO-1	heme oxygenase-1
HPLC	high performance liquid chromatography
HRS	hydroxyl radical scavenging
HSV-1	herpes simplex virus-1
I3C	indole-3-carbinol
IC50	inhibitory concentration 50
IDDM	insulin-dependent diabetes
IDF	International Diabetes Federation
IgA and IgG	immunoglobulin A and immunoglobulin G
IL	interleukin
iNOS	inducible nitric oxide synthase
IPC	International Poultry Council
IR	infrared
IUCN	International Union for Conservation of Nature
IVD	intervertebral disc
Iκβ	inhibitor of κβ
JNK	c-Jun N-terminal kinases
LC	liquid chromatography
LDH	lactate dehydrogenase
LDL	low-density lipoproteins
LLE	liquid–liquid extraction
LLL	combination of linoleic, linoleic, and linoleic
LO	lipooxygenase activity
LPS	lipopolysaccharide
MAbs	monoclonal antibodies
MAE	microwave-assisted extraction
MAP	microwave-assisted processing
MAPK	mitogen-activated protein kinase
MCF-7	Michigan Cancer Foundation-7 (breast cancer cell line)

MCP-1	monocyte chemoattractant protein-1
MDA	malondialdehyde
MDM-LDL	malondialdehyde-modified-LDL
MDR-1	multidrug resistant protein-1
MF	mounting frame
MFRM	mango fruit reject meal
MHG	microwave hydrodiffusion and gravity
MIC	minimum inhibitory concentration
MMC	mitomycin C
MMP 9	matrix metalloproteinase 9
MMP	mitochondrial membrane potential
mRNA	messenger ribonucleic acid
MS	mass spectroscopy
mTOR	mammalian target of rapamycin
MT-α-glucan	maitake α-glucan
NEG	nonenzymatic glycosylation
NF-κB	nuclear factor kappa-light-chain-enhancer of activated B cells
NGF	nerve growth factor
NHPD	Natural Health Product Directorate
NHPs	natural health products
NIDDM	noninsulin dependent diabetes mellitus
NKA	sodium–potassium ATPase activity
NMDA	N-methyl-D-aspartate
NMR	nuclear magnetic resonance
NO	nitric oxide
NOS	nitric oxide synthase
NPQ	nonphotosynthetic quenching
NQO-1	NADPH: quinone oxidoreductase 1
NSCLC	nonsmall cell lung cancer cell
NSS	neurological severity score
ODC	ornithine decarboxylase
OH	Ohmic heating
OLL	combination of oleic, linoleic, and linoleic
ORAC	oxygen radical absorbance capacity
ORCs	olfactory receptor cells
OVA	ovalbumin
p-38	mitogen-activated protein kinases

PABC	pro-oxidant–antioxidant balance
PAL	phenylalanine ammonialyase
PCR	polymerase chain reaction
PE	pulmonary embolism
PEF	pulsed electric field
Phospho-AKT	protein kinase B (PKB) or serine/threonine-specific protein kinase
PLE	pressurized liquid extraction
PLL	combination of palmitic, linoleic, and linoleic
POL	combination of palmitic, oleic, and linoleic
PoMsrA	pleurotusostreatusmethionine sulfoxide reductase A
POP	persistent organic pollutants
PP	phenylpropanoid
PPARy	peroxisome proliferator-activated receptors
PTZ	pentylenetetrazol
PUFA	polyunsaturated fatty acids
qRT-PCR	real-time quantitative reverse transcription-polymerase chain reaction
RAAS	renin–angiotensin aldosterone system
RE	retinol equivalents
RET	rare, endangered, and threatened
RNS	reactive nitrogen species
ROCCs	receptor operated Ca2þ channels
ROS	reactive oxygen species
R-PE	R-phycoerythrin
RSV	respiratory syncytial virus
SAC	S-allyl cysteine
SAMC	S-allyl mercaptocysteine
SF	supercritical fluids
SFA	saturated fatty acids
SFE	supercritical fluid extraction
SI	stimulation index
SIRT	Sirtuin (silent mating type information regulation 2 homolog)
SNPs	silver nanoparticles
SOD	superoxide dismutase
STZ	streptozotocin
SVPs	subviral particles

TAC	total antioxidant capacity
TAG	triacyl glycerol
TBARS	thiobarbituraic acid reactive species
TBI	traumatic brain injury
TC	total cholesterol
TEAC	trolox equivalent antioxidant capacity
TG	triglycerides
TGF	tumor growth factor
Ti	tumor inducing
TLC	thin layer chromatography
TLR	Toll-like receptor
TNF	tumor necrosis factor
TPC	total phenolic content
TPD	Therapeutic Drug Directorate
TrKB	tropomyosin receptor Kinase B
TUNEL	terminal deoxynucleotidyl transferase dUTP nick end labeling
UAE	ultrasound assisted extraction
UCP	uncoupling protein
UFA	unsaturated fatty acids
UNESCO	United Nations Educational, Scientific and Cultural Organization
UPE	ultrahigh pressure extraction
USDA	United States Department of Agriculture
UV	ultraviolet
VDC	voltage-dependent Ca^{2+} channel
VLDL	very low-density lipoproteins
WAT	white adipose tissues
WHO	World Health Organization
XO	xanthine oxidase

PREFACE 1

Medicinal plants have been used in traditional practices of medicine since prehistoric times. These plants contain various bioactive compounds (also referred to as phytochemicals) in leaves, stems, flowers, and fruits that can help to promote human health. Numerous phytochemicals with potential or established biological activity have been identified. However, since a single plant contains widely diverse phytochemicals, the effects of using a whole plant as medicine are uncertain. Further, the phytochemical content and pharmacological actions, if any, of many plants having medicinal potential, remain unassessed by rigorous scientific research to define efficacy and safety. Therefore, these plant products are drawing the attention of researchers and policymakers because of their demonstrated beneficial effects against diseases with high global burdens, such as diabetes, hypertension, cancer, and neurodegenerative diseases.

The side effects associated with conventional medicine have awakened the interest of researchers to explore these plants as an alternative or as complementary medicine. Hence, there is a need for substantial scientific evidence in terms of efficacy, dosage, and safety for traditional herbs to have a place in modern medicine.

This book volume sheds light on the potential of medicinal plants for human health for different technological aspects, and it contributes to the ocean of knowledge on food biochemistry and food science and nutrition. This book would not have been written without the valuable cooperation of these investigators, many of whom are renowned scientists who have worked in the field of food science and nutrition throughout their professional careers. I am glad to introduce my mentor and coeditor of this volume, Prof. Masood Sadiq Butt, who bring his expertise and innovative ideas on functional foods, bioactive compounds, nutraceuticals, and separation sciences in this book.

This book presents scientific reports on therapeutic values of different medicinal plants against diseases. It aims to further encourage the need for the development of plant-based drugs through innovative and groundbreaking research studies and, thus, will help to promote the health and economic well-being of people around the world. The understanding of the therapeutic values of these plants will also help to improve their sustainability, as people and governments will be encouraged to preserve and conserve the plants for future generations. The book covers the phytochemistry and health-promoting potentials of plants against different ailments. This book volume is a treasure house of information and an excellent reference material for researchers, scientists, students, growers, traders, processors, industries, dieticians, medical practitioners, and others. We hope that this compendium will be useful for students and researchers as well as those working in the food, nutraceutical, and herbal industries.

I thank my mentor, Dr. Megh R. Goyal, for his unselfish leadership and educational and editorial qualities, for inviting me to join his team, and for motivating me to fall in love with my profession.

—Hafiz Ansar Rasul Suleria, PhD

PREFACE 2

To be healthy, it is our moral responsibility,
Towards Almighty God, ourselves, our family, and our society;
Eating fruits and vegetables makes us healthy,
Believe and have a faith;
Reduction of food waste can reduce the world hunger and
can make our planet eco-friendly.

—Megh R. Goyal

We introduce this book volume published under the book series Innovations in Plant Science for Better Health: From Soil to Fork. This book mainly covers some new research and case studies and the importance of phytochemicals from plants in therapeutics. The book has three parts: Part I—Bioactive Compounds from Medicinal Plants; Part II—Bioactive Compounds and Health Potentials; and Part III—Scope and Applications of Plant-Based Products.

This book volume sheds light on the potential of plants for human health from different technological aspects, and it contributes to the library of knowledge on food science and technology. We hope that this compendium will be useful for students and researchers as well as for those working in the food, nutraceuticals, and herbal industries.

The contribution of the cooperating authors to this book volume has been most valuable in the compilation. Their names are mentioned in each chapter and in the list of contributors. We appreciate them for having patience with our editorial skills. This book would not have been written without the valuable cooperation of these investigators, many of whom are renowned scientists who have worked in the field of plant science and food science throughout their professional career.

The goal of this book volume is to guide the world science community on how plant-based secondary metabolites can alleviate us from various conditions and diseases.

We will like to thank editorial staff and Ashish Kumar, Publisher and President, at Apple Academic Press, Inc., for making every effort to publish this book when all are concerned with health issues.

We request that readers offer their constructive suggestions that may help to improve future editions.

We express our admiration to our families and colleagues for their understanding and collaboration during the preparation of this book volume.

As an educator, I wish to share a piece of advice to one and all in the world: "Permit that our almighty God, our Creator, provider of all and excellent Teacher, feed our life with Healthy Food Products and His Grace…; and Get married to your profession…"

—Megh R. Goyal, PhD, PE
Senior Editor-in-Chief

PART I
Bioactive Compounds from Medicinal Plants

CHAPTER 1

PHYTOCHEMICALS FROM TRADITIONAL MEDICINAL PLANTS

R. N. RAJI, A. VYSAKH, D. SUMA, M. K. PREETHA, and
M. S. LATHA

ABSTRACT

The evolution and survival of the human race are highly attributed to his dependence on natural resources including plants. Medicinal systems [e.g., Ayurveda, Siddha, and Unani (Hindi)] have extensively used plants to provide health benefits to us. The advent of technology in the field of medicine in drug discovery has led to the decrease in the development of plant-based medicines. However, the increase in the incidence of subsidiary illness in the form of side effects of synthetic medicines combined with changes in lifestyle have prompted mankind to adopt plants as an alternative source of drugs. Therefore, extensive research is being conducted to unveil phytocompounds as natural drugs. This chapter focuses on four traditionally relevant plants (such as, *Amorphophallus campanulatus, Curculigo orchioides, Gardenia gummifera, and Woodfordia fruticosa*) for major phytocompounds and their pharmacological activities.

1.1 INTRODUCTION

Nature facilitated the evolution of mankind in an environment rich in diversity of flora and fauna. Armed with a strategic and logical mind, nature's finest creation has always been plants. From the basic necessities to the formulation of complex utilities, man's dependence on plants is indubitable. The Indian subcontinent is a repository of biodiversity. The growth or evolution of civilization in the Indian subcontinent also witnessed the coevolution of plant-based medicinal systems, such as Ayurveda.[85]

The practice of Ayurveda is India's finest contribution to human civilization. Appropriately called "Science of Life," Ayurveda redefines health as possession of a sound mind, body, and soul and believes in the rejuvenation of all three to surpass an ailment.[86] Ayurveda relies on the phytoconstituents of medicinal plants that act in a singular or a combinational phase. The asavas, arishtas, thailams, choornams, and vatis, used in Ayurvedic formulations, are the result of precise use of selected plant parts in appropriate concentrations and subjected to a meticulous method of preparation. Surgical procedures under this system were widely practiced as mentioned in *Charaka Samhita*, the oldest treatise of Ayurveda.

At a certain point of time, the discovery of antibiotic drugs, the advent of allopathy and presence of quack practitioners of Ayurveda caused a decline in the development, usage, and propagation of the Ayurvedic modality of treatment. However, in due course of time excessive use of synthetic medications, developments in medical research, and an upsurge in a number of diseases prompted the scientific community to trace its steps back to the age-old system of Ayurveda.

Today, Ayurveda is a booming industry with a turnover amassing to millions. Many Ayurvedic treatment centers have developed into research centers, where the efficacy, quality, and mechanism of action of the various plant-based formulations are actively tested and subjected to improvizations with the help of modern technology to bring about oils, pills, and mixtures with easily consumable flavors. The paradigm shift of research companies from synthetic compounds to find more plant-based lead components for drug designing has put medicinal plants back in limelight. Yet, many medicinal plants are left unexplored.

This chapter focuses on four traditionally relevant plants (such as *Amorphophallus campanulatus, Curculigo orchioides, Gardenia gummifera, and Woodfordia fruticosa*) for their pharmacological activities and major phytocompounds.

1.2   HERBAL PLANTS AS SOURCE OF MEDICINES

Medicinal plants refer to plants with certain activities that are used in herbalism. They are considered are as a rich source of phytoconstituents that can be utilized as tools for drug development. These plants can be used as synergic, supportive, or preventive medicine. They can be consumed as concoctions, complex formulation, or extract-based pills.[92]

Plants display a wide variety of chemical entities in them such as alkaloids, glycosides, tannins, etc. with a specific activity. The amount of these precious phytochemicals produced varies according to seasons, habitat, and other climatic conditions. Crude extracts of plants contain a large number of these phytochemicals, which are responsible for various bioactivities. These crude extracts are then thoroughly scrutinized, separated, and purified to isolate their constituents. Some classical examples of such randomly isolated compounds include morphine (isolated from Opium by Freidrich Serturner during 1803–1805) and digitoxin (pure sample first isolated by Oswald Schmiedeberg in 1875).

The year 1897 marked the beginning for the development of plant-based drugs with the discovery of aspirin (acetylsalicylic acid), a derivative of salicylic acid by Arthur Eichengrün and Felix Hoffmann, which led to a revolutionary change in the history of drug development and ushered the pharmaceutical industry. In the years that followed Alexander Flemming's discovery of penicillin from a fungus *Penicillium notatum* in 1928 shifted the focus to microbial-origin based drugs, adding a new dimension to the pharmaceutical industry. An uprising trend in structure activity guided organic synthesis, combinational chemistry, and computational or in-silico drug design has led to a decline in plant-based drug production.[97]

However, a revamp to plant-based drug discovery was seen to address side effects related to drug consumption, such as immune-suppression, metabolic disorders, anti-infective, etc. The approach was based on deciphering a molecular target with definite bioactivity, which could serve as a lead molecule in the development of drug rather than crude extract-based experiments. Industries nowadays make use of powerful new technologies, such as automated separation techniques, high-throughput screening, and combinatorial chemistry for isolation of lead molecules. Even though the process is tedious, and the yield of target molecule could be ridiculously meager, yet the inherent large-scale diversity in structure lures pharmacy companies into this field.[58]

India, a tropical country with enriched biodiversity, is one of the leading exporters of medicinal plants, roughly estimated to about 13% of the global market. It can be rightfully stated that plants can be considered as the largest provider of drugs, due to the research focusing on ailments, such as infectious diseases (mainly viral diseases such as human immunodeficiency virus (HIV)), arthritis, liver diseases, diabetes, cancer, age-related diseases (e.g., memory loss, osteoporosis, and immune disorders), hypertension, sexual dysfunction, and hyperlipidemia; incorporation of techniques, such as plant molecular farming, recombinant DNA technology, micropropogation-based large scale production of phytochemicals, and molecular docking based studies on activities.[59]

1.3   MEDICINAL COMPOUNDS: ISOLATED AND USED TILL 2016

According to the World Health Organization, 80% of people around the world depend on medicinal plants for their primary medical needs and about 21,000 species of plants have been well defined for medicinal properties. Approximately 30% of the world plant population is currently being used as medicinal plants, and of the total drugs used in developed countries, 25% are plant-derived. In developing countries, including China and India, herbal drugs constitute about 80% of the market, and these countries, therefore, are the largest contributors of medicinal plants in the global market.[34]

A large number of phytocompounds have been isolated from various medicinal plants. Many such compounds are used in our daily life as food supplements (e.g., Curcumin from *Curcuma longa* or Turmeric),[75] additives to cosmetics (e.g., methylsalicylate/wintergreen oil from *Gaultheria procumbens*—active ingredient of mouthwashes, such as Listerine),[115] etc., due to their medicinal properties. Preparation of a comprehensive list of all isolated phytocompounds is no less than a mammoth task, yet the following section is an attempt to define, classify, and list few of the phytocompounds isolated and used in drug formulations owing to their bioactivities.

1.3.1   ALKALOIDS

Alkaloid often refers to a class of organic nitrogenous compounds with a pronounced physiological action. They are optically active and exist naturally as salts of organic acids and rarely, sugars. Alkaloids exert a variety of physiological effects; some are poisonous while some are addictive in nature.[23] Among 12,000 or more alkaloids known to mankind, some of the important alkaloids of therapeutic importance are listed in Table 1.1.

TABLE 1.1 Alkaloids: Sources and Their Bioactivities.

Biocompounds	Plant source	Activity	References
Ajmalicine	*Rauvolfia sepentina*	Antihypertensive drug	[63]
Atropine	*Atropa belladonna* (Deadly nightshade)	Anticholinergic, antiemetic	[71]
Berberine	*Berberis vulgaris*—common barberry	Antiarrhythmic, antimicrobial	[41, 62]
Codeine	*Papaver somniferum* (poppy)	Pain killer, treat cough, diarrhea	[89]
Cytisine	*Laburnum anagyroides*	Smoking cessation aid	[122]
Demecolcine, colcemid	*Colchicum autumnale* (Autumn crocus)	Chemotherapy (M phase arrest)	[54]
Galanthamine	*Lycoriss quamigera* (magic lily, naked lady)	Treatment of mild/moderate Alzheimer's	[9]
Hyoscyamine	*Hyoscyamus niger* (black henbane, stinking nightshade, henpin)	Treat peptic ulcers, irritable bowel syndrome (Symax, HyoMax)	[47]
Lobeline	*Lobelia inflate* (Indian tobacco)	Smoking cessation aid	[16]
Morphine	*Papaver somniferum* (poppy)	Pain relief, narcotic	[70]
Nicotine	*Nicotiana tabacum* (tobacco)	Performance enhancer	[38]
Palmatine	*Enantia chlorantha* (African Yellow Wood)	Treating jaundice, hypertension	[31]
Physostigmine	*Physostigma venenosum*—Calabar bean	Cholinesterase inhibitor	[69]
Rotundine	*Stephania sinica*	Anxiolytic, sedative	[27]
Sparteine	*Sarothamnus scoparius*	Antarrhythmic, diuretic	[98]
Strychnine	*Strychnosnux vomica* (poison nut tree)	Sweetener	[14]
Taxol	*Taxus brevifolia* (Pacific yew)	Anticancer treatment	[45]
Theobromine or xantheose	*Theobroma cacao* (cocoa)	Vasodilator, diuretic, heart stimulant	[67]
Tubocurarine	*Chondodendron tomentosum*—Curare vine	Adjunct for clinical anesthesia	[14, 18]
Vinblastine	*Catharanthus roseus*—Madagascar periwinkle	Chemotherapeutic agent	[155]
Vincamine	*Vinca minor* (Lesser Periwinkle)	Nootropic agent	[1]
Vincristine	*Catharanthus roseus*—Madagascar periwinkle	Chemotherapeutic agent	[155]

1.3.2 AMINES, AMINO ACIDS, AND PROTEINS

Amines, amino acids, and proteins are the primary metabolites of plants (Table 1.2). They are nitrogenous compounds and organic in nature. Amines of the plant kingdom can be monoamines, aliphatic diamines, polyamines, and aromatic amines. They are important in the growth, metabolism, and development of plants.[13] Amino acids contain an amine and a carboxyl entity as a functional group along with an organic side chain, which varies according to the amino acid. Although more than 500 amino acids are known, yet only 20 are encoded by the genetic code (canonical amino acids). They are classified in many ways based on polarity, nature of side chain, pH, etc. Apart from being the building blocks of proteins, amino acids serve a variety of other functions such as neurotransmitter biosynthesis, regulation of gene expression, etc.[124]

TABLE 1.2 Amines, Amino Acids, and Proteins of Therapeutic Importance.

Compounds	Plant source	Activity	References
Amines			
Ephedrine	*Ephedra sinica*	Treat asthma	[152]
Kawain	*Piper methysticum*—kava kava	Psychotropic, sedative	[153]
Picrotoxin (Barbiturate)	*Anamirta cocculus*—fish berry	Central nervous system stimulant	[77]
Quinidine	*Cinchona ledgeriana*—quinine tree	Atrial fibrillation	[126]
Quinine	*Cinchona ledgeriana*—quinine tree	Antimalarial, lupus, arthritis	[126]
Amino acid derivatives			
L-Dopa	*Mucuna species* (nescafe, cowage, velvet bean)	Psychoactive drug—atamet	[29]
Proteins			
Bromelain	*Ananas comosus*—pineapple	Meat tenderizer, topical medicine (NexoBrid) for severe skin burns	[154]
Chymopapain	*Carica papaya*—papaya	Proteolytic enzyme, chemonucleolysis	[132]
Papain	*Carica papaya*—papaya	Meat tenderizing, dental caries	[132]
Trichosanthin	*Trichosanthes kirilowii*—snake gourd	Abortifacient	[131]

Proteins are large biomolecules. They are synthesized in accordance with the genetic code under a thoroughly scrutinized procedure in the living organisms to serve a multitude of functions—from formation of muscle fibers to digestion of matter in the body. Biochemically, proteins exist as polypeptide chains that are formed when amino acids link with each other using peptide bonds. The constituent polypeptide chain of a protein undergoes conformational changes enabling the moiety to adopt a specific shape and aid its function.[149]

1.3.3 COUMARINS

Coumarins are a class of fragrant organic compounds found in plants (Table 1.3). They belong to the benzopyrone class and forms colorless crystals in the standard state. The first coumarin was isolated in 1820 from the French Tonka bean "Coumarou" (*Dipteryx odoranta*) and hence the name.[113]

TABLE 1.3 Coumarins of Therapeutic Importance.

Compounds	Plant source	Activity	References
Aesculetin	*Frazinus rhychophylla*	Component of sunscreens, used in varicose treatment	[146]
Bergapten	*Citrus bergamia*	Aids xenobiotic metabolism	[40]
Fraxidin	*Artemisia scotina*	Antioxidant, antiadipogenic	[118]
Khellin	*Ammi visnaga*	Treatment of vitiligo, kidney stones	[60, 116]
Phyllodulcin	*Hydrangea macrophylla*— big leaf hydrangea, French hydrangea	Sweetener	[44]
Umbelliferone	Plants of Umbelliferae family	Component of sunscreens	[148]
Xanthotoxin (Methoxsalen)	*Ammi majus*	Treatment of vitiligo, psoriasis	[101]

1.3.4 FLAVONOIDS

Flavonoids, also called bioflavonoids (from *flavus*—meaning yellow) are a class of plant secondary metabolites with ring structure comprising two phenyl rings and a heterocyclic ring. They are classified on the basis of their structure into flavonoids, isoflavoids, and neoflavonoids. Flavonoids are responsible for the pigmentation/coloration of plant parts, UV filtration, combating oxidative stress, growth, and symbiotic associations as shown in Table 1.4.[56] Rhizobia of soil can sense flavonoids secreted by legumes and are attracted to the roots, thereby forming nodules.[36]

TABLE 1.4   Flavonoids of Therapeutic Importance.

Compounds	Plant source	Activity	References
Astragalin	*Aristolochia indica*	Anti-inflammatory	[24]
Catechin	*Potentilla fragarioides*	Antioxidant	[25]
Rutin	*Citrus species,* e.g., orange, grapefruit	Antioxidant	[6]
Orientin	*Adonis vernalis, Vitex agnus—castus*	Antioxidant	[17, 35]
Luteolin	*Salvia tomentosa*	Anti-inflammatory, antitumor	[64]
Quercetin	Many fruits, vegetables, grains	Anticancer, anti-inflammatory	[6, 73]
Flavanolignan			
Silymarin	*Silybum marianum*—milk thistle	Liver protecting	[117]

1.3.5   GLYCOSIDES

Glycosides are sugars that are chemically bound to any functional group via a glycosidic bond; the sugar group termed as glycone and the nonsugar group termed as aglycone. Plants usually store glycosides in their inactive form and are hydrolyzed to their active forms. Glycosides are classified on the basis of the glycone involved, by the type of glycosidic bond and by the aglycone in the moiety. The most important function carried out by glycosides is signal transduction (Table 1.5).[99]

   Cardiac glycosides are a group of compounds with a steroidal derivative as their aglycone. These compounds affect the cardiac muscles (membranes of cardiac myocytes to achieve faster and more powerful contraction by cross-bridge cycling) and are known to regulate heart rates in arrhythmic conditions.[88]

TABLE 1.5 Glycosides of Therapeutic Importance.

Compounds	Plant source	Activity	References
Bergenin or Cuscutin	*Ardisia japonica*—marlberry	Pashaanbhed of Ayurveda, immunomodulator	[84]
Convallatoxin	*Convallaria majalis*—lily-of-the-valley	Treat acute and chronic heart failure	[127]
Etoposide	*Podophyllum peltatum*—mayapple	Cytotoxic anticancer drug	[156]
Salicin	*Salix alba*—white willow	Anti-inflammatory agent	[115, 157]
Stevioside	*Stevia rebaudiana*—Stevia	Sweetener	[159]
Hesperidin	*Citrus* species, e.g, oranges	Plant defense, chemopreventive	[158]
Cardiac glycoside			
Acetyldigoxin	*Digitalis lanata*—Wolly foxglove	Congestive chronic cardiac failure	[26]
Deslanoside	*Digitalis lanata*	Treat cardiac arrhythmia, congestive heart failure	[39]
Digitoxin	*Digitalis purpurea*—purple foxglove	Anticancer	[26]
Digoxin	*Digitalis purpurea*	Treat atrial fibrillation	[121]
Lanatosides A, B, C	*Digitalis lanata*—Wolly foxglove	Cardiac arrhythmia	
Ouabain or arrow poison	*Strophanthus gratus* (ouabain tree)	Treat arrhythmia, hypertension	[33]

1.3.6 SAPONINS

Saponins belong to the family of glycosides. They are unique in being amphipathic in nature and produce foam when shaken in aqueous solutions. Chemically, they are low molecular weight secondary metabolites tetracyclic steroidal or a pentacyclic triterpenoid aglycone with one or more sugar chains. They are known to protect plants from animals owing to their bitter taste and are also said to aid in (Table 1.6) pest control. They are natural surfactants.[52]

TABLE 1.6 Saponins of Therapeutic Importance.

Compounds	Plant source	Activity	References
Aescin	*Aesculus hippocastanum*	Chronic venous insufficiency	[106]
Asiaticoside	*Centella asiatica*	Antiwrinkle, used in lipsticks	[151]
Avenacosides	*Avena sativa*—Oat	Hypocholesterolemic	[80]
Glycyrrhizin	*Glycyrrhiza glabra*—licorice	Treatment of hepatitis	[129]
Quillaic acid	*Quillaja saponaria*	Anti-inflammatory	[94]
Ruscogenin	*Ruscus aculeatus*	Anti-inflammatory	[42]
Sarasapogenin	*Smilax* sp.	Treatment of Alzheimer's, antitumor	[100]

1.3.7 TANNINS AND TERPENOIDS

Tannins are polyphenolic in nature. The oldest use of tannins was for the preparation of leather from animal hides. Tannins have astringent properties and are bitter in taste, tannin-rich plants were thus protected from animals. Unripe fruits, wine, and tea are rich sources of (Table 1.7) tannins. The pungent, puckering taste is due to a high amount of tannins in them. They are classified into gallotannins, ellagitannins, condensed tannins, and complex tannins on the basis of complexity and constituent subunits. Tannins have potent antioxidant activity and wide industrial use.[50]

Terpenes/terpenoids are a group of organic hydrocarbons derived from isoprene units. Terpenes may be classified as monoterpenes (components of essential oil), diterpenes (a component of plant resins), triterpenes (components of resin, cork, and cutin), sesquiterpenes (aliphatic compounds with three isoprene units). They exert profound physiological effects that make them excellent targets for drug leads.[114]

TABLE 1.7 Tannins and Terpenoids of Therapeutic Value.

Compounds	Plant source	Activity	References
		Tannins	
Castalagin	*Anogeissus leiocarpus*	Antileishmania	[102]
Chebulagic acid	*Terminalia chebula*	Antiviral	[128]
Cyanidin	Red berries	Antioxidant, anti-inflammatory	[37]
Gallic acid	Plants of many families.	Antioxidant, antitumor	[76]
Heterophyllin	*Pseudostellaria heterophylla*	Antitumor	[112]
		Terpenoids	
Artemisinin	*Artemissia annua*	Antimalarial	[150]
Azadirachtin	*Azadirachta indica* (Neem)	Insecticide	[123]
Geraniol	Rose oil, Citonella oil	Antimicrobial	[61]
Taxol	*Taxus breviifolia*	Anticancer	[107]
Vitexin	*Vitex agnus—castus*	Antioxidant	[35]

1.4 PLANT PROFILE

1.4.1 *Amorphophallus campanulatus*

Amorphophallus campanulatus (Roxb.) Bl Blume ex Decne (commonly called the white spot giant arum, elephant foot yam, or stink lily) is a tuber crop[133] grown in areas of Africa, South Asia, South-East Asia, and Pacific Islands. It is widely used as a vegetable.[93,104] It is a perennial herb. After planting the tuber, it germinates to produce a single inflorescence followed by a solitary leaf. After the growing season, the leaf withers leaving behind the tuber underground. At this stage, the tubers are harvested.

1.4.2 *Curculigo orchioides* Gaertn

Curculigo orchioides Gaertn (commonly called "Golden-eye-grass," Weevil wort, Black musli, or Kali Musli) is a flowering plant under the genus Curculigo.[134] It is found widely in parts of Asian subcontinent (including Japan, India, China, Malaysia, etc.) and in Australia.[22] *C. orchioides* Gaertn is a small perennial herb that grows up to about 30 cm in height. The rhizomes are fleshy and tuberous, leaves elongated, and bright yellow flowers seasonally appear on the plant. The rhizome can either be short or elongated with several fleshy and lateral roots, which are blackish to brown in color toward exterior and appear cream in color toward the interior.[43]

1.4.3 *Gardenia gummifera* Linn

Commonly called Dikamali or Cumbi gum tree, *Gardenia gummifera* grows into an approximately 8-m long tree in the wild and bears white scented flowers.[135] The stem is resinous and has renowned medicinal properties. The plant is endemic to peninsular India. It is found in dry forests of Karnataka, Tamil Nadu, Andhra Pradesh, and Kerala. It is a small tree or large woody bush, which grows about 3–7 m tall. The woody part appears yellowish white and is hard.[66] Resin from the leaf buds is used in healing wounds, indigestion, gas trouble, ulcer, and cardiac problems.

1.4.4 *Woodfordia fruticosa* Kurz

Woodfordia fruticosa[136] is one of the two species of the genus *Woodfordia*. *W. fruticosa* is widely distributed in parts of Asia and also included in the IUCN Red List of "Threatened Species" under lower risk category.[144] Commonly called fire flame bush, *W. fruticosa* Kurz is a tree, which grows about 10-feet tall. The bright red flowers bloom in the months of May and June.

1.5 TRADITIONAL USES

1.5.1 *Amorphophallus campanulatus*

The tubers of *A. campanulatus* are very popular as a vegetable in various delicious cuisines. The tuberous roots are reportedly tonic and appetizer in nature. The corm of the plant has been used traditionally for the treatment of bronchitis, anemia, tumors, spleen disorders, piles, abdominal disorders, mild to moderate asthma, and rheumatism.[53] They also find use in the treatment of hemorrhoids, vomiting, arthralgia, inflammations, hemorrhages, cough, anorexia, dyspepsia, metabolic disorders, helminthiasis, liver disorders, menstrual disorders, seminal weakness, and fatigue in general.[30,74]

In traditional Chinese medicine, a gel prepared from the flower of *Amorphophallus paeoniifolius (A. camapanulatus)* has been used for the treatment of asthma, cough, hernia, burns, hematological, and skin disorders. It also finds use in detoxifying the body, tumor suppression, curing blood stagnation, and sputum liquefaction.[12]

1.5.2 *Curculigo orchioides* Gaertn

Curculigo orchioides was mentioned in "Charak Samhita" for the treatment of alleviated cough. It is also used for the treatment of gonorrhea, urinary tract infections, hemorrhoids, jaundice, bronchitis, gastric disorders (such as, heartburn, vomiting, and oligospermia).[22] Rhizomes of *C. orchioides* have been used as Ayurvedic treatment for infertility. Dry powder of rhizome is used in traditional Chinese medicine for the treatment of loss of vigor. Also, the rhizome extract along with the juice of garlic is used for the treatment of various ailments of the eye. The plant is also prescribed for many diseases under the Unani system of medicine.

1.5.3 *Gardenia gummifera* Linn

The powder or paste of the bark of *G. gummifera* is used in the treatment of ageusia (lack of taste), roundworm infections, abdominal distensions, and disorders (such as indigestion, constipation, colic pain), respiratory disorders, and skin diseases. The paste of its tubers is applied to the abdomen of woman to facilitate safe and easy childbirth. It is an anthelmintic, antispasmodic, carminative, diaphoretic, and expectorant. Antiepileptic, antioxidant, and antihyperlipidemic properties have also been attributed to *G. gummifera*.[66]

1.5.4 *Woodfordia fruticosa* Kurz

Woodfordia fruticosa is traditionally used as a fermenter. The flowers with their natural microbiota are added to increase the fermentation of a mixture of medicinal plants during the preparation of asavas and arishtas. According to treatises of Charaka and Sushruta, the sweetened decoction of *W. fruticosa* was used for the treatment of fever, haemothermia, dysentery, etc. The flowers of *W. fruticosa* mixed with honey were known to be prescribed for leucorrhea.[51] The flowers are known to be acrid, cooling, and pungent with alexiteric and sedative properties. The dried flowers are used as an astringent tonic in treating liver disorders and used upon wounds to eliminate discharge and promote granulation.[111]

In Indonesia and Malaysia, crude drug named Sidowava or Sidawayah containing dried flowers of Woodfordia is used in treating sprees, bowel disease, and as an astringent. It is also incorporated into a preparation, which is used to make barren women fertile. According to Yogaratna-kara, the flowers of *W. fruticosa* have been used as a substituent for *Glycyrrhiza glabra*.[55]

1.6 PHYTOCHEMISTRY AND BIOLOGICAL ACTIVITIES OF SELECTED MEDICINAL PLANTS

1.6.1 *Amorphophallus campanulatus*

The corm of *Amorphophallus campanulatus* is widely used part of the plant. The total ash was roughly 6.9% while the acid insoluble ash was about 0.93%. Primary metabolites such as tannins, sugar, phenolics, total protein, and starch were roughly about 0.02%, 1.16%, 0.012%, 1.23%, and 26.93%, respectively. The proximate analysis showed rich carbohydrate content in the corm. The calorific value of the corm was approximately 359.08 Kcal/100 g justifying its use as a nutritionally important food.[110]

The peels of *A. campanulatus* are usually discarded while using only the plant. The standard qualitative tests conducted upon various solvent extracts of the peel indicated the presence of metabolites (such as, carbohydrates, fats), alkaloids, and polyphenolics, such as tannins in the peel.[103]

1.6.1.1 *ANTIBACTERIAL AND ANTIFUNGAL ACTIVITY*

The various extracts of leaf, stem, and root of *A. campanulatus* have been reported to show potent activity against *Escherichia coli, Bacillus subtilis, Staphylococcus aureus, Bacillus cereus, Staphylococcus epidermidis, Pseudomonas aeruginosa, Klebsiella pneumoneae*, and *Proteus vulgaris*. The ethanolic extract of the plant was shown to inhibit the growth of fungi such as *Aspergillus flavus, Candida albicans, A. niger,* and *Rhizopus oryzae.*[81]

1.6.1.2 FREE RADICAL SCAVENGING AND ANTIOXIDANT ACTIVITY

Antioxidant activity and free radical scavenging are important characteristics that impart medicinal properties to a plant. 2,2-diphenyl-1-picrylhydrazyl (DPPH) assay, total antioxidant assay, ferric reducing antioxidant power (FRAP) assay, reducing power and nitric oxide assays conducted on alcoholic and aqueous extracts of *A. campanulatus* have shown promising results.[8,28]

1.6.1.3 LIVER PROTECTION AND ANTICANCER PROPERTIES

Chemicals, such as carbon tetrachloride, N' nitrosodiethylamines, etc., are used widely in animal experiment models to impart liver damage in short-term and long-term studies. These chemicals destroy normal physiology of the liver in such a way that an increase in liver marker enzymes and proteins can easily be assessed in the blood and tissues of the test animal. Preventive and curative model studies conducted using *A. campanulatus* extract as drug control proved the efficacy of the plant as a hepatoprotectant.[5,11] The methanolic extract of the plant also inhibited colon carcinogenesis.[4] The methanolic extract of *A. campanulatus* showed potent anticancer activity against colon cancer in animal models and human cancer cell lines (HCT-15) thereby confirming the claims made by traditional texts. The alcoholic extract of the plant also has immunomodulatory activity.[3]

1.6.2 *Curculigo orchioides*

Phytochemical studies conducted upon Kali musli around the world have provided convincing evidence that the plant is a cocktail of various phytochemicals. Most of the studies conducted showed that the methanolic, ethanolic, and ethyl acetate extracts of the plant are a rich source of tannins, glycosides, saponins, terpenoids, and alkaloids. The mucilage present in the rhizomes is rich in glucuronic acid.[78]

1.6.2.1 HEPATOPROTECTIVE, ANTICANCER, AND IMMUNOMODULATORY ACTIVITIES

Alcoholic extracts of the *C. orchioides* rhizomes showed hepatoprotective activity against carbon tetrachloride and rifampicin-induced hepatotoxicity.[91] Methanolic extract of the rhizomes showed anticancer activity when tested upon MCF-7 cells.[105] The immunomodulatory efficacy of the methanolic extract was well depicted normalized the WBC levels, DTH levels, and antibody titer in a dose-dependent manner when used in both normal and cyclophosphamide-induced immunosuppressed mice.

1.6.2.2 APHRODISIAC AND ANTIDIABETIC ACTIVITY

Experiments conducted on experimental animals showed that administration of *C. orchioides* alcoholic extracts significantly changed their sexual behavior. Studies also showed that the aqueous extract of the plant increased the weight of reproductive organs and, also brought about significant changes in other sexual activities, such as latency of ejaculation.[68]

When compared with a standard antidiabetic drug glimepiride, the ethanolic and aqueous extracts of *C. orchioides* brought about the glucose-loaded and alloxan induced hyperglycemic conditions to normal levels in a dose-dependent manner.[2,21,65]

1.6.2.3 ANTIASTHMATIC, ESTROGENIC, AND ANTIOSTEOPOROTIC ACTIVITY

Studies conducted showed that the ethanolic extract of *C. orchioides* possessed estrogenic properties (normalized uterine functions) and prevented bone loss in ovarectomized experimental animals.[19,119] The ethanolic extract also reduced histamine-induced contraction in experimental animals thereby indicating its antiasthmatic activity.[82]

1.6.3 Gardenia gummifera

Phytochemical evaluation of the roots of *G. gummifera* showed that the ethyl acetate and methanolic extracts of the plant were rich in glycosides, phytosterols, resins, phenols, flavonoids, tannins, and triterpenes.[120]

1.6.3.1   ANTIOXIDANT AND ANTIULCER ACTIVITY

The methanol extract of *G. gummifera* Linn was evaluated for its antioxidant and antiulcer activities in aspirin-induced animal models and was compared with ranitidine standard. The plant extract reduced the complexities of ulcer in a cytoprotective fashion and normalized the antioxidant levels.[96]

1.6.3.2   CARDIOPROTECTIVE AND HEPATOPROTECTIVE ACTIVITIES

Isoproterenol was used to induce myocardial damage in experimental rats. The methanolic extract of *G. gummifera* reversed the damaged myocardial parameters in a dose-dependent manner when compared with ceruloplasmin used as standard.[87] Also, the efficacy of hepatoprotective activity of the various fractions of the methanolic extract of *G. gummifera* was compared with that of Silymarin and revealed compatible results.[95]

1.6.4   *Woodfordia fruticosa* Kurz

The qualitative phytochemicals screening of various solvent extracts of plant showed that the methanolic extract of *Woodfordia fruticosa* leaves contained tannins, flavonoids, anthraquinones, saponins, glycosides, and terpenoids. Metabolites, such as polyphenolics (flavonoids, tannins, terpenoids) and ascorbic acid were identified in organic solvent (ethyl acetate) extract of *W. fruticosa*. Methanol based extract of flowers of *W. fruticosa* was found to contain metabolites,such as polyphenolics, glycosides, reducing sugars, vitamin C, saponins, and anthraquinones.[32]

1.6.4.1   ANTIOXIDANT AND ANTIMICROBIAL ACTIVITIES

DPPH and ABTS assays conducted on *W. fruticosa* flowers showed their antioxidant potential. The antibacterial efficacy of the plant extracts tested upon 14 human pathogens revealed that the petroleum ether extract had exemplary antibacterial activity. However, the methanolic extract was most effective against Gram-negative bacteria and *Pseudomonas pseudoalcaligens* with efficiency as comparable as ciprofloxacin.[57,83]

1.6.4.2   ANTIULCER AND HEPATOPROTECTIVE ACTIVITIES

Ethanol, hydrochloric acid, and dichlofenac sodium-induced ulcer models were studied in experimental animals and *W. fruticosa* chloroform and methanolic extracts were efficient in reversing the damaged parameter levels to normal.[72] Similarly, studies conducted upon experimental animals with carbon tetrachloride, phenytoin, acetaminophen, and N' nitrosodiethyl amine induced liver cancer showed the liver protective activity of *W. fruticosa* plant extracts.[15,20,79]

1.7   PHYTOTHERAPEUTICAL COMPOUNDS: THEIR REPORTED ACTIVITIES

Detailed analysis of the most active extracts of the plants *A. campanulatus, C. orchioides, G. gummifera,* and *W. fruticosa* have been conducted in numerous laboratory studies. The extracts were subjected to a series of fractionation techniques both traditional and automated, to collect fractions. The fractions thus obtained were further scrutinized for their various bioactivities (such as, anticancer, anti-inflammatory, antimicrobial, antifungal, immunoprotective properties, etc.) to isolate active components. These active components could pave the way to the discovery of new drugs.

1.7.1   *Amorphophallus campanulatus*

The extracts of *Amorphophallus campanulatus* were found to have many phytochemicals, which have been studied for their activities.

Amblyone[137] is a triterpenoid isolated from the petroleum ether fraction of the ethanol extract of *A. campanulatus*. The compound showed profound antibacterial, antifungal, and cytotoxic activities.[48]

The 3'-5' diacetylambulin[138] is a flavonoid isolated from the ethanol extract of *A. campanulatus*. The compound inhibited the growth of some selected Gram-positive and Gram-negative bacteria.[49] Some other prominent compounds found in the extracts of *A. campanulatus* include hexadecanoic acid, linoleic acid, oleic acid, stigmasterol, 1,3,5-benzenetriol, vitamin E, and squalene derivatives.[10]

1.7.2   Curculigo orchioides

Curcculigo orchioides extracts have been widely scrutinized for the presence of biocompounds. Curculigosides are natural phenols.[139] Of the many curculigosides isolated from *C. orchioides* Gaertn., curculigoside-A was found to be useful against β-amyloid aggregation in Alzheimer's disease.[125]

Curculigenin-A (a triterpenoid sapogenin and curculigol, a cycloartane triterpene alcohol) are compounds isolated from *C. orchioides* Gaertn. These compounds were tested for their efficacy against thioacetamide and galactosamine-induced hepatotoxicity and were found to possess activity comparable with that of standard liver protectants[90]. curculigenin-B[140] and curculigenin-C are derivatives of curculigenin-A.

1.7.3   Gardenia gummifera

The plant extracts of *G. gummifera* were studied for bioconstituents along with the resin obtained from the plant, which revealed the presence of a large number of biocompounds. Dikamaliartane[141] is a cycloartane isolated from Dikamali resin, an exudate of *Gardenia*. The compound isolated showed significant anticancer activity in in vitro and in vivo systems and was comparable with cisplatin standard. It was also found to reduce locomotor activity and potentiate pentobarbitone- induced sleeping time in mice, thus depicting the ability to affect the central nervous system. The compound also showed potent anticonvulsant activity.[108,109] The major compounds identified in methanol extract of roots of *Gardenia gummifera* include compounds, such asvernolic acid, quininc acid derivatives, lupeol, erythrodiol, Asiatic acid, myricetin, oleanolic aldehyde, chlorogenic acid, and catechin derivatives.

1.7.4 *Woodfordia fruticosa*

Analysis conducted on the various extracts of *Woodfordia fruticosa* Kurz. showed that the plant was rich in tannin compounds. Woodfordins[142] are class of oligomeric hydrolyzable tannins isolated from Woodfordia sp. Woodfordin C (Woodfruticosin C) has been extensively studied and is reported to show antitumor activity. It is said to be a DNA topoisomerase II inhibitor.[46,130]

Ellagic acid[143] in small quantities has been reported in leaves and flowers of *W. fruticosa*.[145] The focus of research on ellagic acid is due to its reported anticancer activity. Ellagic acid is a component of many anticancer foods, such as pomegranate, and studies showed that the compound inhibited DNA binding of carcinogens, such as nitrosamines and polycyclic aromatic hydrocarbons. Some other important compounds isolated from *W. fruticosa* include gallic acid, oenothein B, isochimawalin B, and gentin, etc.

1.8 FUTURE PROSPECTS

From simple decoctions to personalized treatment, the result of the research is reflected clearly in the development and usage of medicines. Yet, the inspirations of many of the path-breaking discoveries have always been plants. And as a result, medicinal plants have become overexploited. This over exploitation of medicinal plants together with after effects of technological upgradation, such as industrialization, pollution, etc., cause destruction of habitats and loss of plant diversity. The most affected are usually endemic species of plants, which can become endangered. The future, therefore, is not just about bringing forth medicines with desirable quality but also about conservation of their source.

1.8.1 *LARGE-SCALE PRODUCTION OF PHYTOCOMPOUNDS*

Thinking beyond bioreactor-based systems and micropropagation, biotechnology has assisted overexpression of desired phytochemicals in plant parts and their isolation could hold the key. Hydroponic cultures can be designed in such a way that desired products can be harvested in a natural way on a large scale. This will also ensure that GM plants used for medicinal purposes do not interfere with the natural flora of an area.

1.8.2 EASY ISOLATION

Another important drawback of the pharmacognosy industry is the tedious procedures of isolation of desired compounds. The research focus should be aimed at devising techniques for easy, one-step isolation of such compounds.

1.8.3 FOOD BE THE MEDICINE

A number of phytochemicals used for medicinal purposes (e.g., gallic acid, ellagic acid, etc.) are naturally present in fruits and vegetables. However, the excessive use of chemical fertilizers, pesticides, etc., mask these properties and cause harm to our body. Organic cultivation needs to be encouraged to ensure the quality of phytochemicals.

1.9 SUMMARY

A large percentage of world population depends on plant-based traditional medicine or alternative medicine. With the increase in population, problems relating to pollution, radiation, etc., taunt man in his day to day life. Lack of effective nutrition, increased consumption of fast foods, and development of sedentary lifestyle has made man prone to variety of diseases. This when coupled with stress and strenuous conditions causes physiological damages to the human body. Also, the usage of synthetic drugs alleviates these conditions through various side effects.

Plants are the richest source of phytocompounds with definite physiological activities and almost no side effects. Utilization of these phytocompounds could hold the key to man's fight for survival. Though a lot of research is being done on plants, not many drugs are being formulated. The low yield of the compounds, seasonal variations, and possible loss of biodiversity may be the reasons for the decline in production and usage of plant-based drugs. Technological advancements in recent years are now being actively incorporated to plant-based research and this may act as a game changer in the field of pharmacognosy. Effective conservation measures for medicinal plants should go hand in hand with large scale production to ensure healthy exploitation of the most precious treasure of plants.

KEYWORDS

- *Amorphophallus campanulatus*
- antioxidant
- *Curculigo orchioides*
- curculigol
- curculigoside
- diacetylambulin

REFERENCES

1. Abdel-Salam, O. M. E.; Hamdy, S. M.; Seadawy, S. A. M.; Galal, A. F.; Abouelfadl, D. M.; Atrees, S. S. Effect of Piracetam, Vincamine, Vinpocetine, and Donezepil on Oxidative Stress and Neurodegeneration Induced by Aluminum Chloride in Rats. *Comp. Clin. Path.* **2016,** *25* (2), 305–318.

2. Anandakirouchanene, E.; Chandiran, I. S.; Kanimozhi, V.; Kadalmani, B. Antioxidant and Protective Effect of *Curculigo orchioides* on Liver, Pancreas and Kidney Tissue in Alloxan Induced Diabetic Experimental Rats. *Drug Invent. Today* **2013,** *5,* 192–200.

3. Ansil, P. N.; Wills, P. J.; Varun, R.; Latha, M. S. Cytotoxic and Apoptotic Activities of *Amorphophallus campanulatus* Tuber Extracts Against Human Colon Carcinoma Cell Lines HCT – 15. *Saudi J. Biol. Sci.* **2014,** *21,* 524–531.

4. Ansil, P. N.; Prabha, S. P.; Nitha, A.; Latha, M. S. Chemopreventive Effect of *Amarphophallus campanulatus* Blume Tuber Against Aberrant Crypt Foci and Cell Proliferation in 1,2–Dimethylhydrazine Induced Colon Carcinogenesis. *Asian Pac. J. Cancer Prev.* **2013,** *14* (9), 5331–5339.

5. Ansil, P. N.; Nitha, A.; Prabha, S. P.; Latha, M. S. Curative Effect of *Amorphophallus campanulatus* (Roxb.) Blume. Tuber on N'-nitrosodiethylamine Induced Hepatocellular Carcinoma in Rats. *J. Environ. Pathol. Toxicol. Oncol.* **2014,** *33* (3), 205–218.

6. Azevedo, M. I.; Pereira, A. F.; Nogueira, R. B.; Rolim, F. E.; Brito, A. C. G.; Wong, D. V. T.; Lima – Junior, R. C. P.; Ribeiro, R. A.; Vale, M. L. The Antioxidant Effects of the Flavonoids Rutin and Quercetin Inhibit Oxaliplatin—Induced Chronic Painful Peripheral Neuropathy. *Mol. Pain.* **2013,** *9,* 53.

7. Bafna, A. R.; Mishra, S. H. Immunostimulatory Effect of Methanol Extract of *Curculigo orchioides* on Immunosuppressed Mice. *J. Ethanopharmacol.* **2006,** *104* (1–2), 1–4.

8. Bais, S.; Singh, K.; Bigoniya, P.; Rana, A. The In Vitro Antioxidant and Free Radical Scavenging Activities of Suran (*Amarphophallus campanulatus,* Araceae) Tuber Extracts. *Int. J. Pharm. Life Sci.* **2011,** *2* (1), 1315–1324.

9. Bar, K. J.; Boettger, M. K.; Siedler, N.; Mentzel, H. J.; Terborg, C.; Sauer, H. Influence of Galanthamine on vasomotor reactivity in Alzheimer's Disease and Vascular Dementia Due to Cerebral Microangiopathy. *Stroke* **2007,** *38,* 3186–3192.

10. Basu, S.; Chaoudhury, U. R.; Das, M.; Dutta, G. Identification of Bioactive Components from the Ethanolic and Aqueous Extract of *Amorphophallus campanulatus* Tuber by GC-MS Analysis. *Int. J. Phytomed.* **2013,** *5,* 243–251.

11. Basu, S.; Das, M.; Datta, G. Protective Activity of Ethanolic Extract of *Amarphophallus campanulatus* Against Ethanol Induced Hepatotoxicity in Rats. *Int. J. Pharm. Pharm. Sci.* **2013,** *5* (2), 411–417.

12. Behera, A.; Kumar, S.; Jena, P. K. Review on *Amorphophallus* species: Important Medicinal Wild Food Crops of Odisha. *Int. J. Pharm. Life Sci.* **2014,** *5* (5), 3512–3516.

13. Bouchereau, A.; Guenot, P.; Larher, F. Analysis of amines of plant materials. *J. Chromatogr. B Biomed. Sci. Appl.* **2000,** *747* (1–2), 49–67.

14. Brams, M.; Pandya, A.; Kuzmin, D.; Elk, R. V.; Krijnen, L.; Yakel, J. L.; Tsetlin, V.; Smit, A. B.; Ulens, C. Structural and Mutagenic Blueprint for Molecular Recognition Of Strychnine and d-Tubocurarine by Different Cys—Loop Receptors. *PLoS Biol.* **2011,** *9* (3), e1001034.

15. Brinda, D.; Geetha, R. Evaluation of the protective efficacy of *Woodfordia fruticosa* on phenytoin induced liver damage in rats. *J Cell Tissue Res.,* **2009,** *9* (3), 1981–1984.

16. Buchhalter, A. R.; Fant, R. V.; Henningfield, J. E. Novel Pharmacological Approaches for Treating Tobacco Dependence and Withdrawal: Current Status. *Drugs* **2008,** *68* (8), 1067–1088.

17. Budzianowski J.; Pakulski, G.; Robak, J. Studies on Antioxidative Activity of Some C-glycosyl Flavones. *Pol. J. Pharmacol. Pharm.* **1991,** *43* (5), 395–401.

18. Burman, M. S. Therapeutic Use of Curare And Erythroidine Hydrochloride for Spastic and Dystonic States. *Arch. Neurol. Psychiatry* **1939,** *41,* 307–327.

19. Cao, D. P.; Zheng, Y. N.; Qin, L. P.; Han, T.; Zhang, H.; Rahman, K.; Zhang, Q. Y. *Curculigo orchioides* Prevents Bone Loss in Ovarectomised Rats. *Maturitas.* **2008,** *59* (4), 373–380.

20. Chandan, B. K.; Saxena, A. K.; Shukla, S.; Sharma, N., Gupta, D. K.; Singh, K.; Suri, J. Hepatoprotective Activity of *Woodfordia fruticosa* Kurz Flowers Against Carbon Tetrachloride Induced Hepatotoxicity. *J. Ethnopharmcol.* **2008,** *119* (2), 218–224.

21. Chauhan, N. S.; Dixit, V. K. Antihyperglycemic Activity of the Ethanolic Extract of *Curculigo orchioides* Gaertn. *Pharmacog. Mag.* **2007,** *3* (12), 236–239.

22. Chauhan, N. S.; Sharma, V.; Thakur, M.; Dixit, V. K. *Curculigo orchioides*: The Black Gold with Numerous Benefits. *J. Chinese Integr. Med.* **2010,** *8* (7), 613–623.

23. Cushnie, T. T. P.; Cushnie, B.; Lamb, A. J. Alkaloids: An Overview of Their Antibacterial, Antibiotic – Enhancing and Antivirulence Activities. *Int. J. Antimicrob. Agents.* **2014,** *44,* 377–386.

24. Desai, D. C.; Jacob, J.; Almeida, A.; Kshirsagar, R.; Manju, S. L. Isolation, Structural Elucidation, and Anti-inflammatory Activity of Astragalin, (-)Hinokinin, Aristolactam I and Aristolochic Acids (I & II) from *Aristolochia indica. Nat. Prod. Res.* **2014,** *28* (17), 1413–1417.

25. Duangyod, T.; Palanuvej, C.; Ruangrungsi, N. The (+)- Catechin and (-)- Epicatechin Contents and Antioxidant Activity of Commercial Black Catechu and Pale Catechu. *J. Chem. Pharm. Res.* **2014,** *6* (7), 2225–2232.

26. Elbaz, H. A.; Steuckle, T. A.; Tse, W.; Rojanasakul, Y.; Dinu, C. Z. Digitoxin and Its Analogs as Novel Cancer Therapeutics. *Exp. Hematol. Oncol.* **2012,** *1,* 4–8.

27. Eldahshan, O. A.; Elsakka, A. M. A.; Singab, A. N. Medicinal Plants and Addiction Treatment. *Med. Aromat. Plants.* **2016,** *5* (4), 260.

28. Firdouse, S.; Alam, P.; Basra, R.; Amreen, A.; Firdouse, N. Antioxidant Activity of *Amarphophallus campanulatus* Tubers (Roxb) Blume. *Int. J. Pharm. Pharm. Sci.* **2012,** *4* (5), 449–451.

29. Foster, H. D.; Hoffer, A. The two faces of L-DOPA: Benefits and Adverse Side Effects in the Treatment of Encephalitis Lethargic, Parkinson's Disease, Multiple Sclerosis and Amyotrophic Lateral Sclerosis. *Med. Hypotheses* **2004,** *62* (2), 177–181.

30. Gajare, M. S. *Amorphophallus campanulatus*: Review of Medicinal Properties. *Pharm. Sci. Monitor.* **2014,** *5* (3), 122–130.

31. Gao, J. M.; Kamnaing, P.; Kiyota, T.; Watchueng, J.; Kubo, T.; Jarussophon, S.; Konishi, Y. One step Purification of Palmatine and its Derivative dl—Tetrahydropalmatine from *Enantia chlorantha* using High Performance Displacement Chromatography (HPDC). *J. Chromatograph A* **2008,** *1208*, 47–53.

32. Grover, N.; Patni, V. Phytochemical Charaterization Using Various Solvent Extracts and GC-MS Analysis of Methanolic Extract of *Woodfordia fruticosa* Leaves. *Int. J. Pharm. Pharm. Sci.* **2013,** *5* (4), 291–295.

33. Gorman, A. L. F.; Marmor, M. F. Long Term Effect of Ouabain and Sodium Pump Inhibition on a Neuronal Membrane. *J. Physiol.* **1974,** *242*(1), 49–60.

34. Gulshan, C.; Manoj, D. K. Role of Herbs in Preventing Cancer. *Int. Res. J. Pharm.* **2014,** *5* (4), 264–266.

35. Hajdu, Z.; Hohman, J.; Forgo, P.; Martinek, T.; Devarics, M.; Zupko, I.; Falkay, G.; Cossuta, D.; Mathe, I. Diterpenoids and Flavonoids from the Fruits of *Vitex agnus— castus* and Antioxidant Activity of the Fruit Extracts and Their Constituents. *Phytother. Res.* **2007,** *21* (4), 391–394.

36. Hassan, S.; Mathesius, U. The Role of Flavonoids in Root Rhizosphere Signaling: Opportunities and Challenges for Improving Plant-Microbe Interactions. *J. Exp. Bot.* **2012,** E-article at: DOI:10.1093/jxb/err430.

37. He, Y. H.; Xiao, C.; Wang, Y. S.; Zhao, L. H.; Zhao, H. Y.; Tong, Y.; Zhou, J.; Jia, H. W.; Lu, C.; Li, X. M.; Lu, A. P. Antioxidant and Anti-inflammatory Effects of Cyanidin from Cherries on Rat Adjuvant-induced Arthritis. *Zhongguo Zhong Yao Za Zhi.* **2005,** *30* (20), 1602–1605.

38. Heishman, S. J.; Kleykamp, B. A.; Singleton, E. G. Meta-analysis of the Acute Effects of Nicotine and Smoking on Human Performance. *Psychopharmacology (Berl).* **2010,** *210* (4), 453–469.

39. Helke, C. J.; Yuhaniak, P. A.; Kellar, K. J.; Gillis, R. A. Effect of Deslanoside on Brain and Spinal Cord Levels of Serotonin and 5-hydroxy-Indoleacetic Acid and Tryptophan Hydroxylase Activity. *Biochem. Pharmacol.* **1978** *27* (20), 2459–2461.

40. Ho, P. C.; Saville, D. J. Inhibition of Human CYP3A4 Activity by Grapefruit Flavonoids, Furanocoumarins and Related Compounds. *J. Pharm. Pharm. Sci.* **2001,** *4* (3), 217–227.

41. Huang, W. W.; Xu, S. Z.; Xu, Y. Q. Study of the Antiarrhythmic Mechanism of Berberine on Delayed Activation Potassium Current by Voltage Clamp. *Zhonghua Xin Xue Guan Bing Za Zhi (Chinese J. Cardiov. Dis.).* **1992,** *20* (5), 310–312.

42. Huang, Y. L.; Kou, J. P.; Ma, L.; Song, J. X.; Yu, B. Y. Possible Mechanism of the Anti-inflammatory Activity of Ruscogenin: Role of Intercellular Adhesion Molecule-1 and Nuclear Factor-kappaB. *J. Pharmacol. Sci.* **2008,** *108* (2), 198–205.

43. Joy, P. P.; Thomas, J.; Samuel, M.; Skaria, B. P. *Curculigo orchioides*: Plant for Health Care. *Ind. J. Areca nut, Spices Med Plants* **2004,** *6* (4), 131–134.

44. Jung, C. H.; Kim, Y.; Kim, M. S.; Lee, S.; Yoo, S. H. The Establishment of Efficient Bioconversion, Extraction, and Isolation Processes for the Production of Phyllodulcin, a Potential High Intensity Sweetener from Sweet Hydrangea Leaves (*Hydrangea macrophylla* Thunbergii). *Phytochem. Anal.* **2016,** *27* (2), 140–147.

45. Kachooei, E.; Moosavi-Movahedi, A. A.; Khodagholi, F.; Mozaffarian, F.; Sadeghi, P.; Hadi-Alijanvand, H.; Ghasemi, A.; Saboury, A. A.; Farhadi, M.; Sheibani, N. Inhibition Study on Insulin Fibrillation and Cytotoxicity by Paclitaxel. *J. Biochem.* **2014,** *155* (6), 361–373.

46. Kadota, S.; Takamori, Y., Nyein, K. N.; Kikuchi, T.; Tanaka, K.; Ekimoto, H. Constituents of the Leaves of *Woodfordia fruticosa*: Isolation, Structure, and Proton and Carbon-13 Nuclear Magnetic Resonance Signal Assignments of Woodfruticosin (woodfordin C), an Inhibitor of Deoxyribonucleic Acid Topoisomerase II. *Chem. Pharm. Bull.* (Tokyo) **1990,** *38* (10), 2687–2697.

47. Keeler, M. H.; Kane, F. J. The Use of Hyoscyamine as a Hallucinogen and Intoxicant. *Psychiatry Online* **1967,** *124* (6), 852–854.

48. Khan, A.; Rahman, M.; Islam, M. Antibacterial, Antifungal and Cytotoxic Activities of Amblyone Isolated from *Amarphophallus campanulatus*. *Ind. J. Pharmcol.* **2008,** *40* (1), 41–44.

49. Khan, A.; Rahman, M.; Islam, M. Antibacterial, Antifungal and Cytotoxic Activities of 3,5-Diacetylambulin Isolated from *Amarphophallus campanulatus* Roxb. Bl Decne. *DARU J. Pharm. Sci.* **2008,** *16* (4), 239–244.

50. Khanbabee, K.; van Ree, T. Tannins: Classification and Definition. *Nat. Prod. Rep.* **2001,** *18*, 641–649.

51. Khare, C. P. *Encyclopaedia of Indian Medicinal Plants.* Springer-Verlag: Berling/Heidelberg, New York, 2007; pp 483–484.

52. Kharkwal, H.; Panthari, P.; Pant, M. K.; Kharkwal, H.; Kharkwal, A. C.; Joshi, D. D. Foaming Glycosides: A Review. *IOSR J. Pharm.* **2012,** *2* (5), 23–28.

53. Kirtikar, K. R.; Basu, B. D. *Indian Medicinal Plants.* Lalit Mohan Basu: Allahabad, India, 1989; Vol. 4; pp 2609–2610.

54. Kiyohiro, H; Takeshi, T. Mitotic Slippage Underlies the Relationship Between p53 Dysfunction and the Induction of Large Micronuclei by Colcemid. *Mutagenesis* **2013,** *28* (4), 457–464.

55. Kumar, D.; Sharma, M.; Sorout, A.; Saroha, K.; Verma, S. *Woodfordia fruticosa* Kurz. Review of its Botany, Chemistry, and Biological Activities. *J. Pharmacogn. Phytochem.* **2016,** *5* (3), 293–298.

56. Kumar, S.; Pandey, A. K. Chemistry and Biological Activities of Flavonoids: An Overview. *Sci. World J.* **2013,** e-article: DOI:10.1155/2013/162750.

57. Kumaraswamy, M. V.; Kavitha, H. S. Antibacterial Extracts of *Woodfordia fruticosa* Kurz on Human Pathogens. *World J. Med. Sci.* **2008,** *3* (2), 93–96.

58. Lahlou, M. The Success of Natural Products in Drug Discovery. *Pharmacol. Pharm.* **2013,** *4*, 17–31.

59. Lahlou, M. Screening of Natural Products for Drug Discovery. *Expert Opin. Drug Discov.* **2007,** *2* (5), 697–705.

60. Leeuw, J-de.; Assen, Y. J.; Bjerring, P.; Neumann, H. A. Treatment of Vitiligo with Khellin Liposomes, UV Light and Blister Roof Transplantaion. *J. Eur. Acad. Dermatol. Venereol.* **2011,** *25,* 74–81.

61. Leite, M. C.; de Brito Bezzera, A. P.; de Souza, J. P.; de Oliveira, L. M. E. Investigating the Antifungal Activity and Mechanism(s) of Geraniol Against *Candida albicans* Strains. *Med. Mycol.* **2015,** *53* (3), 275–284.

62. Li, Y.; Zuo, G. Y. Advances in Studies on Antimicrobial Activities of Alkaloids. *Chin. Tradit. Herb. Drugs* **2010,** *41*(6), 1006–1014.

63. Lobay, D. S. Rauwolfia in the Treatment of Hypertension. *Integr. Med. Clin. J.* **2015,** *14* (3), 40–46.

64. Lopez-Lazaro, M. Distribution and Biological Activities of the Flavonoid Luteolin. *Mini. Rev. Med. Chem.* **2009,** *9* (1), 31–59.

65. Madhavan, V.; Joshi, R.; Murali, A.; Yoganarasimhan, S. N. Antidiabetic Activity of *Curculigo orchioides* Root Tuber. *Pharm. Biol.* **2007,** *45* (1), 18–21.

66. Maitreya, B. B. An Overview of Ethnomedicinal Plants of Family Rubiaceae from Sabarmati River, India. *Int. J. Pharm. Life Sci.* **2015,** *6* (5), 4476–4480.

67. Martinez-Pinilla, E.; Onatibia-Astibia, A.; Franco, R. The Relevance of Theobromine for the Beneficial Effects of Cocoa Consumption. *Front. Pharmacol.* **2015,** *6* (30), 1–5.

68. Mehta, J.; Nama, K. S. A review on Ethanomedicines of *Curculigo orchioides* Gaertn. (*Kali Musli*): Black Gold. *Int. J. Pharm. Biomed. Res.* **2014,** *1* (1), 12–16.

69. Mehta, M.; Abdu, A.; Sabbagh, M. New Acetylcholinesterase Inhibitors for Alzheimer's Disease. *Int. J. Alzheimer's Dis.* **2012,** E-article at: http://dx.doi.org/10.1155/2012/72898.

70. Meine, T. J.; Roe, M. T.; Chen, A. Y.; Patel, M. R.; Washam, J. B.; Ohman, E. M.; Peacock, W. F.; Pollack, C. V.; Gibler, W. B.; Peterson, E. D. Association of Intravenous Morphine Use and Outcomes in Acute Coronary Syndromes: Results from the CRUSADE Quality Improvement Initiative. *Am. Heart J.* **2005,** *149* (6), 1043–1049.

71. Meti, M.; Nandibewoor, S.; Chimatadar, S. Spectroscopic investigation and Oxidation of the Anticholinergic Drug Atropine Sulfate Monohydrate by Hexacyanoferrate (III) in Aqueous Alkaline Media: A Mechanistic Approach. *Turk. J. Chem.* **2014,** *38,* 477–487.

72. Mihira, V.; Ramana, K. V.; Ramakrishna, S.; Rambabu, P. Evaluation of Anti-ulcer Activity of *Woodfordia fruticosa* roots. *Int. J. Advances Pharm. Sci.* **2011,** *2,* 2–3.

73. Miles, S. L.; McFarland, M.; Niles, R. M. Molecular and Physiological Actions of Quercetin: Need for Clinical Trials to Assess its Benefits in Human Disease. *Nutr. Rev.* **2014,** *72* (11), 720–734.

74. Misra, R. S.; Sriram, S. Medicinal Value and Export Potential of Tropical Tuber Crops. *Recent Prog. Med. Plants, Crop Improv. Commer.* **2001,** *5,* 317–325.

75. Namratha, K.; Shenai, P.; Chatra, L.; Rao, P. K.; Veena K. M.; Prabhu, R. V. Antioxidant and Anticancer Effects of Curcumin—A Review. *J. Contemp. Med.* **2013,** *3* (2), 136–143.

76. Nayeem, M.; Asdaq, S. M. B.; Salem, H.; AHEl-Alfqy, S. Gallic Acid: A Promising Lead Molecule for Drug Development. *J. App. Pharm.* **2016**, *8*, 213–217.

77. Newland, C. F.; Cull-Candy, S. G. On the Mechanism of Action of Picrotoxin on GABA Receptor Channels in Dissociated Sympathetic Neurons of the Rat. *J. Physiol.* **1992**, *447*, 191–213.

78. Nie, Y.; Dong, X.; He, Y. J.; Yuan, T. T.; Rahman, K.; Qin, L. P.; Zhang, Q. Y. Medicinal Plants of Genus Curculigo: Traditional Uses and a Phytochemical and Ethnopharmacological Review. *J. Ethnopharmacol.* **2013**, *147*, 547–563.

79. Nitha, A.; Ansil, P. N.; Prabha, S. P.; Latha, M. S. Curative Effect of *Woodfordia fruticosa* Kurz Flowers on N'-Nitrosodiethylamine Induced Hepatocellular Carcinoma. *Int. J. Pharm. Pharm. Sci.* **2014**, *6* (2), 150–155.

80. Onning, G.; Asp, N. G. Effect of Oat Saponins on Plasma and Liver Lipids in Gerbils (*Meriones unguiculatus*) and Rats. *Br. J. Nutr.* **1995**, *73*, 275–286.

81. Pandey, D.; Gupta, A. K. Screening of the Antibacterial Activity of *Amorphophallus campanulatus* from Baster Region of Chhattisgarh, India. *Int. J. App. Bio. Pharm. Technol.* **2013**, *4* (4), 1–6.

82. Pandit, P.; Singh, A.; Bafna, A. R.; Kadam, P. V.; Patil, M. J. Evaluation of Anti-asthmatic Activity of *Curculigo orchioides* Gaertn. Rhizomes. *Indian J. Pharm. Sci.* **2008**, *70* (4), 440–444.

83. Parekh, J.; Chanda, S. In vitro Antibacterial Activity of the Crude Methanol Extract of *Woodfordia fruticosa* Flower. *Braz. J. Microbiol.* **2007**, *38*, 204–207.

84. Patel, D. K.; Patel, K.; Kumar, R.; Gadewar, M.; Tahilyani, V. Pharmacological and Analytical Aspects of Bergenin: A Concise Report. *Asian Pac. J. Trop. Dis.* **2012**, 163–167.

85. Patnaik, N. *The Garden of Life: An Introduction to the healing plants of India.* Doubleday Publishers: Madison **1993**; pages 314.

86. Patwardhan, B.; Vaidya, A. D. B.; Chorghade, M. Ayurveda and Natural Products Drug Discovery. *Curr. Sci.* **2004**, *86* (6), 789–799.

87. Prabha, S. P.; Ansil, P. N.; Nitha, A.; Latha, M. S. Cardioprotective Effect of Methanolic Extract of *Gardenia gummifera* Linn. F Against Isoproterenol Induced Myocardial Infarction in Rats. *Int. J. Pharm. Sci. Res.* **2014**, *5* (9), 3817–3828.

88. Prassas, I.; Diamandis, E. P. Novel Therapeutic Applications of Cardiac Glycosides. *Nat. Rev. Drug Discov.* **2008**, *7*, 926–935.

89. Prommer, E. Role of Codeine in Palliative Care. *J. Opioid Manag.* **2010**, *7* (5), 401–406.

90. Rao K. S; Mishra S. H. Studies on *Curculigo orchioides* Gaertn. For Anti-inflammatory and Hepatoprotective Activities. *Indian Drugs* **1996**, *33*(1), 20–25.

91. Rao, K. S.; Mishra, S. H. Effect of rhizomes of *Curculigo orchioides* Gaertn. on Drug Induced Hepatoxicity. *Indian Drugs* **1996**, *33* (9), 458–461.

92. Rasool Hassan, B. A. Medicinal Plants (importance and uses). *Pharm. Anal. Acta.* **2012**, *3* (10), 1000e139

93. Ravi, V.; Ravindran, C. S.; Suja, G.; George, J.; Nedunchezhiyan, M.; Byju, G.; Naskar, S. K. Crop Physiology of Elephant Foot Yam (*Amorphophallus paeoniifolius*). *Adv. Hort. Sci.* **2011**, *25* (1), 51–63.

94. Rodriguez-Diaz, M.; Delporte, C.; Cartagena, C.; Cassels, B. K.; Gonzalez, P.; Silva, X.; Leon, F.; Wessjohann, L. A. Topical Anti-inflammatory Activity Of Quillaic Acid

from *Quillaja saponaria* and Some Derivatives. *J. Pharm. Pharmacol.* **2011**, *63* (5), 718–724.

95. Sabbani, P. K.; Chityala, P. K.; Gowrishankar, N. L.; Naveen Kumar, G.; Shilpa, K.; Tejaswi C. H. Evaluation of Antioxidant and Antiulcer Activity of Methanolic Extracts of *Gardenia gummifera* L. in Experimental Rats. *Asian J. Pharm. Clin. Res.* **2015**, *8* (4), 41–44.

96. Sabbani, P. K.; Thatipelli, R. C.; Surampalli, G.; Duvvala, P. Evaluation of Hepatoprotective Activity with Different Fractions of *Gardenia gummifera* Linn. on Paracetamol Induced Liver Damage in Rats. *J. Drug Metab. Toxicol.* **2016**, *7* (1), 1–7.

97. Schmidt, B.; Ribnicky, D. M.; Poulev, A.; Logendra, S.; Cefalu, W. T.; Raskin, I. A Natural History of Botanical Therapeutics. *Metab. Clin. Exp.* **2008**, *57* (1), S3–S9.

98. Senges, J.; Ehe, L. Antiarrhythmic Action of Sparteine on Direct and Indirect Models of Cardiac Fibrillation. *Naunyn – Schmiedeberg's Arch. Pharmacol.* **1973**, *280* (3), 265–274.

99. Serafini, G.; Pompili, M.; Innamorati, M.; Giordano, G.; Tatarelli, R.; Lester, D.; Girardi, P.; Dwivedi, Y. Glycosides, Depression and Suicidal Behavior: The Role of Glycoside—Linked Proteins. *Molecules* **2011**, *16*, 2688–2713.

100. Shen, S.; Zhang, Y.; Zhang, R.; Gong, X. Sarsasapogenin Induces Apoptosis via the Reactive Oxygen Species-mediated Mitochondrial Pathway and ER Stress Pathway in HeLa Cells. *Biochem. Biophys Res. Commun.* **2013**, *441* (2), 519–524.

101. Shenoi, S. D.; Prabhu, S. Photochemotherapy (PUVA) in Psoriasis and Vitiligo. *Indian J. Dermatol. Venereol. Leprol.* **2014**, *80* (6), 498–504.

102. Shuaibu, M. N.; Pandey, P. A.; Wuyep, T.; Yanagi, K.; Hirayama, A.; Ichinose, T.; Tanaka, T.; Kouno, I. Castalagin from *Anogeissus leiocarpus* Mediates the Killing of *Leishmania* In Vitro. *Parasitol. Res.* **2008**, *103* (6), 1333–1338.

103. Singh, A.; Srivastava, K. C.; Banerjee, N.; Wadhwa, N. Phytochemical Analysis of the Peel of *Amorphophallus paeniifolius*. *Int. J. Pharm. Bio. Sci.* **2013**, *4* (3), 810–815.

104. Singh, A.; Wadhwa, N. A Review on Multiple Potential of Aroid—*Amorphophallus paeoniifolius*. *Int. J. Pharm. Sci. Rev. Res.* **2014**, *24* (1), 55–60.

105. Singh, R.; Gupta, A. K. Antimicrobial and Antitumor Activity of the Fractionated Extracts of *Kalimusli* (*Curculigo orchioides*). *Int. J. Green Pharm.* **2008**, *2* (1), 34–36.

106. Sirtori, C. R. Aescin–Pharmacology, Pharmacokinetics, and Therapeutic Profile. *Pharmacol. Res.* **2001**, *44* (3), 183–193.

107. Speicher, L. A.; Barone, L.; Tew, K. D. Combined Antimicrotubule Activity of Estramustine and Taxol in Human Prostatic Carcinoma Cell Lines. *Cancer Res.* **1992**, *52*, 4433–4440.

108. Sridhar, G.; Harikiran, L.; Rao, A. V. N.; Reddy, Y. N. Pharmacological Screening of Dikamaliartane A, A Cycloartane Isolated from Gum Resin Dikamali. *Int. J. Appl. Biol. Techn.* **2011**, *2* (4), 1–7.

109. Sridhar, G.; Reddy, N. R. A.; Rao, A. V. N.; Reddy, Y. N. Evaluation of Anticancer Activity of Dikamaliartane – A Cycloartane Isolated from Dikamali, a Gum Resin. *Int, J. Pharm. Pharm. Sci.* **2012**, *4* (4), 501–504.

110. Srivastava, S.; Verma, D.; Srivastava, A.; Tiwari, S. S.; Dixit, B.; Singh, R. S.; Rawat, A. K. S. Phytochemical and Nutritional Evaluation of *Amorphophallus campanulatus* (Roxb.) Blume Corm. *J. Nutr. Food Sci.* **2014**, *4*, 274; E-article at: DOI: 10.4172/2155-9600.1000274.

111. Syed, Y. H.; Khan, M.; Bhuvaneshwari, J. J. A. Phytochemical Investigation and Standardization of Extracts of Flowers of *Woodfordia fruticosa*; A Preliminary Study. *J. Pharm. Biosci.* **2013**, *4*, 134–140.

112. Tantai, J. C.; Zhang, Y.; Zhao, H. Heterophyllin B Inhibits the Adhesion and Invasion of ECA-109 Human Esophageal Carcinoma cells by Targeting PI3K/AKT/β-Catenin Signaling. *Mol. Med. Reports.* **2016**, *13*, 1097–1104.

113. Thakur, A.; Singla, R.; Jaitak, V. Coumarins as Anticancer Agents: A Review on Synthetic Strategies, Mechanism of Action and SAR Studies. *Eur. J. Med. Chem.* **2015**, *101*, 476–495.

114. Thoppil, R. J.; Bishayee, A. Terpenoids as Potential Chemopreventive and Therapeutic Agents in Liver Cancer. *World J. Hepatol.* **2011**, *3* (9), 228–249.

115. Thompson, T. M.; Toerne, T.; Erickson, T. B. Salicylate Toxicity from Genital Exposure to a Methylsalicylate–containing Rubefacient. *West J. Emerg. Med.* **2016**, *17* (2), 181–183.

116. Vanachayangkul, P.; Chow, N.; Khan, S. R.; Butterweck, V. Prevention of Renal Crystal Deposition by an Extract of Ammi visnaga L. and its Constituents Khellin and Visnagin in Hyperoxaluric Rats. *Urol. Res.* **2011**, *39* (3), 189–195.

117. Vargas-Mendoza, N.; Madrigal-Santillan, E.; Moralez-Gonzalez, A.; Esquivel-Soto, J.; Esquivel-Chirino, C.; Garcia-Luna, M.; Gayosso–de-Lucio, J. A.; Morales-Gonzalez, J. A. Hepatoprotective Effect of Silymarin. *World J. Hepatol.* **2014**, *6* (3), 144–49.

118. Venugopala, K. N.; Rashmi, V.; Odhav, B. Review on Natural Coumarin Lead Compounds for their Pharmacological Activity. *BioMed Res. Int.* **2013**, 963248.

119. Vijayanarayana, K.; Rodrigues, R. S.; Chandrashekhar, K. S.; Subramanyam, E. V. Evaluation of Estrogenic Activity of Alcoholic Extract of Rhizomes of *Curculigo orchioides.* *J. Ethanopharmacol.* **2007**, *114* (2), 241–245.

120. Vindhya, K.; Sampath Kumara, K. K.; Neelambika, H. S.; Leelavathi, S. Preliminary Phytochemical Screening of *Gardenia latifolia* Ait. and *Gardenia gummifera* Linn. *Res. J. Pharm. Biol. Chem. Sci.* **2014**, *5*(2), 527–532.

121. Virgadamo, S.; Chamigo, R.; Darrat, Y.; Morales, G.; Elayi, C. S. Digoxin: A Systematic Review in Atrial Fibrillation, Congestive Heart Failure and Post Myocardial Infarction. *World J. Cardiol.* **2015**, *7* (11), 808–816.

122. Walker, N.; Howe, C.; Glover, M.; McRobbie, H.; Barnes, J.; Nosa, V.; Parag, V.; Bassett, B.; Bullen, C. Cytisine versus Nicotine for Smoking Cessation. *New Eng. J. Med.* **2014**, *371* (25), 2353–2362.

123. Wilps, H.; Kirkilionis, E.; Muschenich, K. The Effects of Neem Oil and Azadirachtin on Mortality, Flight Activity, and Energy Metabolism of Schistocerca gregaria forskal—A Comparison Between Laboratory and Field Locusts. *Compar. Pharmacol.* **1992**, *102* (1), 67–71.

124. Wu, G. Functional Amino Acids in Growth, Reproduction and Health. *Adv. Nutr.* **2010**, *1*, 31–37.

125. Xiu-Ying, W.; Li, J. Z.; Guo, J. Z.; Hou, B. Y. Ameliorative Effects of Curculigoside from *Curculigo orchioides* Gaertn. on Learning and Memory in Aged Rats. *Molecules* **2012**, *17*, 10108–118.

126. Yan, M.; Fan, P.; Shi, Y.; Feng, L.; Wang, J.; Zhan, G.; Li, B. Stereoselective Blockage of Quinidine and Quinine in the hERG Channel and the Effect of Their

Rescue Potency on Drug Induced hERG Trafficking Defect. *Int. J. Mol. Sci.* **2016,** *17* (10), 1648–1652.

127. Yang, S. Y.; Kim, N. H.; Cho, Y. S.; Lee, H.; Kwon, H. J. Convallotoxin, a Dual Inducer of Autophagy and Apoptosis, Inhibits Angiogenesis In Vitro and In Vivo. *PLOS One* **2014,** *9* (3), E-article: 91904.

128. Yang, Y.; Xiu, J.; Liu, J.; Zhang, L.; Li, X.; Xu, Y.; Qin, C.; Zhang, L. Chebulagic Acid, a Hydrolyzable Tannin, Exhibited Antiviral Activity In Vitro and In Vivo Against Human Enterovirus 71. *Int. J. Mol. Sci.* **2013,** *14*, 9618–9627.

129. Yasui, S.; Fujiwara, K.; Tawada, A.; Fukuda, Y.; Nakano, M.; Yokosuka, O. Efficacy of Intravenous Glycyrrhizin in the Early Stage of Acute Onset Autoimmune Hepatitis. *Dig. Dis. and Sci.* **2011,** *56* (12), 3638–47.

130. Yoshida, T.; Chou, T.; Nitta, A.; Miyamoto, K. I.; Koshiura, R.; Okuda, T. Wood-fordin-C, a Macro-ring Hydrolyzable Tannin Dimer with Antitumor Activity, and Accompanying Dimers from *Woodfordia fruticosa* Flowers. *Chem. Pharm. Bull.* **1990,** *38*, 1211–1217.

131. Zhang, J. S.; Liu, W. Y. The Mechanism of Action of Trichosanthin on Eukaryotic Ribosomes – RNA N Glycosidase Activity of the Cytotoxin. *Nucleic Acids Res.* **1992,** *20* (6), 1271–75.

132. Zucker, S.; Buttle, D. J.; Nicklin, M. J.; Barrett, A. J. The proteolytic activities of Chymopapain, Papain and Papaya Proteinase III. *Biochim Biophys Acta.* **1985,** *828* (2), 196–204.

CHAPTER 2

PLANT-BASED SECONDARY METABOLITES FOR HEALTH BENEFITS: CLASSIFICATION AND PROCESSING

MONIKA SHARMA, JYOTSANA DWIVEDI, BHANU KUMAR, BRAMHANAND SINGH, and A. K. S. RAWAT

ABSTRACT

This chapter presents the classification of che mically diverse and medicinally potential bioactive compounds. Plants are sources of a wide range of natural products in the form of health-promoting secondary metabolites with attributes contributing to preventive and therapeutic medication for diseases. Quality control of herbal drugs is also an important requirement to ensure a high yield of secondary metabolites from plants along with accurate identification and standardization of potential biocompounds. Based on their chemistry, bioactive compounds are categorized as alkaloids, glycosides, lignans, flavones, flavan-3-ols, isoflavones, flavanones, flavonols, anthocyanidins, salicylates, glucosinolates, stanols, and sterols. These secondary metabolites are absorbed and metabolized in the biological system to exert a specific pharmacological effect. Thus, various studies on the pharmacological activity of these bioactive compounds have been conducted at preclinical level (in vitro and in vivo) for establishing a detailed pharmacological profile. Potential mechanisms of action have also been explored for ROS scavenging potential, a hypolipidemic effect for reduction of circulating LDL, antiplatelet aggregation effect antitumor effect and anti-inflammatory effects.

2.1 INTRODUCTION

Research based on natural products focuses on the central idea of investigating chemical and biological aspects including biosynthetic pathway, chemical properties, and biological screening of secondary metabolites from natural resources. Thus, natural product can be redefined as a biologically active chemical entity of natural origin either manufactured as a by-product through living organisms or derived from the plants as plant bioactive.[52,76] "Natural product" is generally more related to "secondary metabolite" that is produced by the plants. Now, secondary metabolites are such chemical entities having the exact chiral configuration required to exercise the pharmacological effect in biological systems, irrespective of their contribution in any "primary" function related to the reproduction, growth, and development of an organism. The natural products have been extensively exploited as pharmaceutical drugs, cosmeceutical ingredients, functional foods, and dietary supplements thereby utilized as nutraceuticals.[67,68]

For developing these products based on natural sources: Identification of source, isolation of bioactive compounds, and characterization of natural products are important aspects based on scientific details of pharmacognosy. On broad perspective, natural source of the bioactive compounds can be from all biological kingdoms of plants and animals, and markedly they can be plants, fungi, marine invertebrates, and bacteria.[18] On the basis of source of natural products, these are broadly grouped in four categories,[18] such as

- natural products from plant origin,
- natural products from microorganisms,
- natural products from marines, and
- natural products from animal sources.

This chapter presents detailed classification of chemically diverse and medicinally potential bioactive compounds. It also delivers an insight into the processing of plant material and a brief discussion on the associated toxicity concerns with their use and present status of phytochemicals in health regulation. The role of these bioactive compounds in human health modulation has also been detailed eliciting potential pharmacological and therapeutic benefits.

2.2 NATURAL PRODUCTS FROM MICROORGANISMS

At the advent of 1930, microorganisms were investigated as a source of prospective drug candidates. Discovery of penicillin in 1929 by Dr. Alexander Fleming was the feature story for these investigations. This lead from fungi *Penicillium* paved the way for exploring other terrestrial and marine microorganisms for drug discovery. Since then, number of microorganisms has been screened and new potential bioactive compounds from antibacterial agents have been discovered such as antidiabetic agents, like acarbose, anticancer agents, like epirubicin and cephalosporins.[18,68]

2.3 NATURAL PRODUCTS FROM MARINE ORGANISMS

Marine organisms are a rich source of bioactive compounds. In the 1950s, the first active compound was identified and isolated from Caribbean sponge *Cryptothecacrypta*: spongouridine and spongothymidine. These potential bioactive compounds are nucleotides and have a tremendous scope as antiviral agents and cancer chemotherapeutic agents. This discovery attracted investigators to research on marine organisms for identification of more novel drug candidates. Since 70% of the earth consists of water bodies (oceans), thereby they provide huge biodiversity for exploring novel drug sources. The marine organisms synthesize various complex chemical compounds, which have the possibility to serve as remedies for various critical diseases, such as cancer.

For example, Discodermolide is such a chemical compound isolated from *Discodermia dissolute*, a marine sponge, having strong antitumor potential and mechanism of action is similar to paclitaxel. The physiochemical profile of this compound is better than paclitaxel and it is more hydrophilic compared to paclitaxel. A synergistic antitumor response was positive from the combination therapy of these two drugs.[67,68]

2.4 NATURAL PRODUCTS FROM ANIMAL SOURCES

Research has unveiled that animals are also a source of some interesting novel drugs. Venoms and toxins from animal sources are in use for curing various diseases. Epibatidine is 200-folds more effective in comparison with morphine and it is extracted from Ecuadorian poison frog skin.

Teprotide is obtained from a Brazilian viper that has been used for developing hypertension drugs cilazapril and captopril.[67,68]

2.5 PLANT-BASED MEDICINES FOR HUMAN HEALTH

Plant based medicines are involved actively from ancient times in treatment of number of diseases. From earlier times (2600 B.C.), plants have been utilized as herbal remedies in traditional medicine.[76] Till now, 35,000–70,000 plant species have been investigated successfully for their medicinal attributes. Herbal medicines include various drugs, such as chemotherapeutic agents' camptothecin and Taxol®, antimalarial drug quinine and analgesic drug morphine, etc. Plants are widely used in several indigenous systems of medicines and also they have gained attention of modern practitioners and now have been established as alternate therapy to synthetic medicines. Economical cost, availability of product, and low associated side effects contribute to the growth of herbal medicines in market.[99] Recent research developments for discovery of medicinal plant-based drugs have exposed multifaceted tactics via an amalgamation of phytochemicals, their bioactivity and corresponding molecular mechanism.[7]

The bioactive principles of herbal medicines are key principles for therapeutic activities of these medicinal plants. Various plant bioactive compounds have potential to cure diabetes, malaria, cancer, arthritis, fever, wound healing, etc.[11,12,95] Thus, there are thriving opportunities for plant-derived bioactive compounds to develop into new drugs.[95]

The recent research developments in natural products have introduced many new phytoconstituents, such as the popular Paclitaxel commonly known as Taxol obtained from *Taxus brevifolia* Nutt. Furthermore, *Catharanthus roseus* (L.) G. Don had contributed vinca alkaloids vinblastine and vincristine. Some other important anticancer drugs, like etoposide and teniposide, were derived from *Podophyllum roots*. Also, antimalarial drug quinine was isolated from the bark of *Cinchona officinalis*. World Health Organization (WHO) recommended antimalarial drug artemisinin that has its source from *Artemisia annua*. In addition to drugs, plant bioactive compounds are also flavoring agents, food additives, pesticides, and health supplements.[25,104] According to a recent report by WHO, traditional medicines are extensively used by nearly 80% of the population throughout the world. Furthermore, 47% of anticancer drugs available in the market are from natural plants.[94]

2.6 MAJOR GROUPS OF PHYTOCONSTITUENTS

Most of the bioactive phytoconstituents are classified according to their clinical function (therapeutic potential), biological effect (chemical and their clinical benefits) and on the basis of family and genera (plant species producing chemically similar or dissimilar bioactive compounds).[14,98] Since this chapter deals with plant products, therefore, bioactive compounds derived from plants are endowed as major components. Hence, detailed classification of the plant bioactive compounds has been discussed in this chapter according to biochemical pathways and chemical classes.

2.6.1 GLYCOSIDES

Glycosides are organic compounds obtained from plant or animal sources, which give one or more moieties along with nonsugar moiety, on enzymatic or acid hydrolysis. The former is known as glycone (saccharide or uronic acid part) and the latter is aglycone or genin. According to the chemical nature of aglycone moiety, glycosides are grouped into categories that are described in Table 2.1.

TABLE 2.1 Classification of Glycosides with Examples.

Classes of glycosides	Examples	Uses
Aldehyde glycoside	Vanillin	Flavoring agent
Anthraquinone or anthracene glycosides	Cascara, senna, aloe, rhubarb	Purgative
Cardioactive or sterol glycosides	Digitalis, Thevetia, Indian squill, European squill	Cardiotonic
Coumarin glycoside	Psoralen cantharidin	Leucoderma counterirritant, rubefacient
Cyanophore glycosides	Amygdalin, prunisin	Demulcent, sedative
Flavonoid glycoside	Ginko	Vascular disorders
Isothiocyanate glycosides	Wild mustard	Counter-irritant, rubefacient
Phenolic glycoside	Bearberry (*Arctostaphylosuvaursi*)	Diuretic
Saponin glycosides	Dioscorea, Glycyrrhiza, Shatavari	Synthesis of steroids, expectorant, peptic ulcer galactagogue

The aglycone of cardiac glycoside possesses a steroidal structure. They hinder sodium/potassium ATPase-pumps within cell membranes. Also, they exert a potential effect on the heart by increasing contractility and reducing heart rate. The anthraquinone glycosides are responsible for the secretion of water and electrolyte, as well as peristalsis motion occurring inside the colon. The aglycone part of cyanogenic glycosides is derived from amino acids. Furthermore, cyanogenic glycosides are responsible for hypothyroidism as they hamper the iodine utilization. Sometimes they release highly toxic hydrogen cyanide being lethal at high dosage. The saponin glycosides are usually soap forming compounds that display immune modulating and antineoplastic effects in humans. Their aglycone part is hydrophobic and comprises of pentacyclic triterpenoids or tetracyclic steroids, whereas the glycone part is hydrophilic.[36]

2.6.2   ALKALOIDS

These are a heterogeneous group of natural constituents and comprise more than 6000 basic nitrogen-containing organic compounds. Due to the availability of a lone pair of electrons on nitrogen, they are basic in nature. All alkaloids are colorless, crystalline solids with a sharp melting point or decomposition range. They are bitter in taste[1] and can be classified based on pharmacological action, taxonomy, biosynthetic pathway, and chemical constituents (Table 2.2). Alkaloids are reserve substances that supply nitrogen and also contribute the to a defensive mechanism for plant dwellings in dry regions.[53] They possess specific physiological actions on human body when used in minute quantity.

2.6.3   LIGNANS

These are tremendously broad groups of natural compounds, which are 18 C skeleton made of two phenylpropanoid components having numerous functional groups. Lignans possess lipophilic nature and own structural functions in cell membranes. They exist in oilseeds in large volume and are also widely present in different plant species. Due to the huge structural variation of these phytoconstituents, they have uncommon range of therapeutic properties thus leading to fruitful investigations.[74]

TABLE 2.2 Alkaloidal Classification Based on Basic Chemical Structure.

Types of alkaloids	Examples	Uses
Amino alkaloids	Ephedrine, pseudoephedrine	Sympathomimetic, antiasthmatic
Diterpene	Aconitine, aconine, hypoaconitine	Rheumatism
Imidazole	Pilocarpine, isopilocarpine, pilosine	Glaucoma
Indole (Benz pyrrole)	Ergometrine, ergotamine, reserpine, vincristine, vinblastine	Oxytocic, hypotensive, anticancer
Isoquinoline	Morphine, codeine, papaverine, narcotine, d-tubocurarine, berberine, emetin	Analgesic, antitussive, skeletal muscle relaxant, astringent, emetics
Purine	Caffeine, theobromine, theophylline	CNS stimulant
Pyrrole and pyrrolidine; Pyridine and piperidine	Arecoline, anabasine, coniine, lobeline, trigonelline	Respiratory stimulant
Quinoline	Quinine, quinidine, cinchonine, cinchonidine, cupriene, camptothecin	Antimalarial agent
Steroidal	Protoveratine, conessine, withanolide	Hypotensive, antiamoebic, CNS disorders, antirheumatic
Tropane	Atropine, hyoscine, hyoscyamine	Anticholinergic, antispasmodic

2.6.4 FLAVONOIDS AND PROANTHOCYANIDINS

Amongst secondary metabolites, flavonoids are categorized on the basis of the flavan nucleus and C6-C8-C6 carbon skeleton. The major categories of flavonoids depending on the molecular structures are: flavones, flavonols, flavanones, anthocyanidin, catechin, isoflavone, and chalcones.[2] Proanthocyanidins are oligomers of flavonoids. They show potential antioxidant effects due to the presence of phenol. These are pigment containing groups present widely in plant families.[26,43]

2.6.5 TANNINS

These are complex polyphenolic phytoconstituents present widely in different plants. Tannins can be explained as "any high molecular weight phenolic compound with adequate hydroxyls and additional appropriate groups (i.e., carboxyls) so that they can interact indefinite environmental condition with protein as well as macromolecules to form effectively strong complexes." They are efficient to precipitate, collapse, or bind protein due to their astringent nature.[21] Tannins are divided into two categories: the hydrolyzable and the condensed. Both possess common properties but hydrolyzable tannins have more capability to cause toxicity and are less stable.[21]

2.6.6 TERPENOIDS

These are volatile components, which are responsible for the fragrance of plants and flowers. Terpenoids on thermal decomposition break into isoprene. Isoprene itself is considered the only hemiterpene, but oxygen-containing derivatives, such as prenol and isovaleric acids, are hemiterpenoids.[10,65,102]

2.6.6.1 CLASSIFICATION OF TERPENOIDS BASED ON HYDROCARBON FORMULA

Naturally occurring terpenoid hydrocarbons have the general formula $(C_5H_8)_n$. They are classified on the basis of the value of "n" or number of C atoms in the compound[102] as presented in Table 2.3.

2.6.6.2 CLASSIFICATION OF TERPENOIDS BASED ON NUMBER OF RINGS IN STRUCTURE

Each terpenoid class can be subclassified based on a number of rings existing in the molecule.[102]

- *Acyclic terpenoids*: They have an open structure.
- *Bicyclic terpenoids*: They hold two rings in the molecule.
- *Monocyclic terpenoids*: They possess one ring in the molecule.
- *Tetracyclic terpenoids*: They have four rings in the molecule.
- *Tricyclic terpenoids*: They have three rings in the molecule.

TABLE 2.3 Classification of Terpenoids on the Basis of Hydrocarbon Formula.

Number of C atoms	Value of n	Class and formula	Examples
10	2	Monoterpenoids $(C_{10}H_{16})$	Vetiver, eucalyptus oil, peppermint, caraway
15	3	Sesquiterpenoids $(C_{15}H_{24})$	Artimisia, arnica, clove, valerian
20	4	Diterpenoids $(C_{20}H_{32})$	Taxus, coleus
30	6	Triterpenoids $(C_{30}H_{48})$	Ambergris
40	8	Tetraterpenoids $(C_{40}H_{64})$	Carotene, annatto
>40	>8	Polyterpenoids $(C_5H_8)_n$	Rubber

2.6.7 RESINS

Resins may be defined as solid, semisolid, or amorphous products, which first softens and then melts on heating. Resins form an ill-defined group of complicated lipid-soluble plant products typically nonvolatile as well as volatile, which are produced normally during growth or secreted as a result of injury to the plant. The diterpenoid and triterpenoid components come under nonvolatile fraction, while mono- and sequiterpenoids dominate in the volatile fraction. These all are sticky in nature and their fluidity is related to the volume of volatile components.

2.7 PROCESSING OF PLANTS FOR DEVELOPMENT OF PHYTOMEDICINES

Medicinal plants collected for scientific investigations are selected on the basis of indigenous knowledge, inherited by traditional practitioners.[15] Use of plant extracts for specific pathological conditions is based on a long hit and trial basis from which biologically active principles of medicinal interest can now be identified with the aid of modern tools.[15,33,90] More or less, the same general strategy is followed by most of the workers to investigate plants for their therapeutic potential.[4,90] Field observations can help in the selection of target plants. For example, plants growing in stressful environmental conditions tend to accumulate more secondary metabolites and synthesize defensive natural products in order to thrive in tough habitats. Further, screening of plants on the basis of the chemotaxonomy is also helpful in selecting desired plant material.[90] Another method recognized as data-driven methodology develops an amalgamation of ethnobotanical, chemotaxonomic, and unsystematic tactics collected with a database that covers all applicable evidence about a certain plant species.[77,88]

2.7.1 IDENTIFICATION AND COLLECTION OF PLANT MATERIALS

The desired plant must be identified scientifically before the collection in order to ensure the collection of adequate plant material. This can be done with the help of monographs, flora, plant databases on the internet and other scientific literature available. The plants must also be authenticated post collection by a taxonomist or a botanist at any herbarium near the locality of plant collection.[88]

The plant part, required for the extraction of bioactive components should also be identified in advance so that only the desired part is collected. This is crucial to conserve our natural wealth and also plants that are in the rare, endangered, and threatened category.[42]

Every aspect pertaining to the plant collection must be documented. The basic details are important to record are scientific and vernacular names of the plant collected, location, date, and season of collection. A herbarium sheet of the plant should be prepared and deposited in a repository. GPS information should also be taken for each collection to make feasible the repeat collections from exactly same location. The altitude, temperature, rainfall, sunlight, soil characteristics, etc., are factors, which influence the accumulation of bioactive compounds/secondary metabolites in plants. Keeping these points in mind is helpful in repeating and validating the chemoprofiling studies of the medicinal plants.[42]

2.7.2 DRYING AND GRINDING OF PLANT MATERIAL

After collecting plant samples, it should be chopped and kept for drying in shade at ambient temperature with adequate ventilation. It is necessary to make sample free of moisture to avoid microbial infections and subsequent degradation of the plant metabolites. Plant samples after drying should be stored in airtight containers away from moisture. For extraction purpose, the plant material is ground using electric grinders.[45]

2.7.3 EXTRACTION OF PLANT MATERIALS

Plants contain numerous biocompounds of diverse chemical nature within them, which have different solubility profiles. Thereby, it is important to select the suitable extraction solvent[45] (Fig. 2.1). Therefore, prior to extraction, it is important to work out which kind of chemical compound is targeted.[11] The extraction protocol should be designed keeping in mind the most ability of the compounds targeted. We can choose from hot and cold extraction procedures for maximum yield. A suitable extraction solvent can be identified from a range of polar and nonpolar solvents.[20] Extraction can be done in several ways starting from the very basic maceration to advanced techniques such as supercritical fluid extraction.[89]

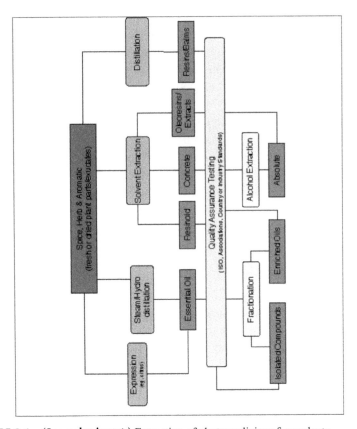

FIGURE 2.1   (See color insert.) Extraction of phytomedicines from plants.

Nowadays, there are various sophisticated instruments available for highly efficient extraction that utilizes pressure, ultrasound, microwaves, and surfactants, etc.[20,83,89,101]

2.7.4　ISOLATION, IDENTIFICATION, AND CHARACTERIZATION OF BIOACTIVE PRINCIPLES

As the plants harboring numerous phytochemicals are extracted, the extracts obtained obviously contain a high concentration of those phyto-constituents. Since the chemical moieties are diverse and numerous, it is a tough task to get them separated from each other.[89] However, modern science has techniques to resolve and identify the different constituents. Chromatography techniques have played a vital role in this task.

Different versions of chromatography such as HPTLC, HPLC, GC, etc. have led to separation and identification of pure compounds (Fig. 2.1). Few other techniques have also played a crucial role in the identification and structure elucidation of compounds, such as immunoassays, FTIR, GC-MS, etc.[13,83,89,101]

Several hyphenated techniques have played an important role in herbal drug research. Liquid chromatography-ultraviolet, liquid chromatography-mass spectroscopy, and liquid chromatography-nuclear magnetic resonance are very helpful in identifying the novel molecules. Other techniques such as ultraviolet/visible and infrared spectroscopy, Nuclear magnetic resonance, mass spectroscopy, X-ray diffraction, etc., along with different chemical reaction are used for depicting the structure and identity of phytoconstituents.[7,45]

2.7.5　BIOLOGICAL AND PHARMACOLOGICAL SCREENING

Understanding the chemical constituents profile gives us an idea about the nature of the biological activity the plant. The pharmacological attributes of plant extracts can be studied using either in vitro or in vivo or using both kinds of studies.[24,49,60,100] Since, plant extracts contain a vast number of biomolecules, it is not easy to obtain absolutely pure bioactive compound. Therefore, studying the therapeutic activity is more fruitful using bioassay-guided approach. In this way toxicity of the phytoconstituents can also be monitored. In vitro biochemical assays are very helpful in identifying different biological activities, for example, antioxidant activity, antidiabetic activity, etc. Apart from this, the in vivo studies are very crucial to give us a better insight into how the molecule or plant extracts work in the living system.[9,44]

Plant secondary metabolites, which also act as the bioactive molecules, are derived from the intermediated and final products of primary metabolism. Secondary metabolites seem to have no direct involvement in the regular metabolism and development of the plants. However, they are directly related to the processes involved in the interaction with the environment. They play a definite role in allowing the plant in surviving and sustaining in stressful environmental conditions.[9] Secondary metabolites have a potential role in protecting the plants from microbial attacks and also from grazers.[79,80]

2.8 QUALITY CONTROL FOR THE DEVELOPMENT OF HERBAL DRUGS/PRODUCTS

Quality control is an important aspect for the development of herbal drugs/ products. Herbs/plants traditionally in use are required to be identified taxonomically before evaluating their traditional claims in respect of pharmacological effects (Fig. 2.2). Further, raw material collected from wild/ cultivated or either procured from market, are mandatory to ensure for their quality by means of following guidelines given in good agriculture practices, good collection practices, good procurement practices, good ethical practices, and good storage practices.[89] In India, looking at the urgent need for the quality and regulatory requirement for herbal drugs, documentation of scientific data has been started. Ministry of AYUSH, Govt. of India, published more than 500 monographs of single drugs and 101 classical Ayurvedic formulations in six volumes. Similarly, such efforts for developing quality control parameters have also attempted in other countries. Furthermore, WHO has also developed standards for quality control of herbal drugs.[89]

2.8.1 STANDARDIZATION OF HERBAL PRODUCTS

In search of novel bioactive compounds, new technologies are under continuous development for the purpose of isolation and identification of bioactive components in medicinal plants. Since plant extract is compoed of large number bioactive components, thereby the biological response exerted by plant extracts is a cumulative or synergistic effect of multiple bioactive compounds present in the extract. Therefore, qualitative and quantitative determinations of varied phytocomponents are required to eradicate the possible limitations in the use of medicinal plants.[93]

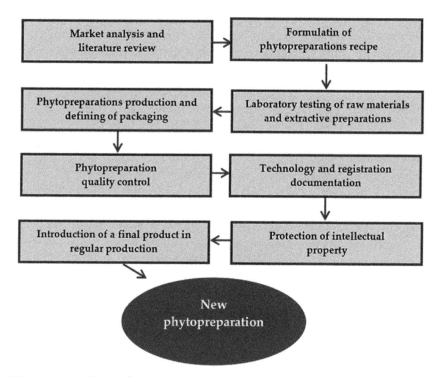

FIGURE 2.2 (See color insert.) Schematic representation of the planning and development of new herbal preparation.

Source: Reprinted from Sofija M. Djordjevic (2017). From Medicinal Plant Raw Material to Herbal Remedies. Chapter 16; In: *Aromatic and Medicinal Plants - Back to Nature*; Hany El-Shemy (Ed.); Online; InTech; Open Access; DOI: 10.5772/66618; Available at: https://www.intechopen.com/books/aromatic-and-medicinal-plants-back-to-nature/from-medicinal-plant-raw-material-to-herbal-remedies.

Secondary metabolites are altered qualitatively and quantitatively by the influence of various elicitors including stress conditions,[82] chemotype variations due to microenvironments, locations, physical, and chemical stimuli, and climates.[64,82,92] Such variations are very common and well documented in phytochemicals such as alkaloids,[64] terpenoids,[47] and phenylpropanoids[47]; and the content level of these phytochemicals is increased by two to three folds under stress or elicitation.[89,92] Moreover, increased level of phytochemicals due to elicitation may increase the biological output of the plant, which is needed to recognize during the screening process for improving the reliability and efficiency of plant extracts in drug discovery significantly. Thus, standardization, optimization, and control of growth conditions for cultivation ensure the quality production of many plant-derived compounds.[43,92]

Quality, safety, and efficacy are three main factors behind an ideal herbal drug and its rational use. The maintenance of quality standards and batch to batch consistency of herbal drugs are major challenges for the herbal drug industries. The composition and specifications of herbal drugs are not well defined and characterized. Therefore, they do not stand along three pillars of standard herbal drugs.

The traditional medicinal practitioners had a vast knowledge of indigenous plants and they were able to identify medicinal important plants from specific habitats. They were also aware of the fact that at what stage of growth and in which season, a plant will be more useful. On the basis of this expertise, they used to maintain the quality of their drugs or formulations. However, in past decades it has been observed that there had been a significant decrease in the number of such experienced plant collectors. The transformation of the individualized system of traditional medicine to industrial manufacturing system is one of the causes behind this situation. The industrialization process has resulted in vast enfeeblement of the whole process of traditional medicine systems.[77]

Unscientific and indiscriminate collection of medicinal plants from wild habitats has led to huge destruction in the population of medicinal plants and the random collection has added to the decreased quality of herbal drugs. The escalating demand for the raw material for industrial applications has encouraged people to use low-quality material or some adulterants, which have further degraded the quality standards of herbal drugs.[96] The traditional Indian systems of medicine have emphasized the role of appropriate plant collection practices. In systems, like Ayurveda, Siddha, Unani, Amchi, etc., it has been instructed about the season, developmental stage of the plant, edaphic factors, environmental conditions that regulate the presence of specific biologically active metabolites. This has now been scientifically proven that there arise fluxes of variations in the metabolite profile in different growing conditions since the chemical constituents present in the medicinal plants are governed by several factors. Therefore, it is of prime concern to standardize and establish quality parameters through scientific studies using modern tools and techniques including traditional knowledge.[40,96]

2.8.2   HEALTH ORGANIZATIONS AND REGULATORY BODIES GOVERNING NATURAL HEALTH PRODUCTS

The tenacity of scientific validation is to spread awareness about natural products and the natural health products (NHPs) by standardizing and establishing their efficacy and safety. In the USA, the NHPs are produced and provisioned to the public via the Food and Drug Administration. NHPs are classified into five major types: (1) Ayurveda, homeopathy; (2) mind body medicine; (3) dietary supplements; (4) body-centered practices like a massage; and (5) energy medicine.

In Canada, there is a distinct system for regulation of the introduction of NHPs and drugs to the public such as Natural Health Product Directorate (NHPD) and Therapeutic Drug Directorate. Health Canada is the parent governing agency that assists and ensures the supply of "safe, effective, and of high quality" NHPs to the citizens. In 2012, NHPD published guidelines regarding the use of NHPs singly, as well as in combination, risk information, how to assess NHPs, their health claims, etc. Apart from this, Health Canada also collects information about adverse reaction reports for NHPs. Using this data, the agency also contributed in the "quality of NHPs guide" and "the pathway for licensing NHPs making modern health claims" and "the pathway for licensing NHPs making traditional health claims." In this way, Health Canada continuously monitors the quality of NHPs and ensures continuous efficacy and safety.[23,27]

There is an urgent need for the development of effective, novel pharmaceutical agent or lead structures from natural products using the novel screening systems based on genetic information. To date, the increased access to the plants has increased the discovery of many vital phytomolecules and still a lot of phytomolecules are to be discovered from the natural products.[27] There are certain areas that are still not well explored, such as the marine flora, especially marine-sourced fungi, which possibly will play a significant role in the future. In the future, researches on plant endophytes have huge potential in the field of therapeutics as well as nutraceuticals. The introduction of nanotechnology and synthetic chemists has also provided strong support for developing novel molecules, drugs, and nutraceuticals eliminating challenges, such as poor bioavailability.[23,27,93]

2.9 ROLE OF PHYTOCHEMICALS IN REGULATION OF HUMAN HEALTH

Addressing the urgent need of new drug candidates to surpass the limitations of the existing drugs and synthetic drugs, phytochemicals have continuously played a significant role. Along with therapeutic effects, phytochemicals also exerted a protective role on human health against various diseases.[80] In recent studies, nutritionists investigated that there exists a strong correlation in dietary intake of fruit and vegetables and immunomodulation in the human system, which is further related to the protective mechanism of phytochemicals.[30] Moreover, analysis of cellular mechanisms of phytochemicals amplifies the growth of health foods in the market as nutraceuticals.[30,31]

Biochemical compounds of plant origin are crucial for maintaining human health in various ways. These molecules are helpful in the treatment of several pathological conditions and also for maintaining vigor and vitality, prevent aging, cure several lifestyle diseases as diabetes, heart diseases, etc. They also show diuretic, immunomodulatory, and carminative effect and protects from UVB-induced carcinogenesis.[28,63] Among numerous natural products, nutraceuticals including functional foods, micronutrient fortified foods, certain bioactive phytochemicals exert a number of health benefits and other health claims based on science and ethics. They are obtained from plants, which may be well known or may also be less known and underutilized.[28]

There are number of examples of phytochemicals, which have scientifically proven biological activities, such as capsaicin for anticarcinogenic and antimutagenic effect, curcumin for anti-inflammatory and cancer prevention, genistein for antiproliferative effects on mitogen-stimulated growth,[20,63] etc. Scientific investigations have proved that the consumption of isoflavonoids rich food is able to reduce the risk of cancer. It is also found that these isoflavonoids also known as phytoestrogens are cardio protective and can reduce the cardiovascular diseases in vegetarians.[2,17] In

Asian countries, the cases of cancer (hormone-dependent) are found less in the population taking phytoestrogen-rich diet.[2,98]

2.9.1 NATURAL PRODUCTS AS ANTIOXIDANTS

Antioxidants are free radical scavenging molecules, which ensure minimum risk of oxidative damage and other problems associated with the oxidation process. Examples of natural antioxidant molecules having free-radical scavenging activity are carotenoids, tocopherols, ascorbates, lipoic acids, and polyphenols. Food material rich in antioxidant activity are of great significance by virtue of their health benefits.[87,97] Many studies on animals have suggested that optimum consumption of whole grains, fruits, especially citrus, and green vegetables help in preventing complications arising due to oxidative stress.[50,51] Synthetic antioxidants, such as butylated hydroxyl anisole and butylated hydroxyl toluene play a vital role in the food and drug sector.[32,78,103]

Polyphenols constitute a large family of plant products abundantly available in fruits and vegetables, legumes, berries, spices, tea, cocoa, etc.[69,103] The active constituents in the dietary phytochemicals (e.g., vitamins, curcumin, capsaicin, catechins, β-carotene, etc.) are involved in protection against cancer, cardiovascular diseases, and also in neurodegenerative disorders by mediating several biological processes, like redox balance, suppressing the inflammatory processes, and moderating cell signaling pathways, etc.[1,16] The polyphenols work by activating at different stages through various mechanisms, like modulation of mitogen-activated protein kinase, protein kinase B (Akt), and nuclear factor kappa-light-chain-enhancer of activated B cells signaling ways to inhibit cancer cell growth, obstructing the generation of inflammatory cytokines and chemokines, overpowering the activity of cyclooxygenase, and inducible nitric oxide synthase; and decreasing the production of free radicals.[75,103] Polyphenols also work by some other mechanisms, like maintaining the levels of enzymes, like catalase, superoxide dismutase, glutathione peroxidase, etc. which show significant title role in antioxidant processes. The anti-inflammatory property of polyphenols is imparted by the phenolic acids and aldehydes produced after the breakdown of polyphenols. These days, there is great awareness about foods fortified with polyphenols due to their health benefits, like antioxidant potential, anti-inflammatory activity, anti-cancer activity, skin photoprotective activity, and anti-aging effect, etc.[84]

2.9.1.1 *BIOLOGICAL IMPORTANCE AND APPLICATION*

Phenolic acids scavenge reactive oxygen species (free radicals), preventing oxidation of vital biomolecules, cellular components[6] and food matrices.[19] Dietary intake of phenolic acids, their uptake, and metabolism in the brain have been explored extensively. Natural antioxidants have considerable effects in neurodegenerative disorders. Phenolic acids exert their effect through natural antioxidant property after getting metabolized as conjugated derivatives and get absorbed mainly in the colon as a result of hydrolysis by the intestinal microflora. A new area of research has been emerged to control diseases caused by oxidative damage by the natural antioxidants, like ferulic acid ethyl ester by improving its capability to pass through lipid-rich cell membranes as in neurons. Thus, ferulic acid ethyl ester can act as a good nutraceutical agent. This will be helpful in the treatment of neurodegenerative diseases.[19,85,103]

2.9.1.2 *SOURCES OF ANTIOXIDANT PHYTOCHEMICALS*

Antioxidants are primarily of herbal origin. Antioxidant molecules are present in different fruits (especially citrus fruits), vegetables, nuts, grains, tea, coffee, etc. The nuts and grains are found to contain a significant amount of polyphenolic compounds.[86] Phytochemicals rich in antioxidant are found in both dietary and nondietary compounds. Higher content of phenolic compounds is found to be present in vegetarians.[34,88]

2.9.2 *NATURAL PRODUCTS IN ADJUVANT THERAPY FOR CHRONIC RENAL FAILURE*

Antioxidant-rich phytoformulations may play a significant role in reducing the occurrence and development of chronic renal failure (CKD). Scientific investigations followed by clinical trials may lead to the identification of such phytomolecules to minimize the progression of CKD. The effect of phytomolecules may be through modulation in the causative mechanisms of the disease, primarily the oxidative stress. Toxicity studies, preliminary in vivo preclinical studies and clinical trials are necessary to draw decision regarding the selection of combinations of phytotherapies alone or in combination with conventional therapies. Such experiments are currently being performed in the laboratories and should be encouraged and published by the scientific community.[78]

2.9.2.1 MODULATION OF CKD WITH PHYTOCHEMICALS

Nowadays, it has become the area of prime concern for the researchers to prevent or delay the onset of CKD by identification of natural phyto-molecules with high efficacy and safety. Phytochemicals have shown promising results against many diseases like diabetes mellitus type II, which have close links with CKD.[55] Clinical and preclinical trials using phytomolecules based drugs are increasing and have shown huge market potential as these are well tolerated as a therapy.

2.9.2.2 SELECTED PHYTOCHEMICALS FOR CKD THERAPY

2.9.2.2.1 Curcumin

Curcumin is the key phytomolecule present in turmeric, which is effective as an anti-inflammatory and antioxidant agent.[71] Studies have shown that curcumin in the form of tetrahydro curcumin exhibited the highest radical scavenging potential. It was found that curcumin exerted its antioxidant activity by inducing antioxidative enzymes and detoxification enzymes. In cases of chronic renal pathologies, curcumin works in creatinine and urea clearance. Curcumin and its derivatives protect from chronic renal allograft nephropathy. Chronic renal failure conditions have been studied with the help of experimental models, like 5/6th nephrectomy model of CKD in rats.[37] To reduce the development of CKD, studies were conducted to estimate the effectiveness of curcumin (75 mg/kg) against angiotensin converting enzyme inhibitorenalapril (10 mg/kg) and expression of inflammatory agents.[37,71]

2.9.2.2.2 Resveratrol

Resveratrol is a polyphenol having high antioxidant and anti-inflamma-tory benefits. It is commonly found in grapes.[22] Consumption of red wine has been found to decrease the progression of atherosclerosis and helpful in aging. Although it is a highly effective molecule, yet its bioavailability is not very good since it is excreted after rapid metabolism. It is a challenge to stabilize resveratrol in any drug delivery system to increase its bioavailability. The oxidative stress and endothelial dysfunction (atherosclerosis) conditions common in CKD patients can be managed by resveratrol intake.[66]

2.9.2.2.3 *Quercetin*

In rats in a model of CKD, quercetin (a flavanol) has been found to be most effective phytomolecule to control CKD by indicating improved serum creatinine among other effects.[54] Quercetin has also been found to be highly effective in conditions of hypertension and cardiac pathologies. It is supposed to prevent hematocrit and erythrocyte depletion in bone marrow as well as aortic calcification. Apart from this, quercetin has shown to be completely effective in reducing caspase-3 cleavage, hence ameliorating apoptosis but partially effective in suppressing proapoptotic proteins B-cell lymphoma 2 (BCL2) associated X protein, BCL2-associated death promoter, and restoring antiapoptotic BCL2).[38]

2.9.2.2.4 *Genistein*

The anti-inflammatory action of isoflavo-genistein is exhibited by selective inhibition of monocyte Tumor necrosis factor alpha (TNFα) generation in experimental animals. A robust TNFα response was observed when whole blood and isolated mononuclear cells from hemodialysis patients and healthy control subjects were incubated with genistein and stimulated with lipopolysaccharides. Genistein blocked lipopolysaccharide-induced TNFα formation but did not inhibit interleukin-6 formation and did not alter basal TNFα.[73]

2.9.3 *COMBINATORIAL THERAPY WITH NATURAL HEALTH PRODUCTS*

NHPs contain multiple components, which enable them to have selective efficacy contrary to cancer cells, in vitro and in xenograft models. Several studies have illustrated that the effect of the whole extract is greater than the effect of a single molecule. This might be one of the reasons behind the use of multiple NHPs in combination to achieve better results. The NHPs having multiple phytoconstituents are associated with targeting multiple signaling pathways and are effective even at lower treatment doses. The increased efficacy is attributed to the synergistic effect of the multiple components present within the extract. Due to these benefits, the NHPs may serve as an alternative to chemotherapy in cancer treatment.[72]

2.9.3.1   NATURAL PRODUCTS FOR CHEMOTHERAPY

Plants are opulently consecrated with secondary metabolites having anticancer properties.[52] These phytocompounds covers a vast category of secondary metabolites comprising monoterpene indole alkaloids vinblastine and vincristine (from *Catharanthus roseus*). Vinblastine effectively manages Hodgkin's lymphoma.[58] Another important diterpene alkaloid is taxol (pacli-taxel), which has earned tremendous success as an antitumor molecule. It is commonly derived from the bark of gymnosperm *Taxus brevifolia*. Taxol is also found in nature as a fungal metabolite. It has been a drug of choice for clinical therapy of different cancers, mainly breast cancer, ovarian cancer, and liver cancer. Taxol inhibits the initiation and propagation of cancer cells by hindering depolymerization of microtubules. Taxol also endorses tubulin polymerization that interplays for inhibition of proliferation of mammalian cancer cells. However, progress in natural chemistry had extended these potential secondary metabolites by developing the analogs of the active mole-cule.[56] Through this strategy, new drug candidates for cancer therapy have been developed with huge potential. A reformed monoterpene indole alkaloid from assured plants (angiosperms) called camptothecin is also used as an anti-tumor agent for colon, lung, ovarian, and uterine cancer.[3] Another anticancer agent etoposide exhibited topoisomerase II inhibitory effect. Topoisomerase II is an essential enzyme that regulates DNA supercoiling.[8] Etoposide has been used for lung cancer, ovarian and testicular cancer, lymphoma, and leukemia.[8] Moreover, Teniposide has a remarkable effect on central nervous system tumors, lymphomas, and bladder cancer. Some other important natural compounds such as curcumin, (-)-epigallocatechin-3-gallate, lycopene, genistein, indole-3-carbinol, resveratrol and 3,3'-di-indolylmethane are well reported anticancer agents.[48] These phytocompounds work on multiple cell signaling pathways to activate cell apoptosis. Till now, several molecules of plant origin had been examined for their anticancer potential to achieve safe, efficacious, and therapeutic benefits. Some triterpenes are reported to possess considerable anticancer effects exhibiting chemotherapeutic potential against different types of cancers.[70]

2.9.3.2 NATURAL ANTIOXIDANTS FOR CHEMOTHERAPY

Concerning the toxicity issues of the cancer chemotherapy, the phytocompounds are more preferred over synthetic drugs. They are safe and potentially active at low doses, favorable to oral administration, economical, and naturally available. These phytocompounds are also effective antioxidants and interfere in the free radical formation in normal cells. Free radical species are accountable for chronic diseases like diabetes, chronic inflammation, and cancer. Thus, a new regimen of therapy has developed owing to the chemopreventive nature of the antioxidant properties of phytocompounds.[51] Current chemotherapy encountered disquieting issue that is drug resistance by developing confrontation for anticancer drugs through various mechanisms. Some of the important mechanisms are:

- Alteration in the target sites or enzyme or receptor.
- Drug efflux mechanism can limit the therapeutic effect of the anticancer drugs due to the low intracellular concentration of drugs, like vinca alkaloids paclitaxel, anthracyclines, epipodophyllotoxins, and dactinomycin.
- Interfering with active drug through alkylation or other agents, as in antimetabolites and bleomycin.

Phytocompounds act through various mechanisms to interfere with cancer events. Moreover, simultaneous administration of these phytocompounds having different mechanisms can result in an exceptional synergistic effect. Thus codelivery of two or more phytocompounds results in enhancement of therapeutic efficacy and opens multiple apoptotic pathways. The ultimate result will be a reduction of dose and related side effects too. In early reports on phytocompounds in cancer therapy, a high-effective dose of phytocompounds was discussed as a major limitation of phytotherapy in cancer. This may induce a metabolic burden on the body system with undue consequences.[39]

2.9.4 PHYTOCHEMICALS AS NUTRACEUTICALS

Several medicines and food supplements have been developed based on traditional knowledge. In broad term, nutraceuticals include various products ranging from isolated nutrients, food supplements, herbal products, genetically engineered designer foods, processed foods, and beverages. Moreover, "vitamin-enriched" fresh foods, like vegetables and fruits are a major component of functional foods. Thus, nutraceuticals from plant origin are proactive healthcare system impending their tremendous health benefits.[5,28] The "novel" herbal nutraceuticals may develop as vigorous aspect of dietary disease-preventive food components. Careful studies are necessary on various phytoconstituents for their role in the inhibition of chronic degenerative diseases. Further, different phytochemicals also act synergistically to give the desired action that should also be taken into account. The revival of importance in these phytoconstituents will ultimately lead to desirable data on structure–function relationships.

Nutraceuticals are designed for the delivery of food bioactive compounds in various dosage forms (injectables, tablets, capsules, parenteral, etc.).[35,61] The food bioactive constituents are phytochemicals with health-promoting attributes along with some of them having specific pharmacological properties. Some phytochemicals have been exploited by the nutraceutical industry for intended health purpose, such as terpenoids, phytoestrogens, glucosinolatesphytosterols, polyphenols, limonoids, flavonoids, carotenoids, isoflavonoids, and anthocyanidins. These phytochemicals exerted huge biological activities.[41,46] Thereby, nutraceuticals are promising aspects of developing phytochemicals for public health. The herbal formulation well standardized and characterized by herbal monographs, bioactive fraction of herbal extracts are key ingredients of functionally active nutraceuticals used in food and pharmaceutical preparations.[28]

2.10 FUTURE OF NUTRACEUTICALS

The future of nutraceuticals from different sources grasps exciting prospects for the newer food products in food industries. Thus, the requirement of the food industries is to encourage investors with economical rewards that can be increased by capitalizing in the cost of nutraceuticals, and to gain the interest marketing should be done perfectly, and most importantly, to satisfy the tastes of buyers.[39]

2.11 SUMMARY

In this chapter, plant attributes for health promotion have been discussed with special focus on the secondary metabolites. This chapter gives details on processing techniques for proper extraction and isolation of plant secondary metabolites including identification and collection of plants, grinding, and drying, etc., various techniques of extraction and finally characterization of plant bioactive compounds. Furthermore, the classification of plant bioactive compounds responsible for the pharmacological outcomes of the plants has been elucidated. Quality control parameters being an important consideration for the development of natural products are also discussed. The role and expected mechanism of phytochemicals in modulating health emphasizing the profound effect as an antioxidant and in chemotherapy and chronic kidney disease are also explained. Further, new extension of plant product in nutraceuticals is also highlighted.

KEYWORDS

- alkaloids
- glycoside
- lignans
- nutraceuticals
- polyphenols
- secondary metabolites

REFERENCES

1. Aggarwal, B. B.; Shishodia, S. Molecular Targets of Dietary Agents for Prevention and Therapy of Cancer. *Biochem. Pharmacol.* **2006**, *71* (10), 1397–1421.
2. Al-Azzawi, F.; Wahab, M. Effectiveness of Phytoestrogens in Climacteric Medicine. *Ann. New York Acad. Sci.* **2010**, *1205*, 262–267.
3. Amna, T.; Puri, S. C.; Verma, V.; Sharma, J. P.; Khajuria, R. K.; Musarrat, J.; Spiteller, M.; Qazi, G. N. Bioreactor studies on the Endophytic Fungus *Entrophospora infrequens* for the Production of an Anticancer Alkaloid Camptothecin. *Can. J. Microbiol.* **2006**, *52* (3), 189–196.
4. Arnason, J. T.; Mata, R. *Phytochemistry of Medicinal Plants.* Springer: New York,; 2013; pages 356.
5. Bagchi, D.; Preuss, H. G.; Swaroop, A. *Nutraceuticals and Functional Foods in Human Health and Disease Prevention.* CRC Press: Boca Raton, USA, 2015; 700 pp.
6. Bartosz, G. *Food Oxidants and Antioxidants: Chemical, Biological, and Functional Properties.* CRC Press: Boca Raton, USA; 2013; 568 pp.
7. Balunas, M. J.; Kinghorn, A. D. Drug Discovery from Medicinal Plants. *Life Sci.* **2005**, *78* (5), 431–441.
8. Bender, R. P.; Jablonksy, M. J.; Shadid, M.; Romaine, I.; Dunlap, N.; Anklin, C.; Graves, D. E.; Osheroff, N. Substituents on Etoposide that Interact with Human Topoisomerase II alpha in the Binary Enzyme–Drug Complex: Contributions to Etoposide Binding and Activity. *Biochemistry* **2008**, *47* (15), 4501–4509.
9. Bernhoft, A.; Clasen, P. E.; Kristoffersen, A. B.; Torp, M. Less Fusarium Infestation and Mycotoxin Contamination in Organic than in Conventional Cereals. *Food Addit. Contam. Part A, Chem. Anal. Control. Expo. Risk Assess.* **2010**, *27* (6), 842–852.
10. Bonora, A.; Pancaldi, S.; Gualandri, R.; Fasulo, M. P. Carotenoid and Ultrastructure Variations in Plastids of Arum Italicum Miller Fruit During Maturation and Ripening. *J. Exp. Botany,* **2000**, *51* (346), 873–884.
11. Brahmachari, G. *Bioactive Natural Products: Opportunities and Challenges in Medicinal Chemistry.* World Scientific Publishing Company: Singapore, 2012; 696 pp.
12. Brahmachari, G. *Bioactive Natural Products: Chemistry and Biology.* Wiley: New York, 2015; 544 pp.
13. Brusotti, G.; Cesari, I.; Dentamaro, A.; Caccialanza, G.; Massolini, G. Isolation and Characterization of Bioactive Compounds from Plant Resources: The Role of Analysis in the Ethnopharmacological Approach. *J. Pharm. Biomed. Anal.* **2014**, *87*, 218–228.
14. Carkeet, C.; Grann, K.; Randolph, R. K.; Venzon, D. S.; Izzy, S. *Phytochemicals: Health Promotion and Therapeutic Potential.* Taylor & Francis: Boca Raton, FL, 2012; 270 pp.
15. Carlson, L. K. Reimbursement of Complementary and Alternative Medicine in the Context of The Future of Healthcare. *Altern. Ther. Health Med.* **2002**, *8* (1), 36–42.
16. Chandra, D. *Mitochondria as Targets for Phytochemicals in Cancer Prevention and Therapy.* Springer: New York, 2013; 243 pp.
17. Cherdshewasart, W.; Sutjit, W.; Pulcharoen, K.; Chulasiri, M. The Mutagenic and Antimutagenic Effects of the Traditional Phytoestrogen-Rich Herbs, *Pueraria mirifica* and *Pueraria lobata. Braz. J. Med. Biol. Res.* **2009**, *42* (9), 816–823.

18. Chin, Y. W.; Balunas, M. J.; Chai, H. B.; Kinghorn, A. D. Drug discovery From Natural Sources. *AAPS J.* **2006,** *8* (2), E239–E253.
19. Choe, E.; Min, D. B. Mechanisms of Antioxidants in the Oxidation of Foods. *Compr. Rev. Food Sci. Food Safety* **2009,** *8* (4), 345–358.
20. Chua, L. S. Review on Plant-based Rutin Extraction Methods and its Pharmacological Activities. *J. Ethnopharmacol.* **2013,** *150* (3), 805–817.
21. Chung, K. T.; Wong, T. Y.; Wei, C. I.; Huang, Y. W.; Lin, Y. Tannins and Human Health: A Review. *Crit. Rev. Food Sci. Nutr.* **1998,** *38* (6), 421–464.
22. Caimi, G.; Carollo, C.; Lo Presti, R. Chronic Renal Failure: Oxidative stress, Endothelial Dysfunction and Wine. *Clin. Nephrol.* **2004,** *62* (5), 331–335.
23. Cordell, G. A. Sustainable Medicines and Global Health Care. *Planta Medica* **2011,** *77* (11), 1129–1138.
24. Crawford, A. D.; Liekens, S.; Kamuhabwa, A. R.; Maes, J.; Munck, S.; Busson, R.; Rozenski, J.; Esguerra, C. V.; de Witte, P. A. Zebrafish Bioassay-Guided Natural Product Discovery: Isolation of Angiogenesis Inhibitors from East African Medicinal Plants. *PloS One* **2011,** *6* (2), E-14694.
25. Cseke, L. J.; Kirakosyan, A.; Kaufman, P. B.; Warber, S.; Duke, J. A.; Brielmann, H. L. *Natural Products from Plants.* 2nd ed; CRC Press: Boca Raton, FL, **2016;** 632 pp.
26. Cushnie, T. P.; Lamb, A. J. Antimicrobial Activity of Flavonoids. *Int. J. Antimicrob. Agents* **2005,** *26* (5), 343–356.
27. Daliri, E. B.-M.; Lee, B. H. Current Trends And Future Perspectives On Functional Foods and Nutraceuticals. In *Beneficial Microorganisms in Food and Nutraceuticals*; Liong, M. T., Ed.; Springer International Publishing: New York, **2015;** pp 221–244.
28. Das, L.; Bhaumik, E.; Raychaudhuri, U.; Chakraborty, R. Role of Nutraceuticals in Human Health. *J. Food Sci. Technol.* **2012,** *49* (2), 173–183.
29. Dixon, R. A.; Ferreira, D. Genistein. *Phytochemistry* **2002,** *60* (3), 205–211.
30. Dreosti, I. E. Recommended Dietary Intake Levels for Phytochemicals: Feasible or Fanciful? *Asia Pac. J. Clin. Nutr.* **2000,** *9* (Suppl 1), S119–S122.
31. Dreosti, I. E. Antioxidant Polyphenols in Tea, Cocoa, and Wine. *Nutrition* **2000,** *16* (7–8), 692–694.
32. Escarpa, A.; González, M. C. Total Extractable Phenolic Chromatographic Index: An Overview of the Phenolic Class Contents from Different Sources of Foods. *Euro. Res. Technol.* **2001,** *212* (4), 439–444.
33. Fabricant, D. S.; Farnsworth, N. R. The Value of Plants Used in Traditional Medicine for Drug Discovery. *Environ. Health Perspect.* **2001,** *109*, 69–75.
34. Felgines, C.; Texier, O.; Morand, C.; Manach, C.; Scalbert, A.; Regerat, F.; Remesy, C. Bioavailability of the Flavanone Naringenin and its Glycosides in Rats. *Am. J. Physiol. Gastroint. Liver Physiol.* **2000,** *279* (6), G1148–G1154.
35. Folkerts, G.; Garssen, J. *Pharma-nutrition: An Overview.* Springer: New York, **2014;** 479 pp.
36. Fozzard, H. A.; Sheets, M. F. Cellular Mechanism of Action of Cardiac Glycosides. *J. Am. Coll. Cardiol.* **1985,** *5* (5 Suppl A), 10A–15A.
37. Ghosh, S. S.; Massey, H. D.; Krieg, R.; Fazelbhoy, Z. A.; Ghosh, S.; Sica, D. A.; Fakhry, I.; Gehr, T. W. Curcumin ameliorates Renal Failure in 5/6 Nephrectomized Rats: Role of Inflammation. *Am. J. Physiol. Renal Physiol.* **2009,** *296* (5), F1146–F1157.

38. Granata, S.; Dalla Gassa, A.; Tomei, P.; Lupo, A.; Zaza, G. Mitochondria: A New Therapeutic Target in Chronic Kidney Disease. *Nutr. Metabol.* **2015,** *12*, 49.

39. Grumezescu, A. *Nutraceuticals.* Elsevier Science: New York, **2016;** 890 pp.

40. Guo, L.; Duan, L.; Dou, L. L.; Liu, L. L.; Yang, H.; Liu, E. H.; Li, P. Quality Standardization of Herbal Medicines Using Effective Compounds Combination as Labeled Constituents. *J. Pharm. Biomed. Anal.* **2016,** *129*, 320–331.

41. Gupta, C.; Prakash, D. Phytonutrients as Therapeutic Agents. *J. Complement. Integr. Med.* **2014,** *1* (3), 151–169.

42. Harborne, A. J., *Phytochemical Methods A Guide to Modern Techniques of Plant Analysis.* Springer: Netherlands, **1998;** 288 pp.

43. Havsteen, B. H. The biochemistry and Medical Significance of the Flavonoids. *Pharmacol. Therapeut..* **2002,** *96* (2–3), 167–202.

44. Jager, A. K.; Eldeen, I. M.; van Staden, J. COX-1 and COX-2 Activity of Rose Hip. *Phytother. Res.* **2007,** *21* (12), 1251–1252.

45. Jones, W. P.; Chin, Y. W.; Kinghorn, A. D. The Role of Pharmacognosy in Modern Medicine and Pharmacy. *Curr. Drug Targets* **2006,** *7* (3), 247–264.

46. Karwande, V.; Borade, R. *Phytochemicals of Nutraceutical Importance.* Scitus Academics LLC: New York, **2015;** 275 pp.

47. Keefover-Ring, K. M. *One Chemistry, Two Continents: Function and Maintenance of Chemical Polymorphism in the Mint Family (Lamiaceae).* University of Colorado: Boulder, **2008;** 230 pp.

48. Ketron, A. C.; Osheroff, N. Phytochemicals as Anticancer and Chemopreventive Topoisomerase II Poisons. *Phytochem Rev.* **2014,** *13* (1), 19–35.

49. Kinghorn, A. D.; Chai, H. B.; Sung, C. K.; Keller, W. J. The Classical Drug Discovery Approach to Defining Bioactive Constituents of Botanicals. *Fitoterapia* **2011,** *82* (1), 71–79.

50. Kris-Etherton, P. M.; Keen, C. L. Evidence that the Antioxidant Flavonoids in Tea and Cocoa are Beneficial for Cardiovascular Health. *Curr. Opin. Lipidol.* **2002,** *13* (1), 41–49.

51. Kris-Etherton, P. M.; Hecker, K. D.; Bonanome, A.; Coval, S. M.; Binkoski, A. E.; Hilpert, K. F.; Griel, A. E. Bioactive Compounds in Foods: Their Role In The Prevention of Cardiovascular Disease and Cancer. *Am. J. Med.* **2002,** *113* (Suppl 9B), S71–S88.

52. Kong, J. M.; Goh, N. K.; Chia, L. S.; Chia, T. F. Recent Advances in Traditional Plant Drugs and Orchids. *Acta Pharmacol. Sin.* **2003,** *24* (1), 7–21.

53. Kutchan, T. M. Alkaloid Biosynthesis: The basis for Metabolic Engineering of Medicinal Plants. *The Plant Cell* **1995,** *7* (7), 1059–1070.

54. Lasek-Bal, A.; Holecki, M.; Kret, B.; Hawrot-Kawecka, A.; Duława, J. Evaluation of Influence Of Chronic Kidney Disease and Sodium Disturbances on Clinical Course of Acute and Sub-acute Stage First-ever Ischemic Stroke. *Med. Sci. Monitor Int. Med. J. Exp. Clin. Res.* **2014,** *20*, 1389–1394.

55. Leiherer, A.; Mundlein, A.; Drexel, H. Phytochemicals and their Impact on Adipose Tissue inflammation and Diabetes. *Vasc. Pharmacol.* **2013,** *58* (1–2), 3–20.

56. Lobert, S.; Fahy, J.; Hill, B. T.; Duflos, A.; Etievant, C.; Correia, J. J. Vinca Alkaloid-Induced Tubulin Spiral Formation Correlates with Cytotoxicity in The Leukemic L1210 Cell Line. *Biochemistry* **2000,** *39*, 12053–12062.

57. Makkar, H. P.; Becker, K. Vanillin-HCl Method for Condensed Tannins: Effect of Organic Solvents Used for Extraction of Tannins. *J. Chem. Ecol.* **1993**, *19* (4), 613–621.

58. Manfredi, J. J.; Horwitz, S. B. Vinblastine Paracrystals from Cultured Cells are Calcium-stable. *Exp. Cell Res.* **1984**, *150* (1), 205–212.

59. Manfredi, J. J.; Horwitz, S. B. Taxol: An Antimitotic Agent with a New Mechanism of Action. *Pharmacol. Ther.* **1984**, *25* (1), 83–125.

60. Marles, R. J.; Farnsworth, N. R. Antidiabetic Plants and Their Active Constituents. *Phytomed. Int. J. Phytother. Phytopharmacol.* **1995**, *2* (2), 137–189.

61. Martinez, M. G. *Open Innovation in the Food and Beverage Industry.* Elsevier Science: New York, **2013**; 448 pp.

62. Máthé, A. *Medicinal and Aromatic Plants of the World: Scientific, Production, Commercial and Utilization Aspects.* Springer: Netherlands, **2015**; 460 pp.

63. Meskin, M. S.; Bidlack, W. R.; Davies, A. J.; Omaye, S. T. *Phytochemicals in Nutrition and Health.* CRC Press: Boca Raton, FL; **2002**; 224 pp.

64. Milanowski, D. J.; Winter, R. E.; Elvin-Lewis, M. P.; Lewis, W. H. Geographic Distribution of Three Alkaloid Chemotypes of *Croton lechleri. J. Nat. Prod.* **2002**, *65* (6), 814–819.

65. Nagegowda, D. A.; Gutensohn, M.; Wilkerson, C. G.; Dudareva, N. Two Nearly Identical Terpene Synthases Catalyze the Formation of Nerolidol and Linalool in Snapdragon Flowers. *Plant J, For Cell Mol. Biol.* **2008**, *55* (2), 224–239.

66. Neves, A. R.; Lucio, M.; Lima, J. L.; Reis, S. Resveratrol in Medicinal Chemistry: A Critical Review of its Pharmacokinetics, Drug-Delivery, and Membrane Interactions. *Curr. Med. Chem.* **2012**, *19* (11), 1663–1681.

67. Newman, D. J.; Cragg, G. M. Natural Products as Sources of New Drugs Over the Last 25 Years. *J. Nat. Prod.* **2007**, *70* (3), 461–477.

68. Newman, D. J.; Cragg, G. M.; Snader, K. M. Natural Products as Sources of New Drugs over the Period, 1981–2002. *J. Nat. Prod.* **2003**, *66* (7), 1022–1037.

69. Nichenametla, S. N.; Taruscio, T. G.; Barney, D. L.; Exon, J. H. Review of the Effects and Mechanisms of Polyphenolics in Cancer. *Critic. Rev. Food Sci. Nutr.* **2006**, *46* (2), 161–183.

70. Oh, J.; Hlatky, L.; Jeong, Y.-S.; Kim, D. Therapeutic Effectiveness of Anticancer Phytochemicals on Cancer Stem Cells. *Toxins.* **2016**, *8* (7), 199–202.

71. Osawa, T. Nephroprotective and Hepatoprotective Effects of Curcuminoids. *Adv. Exp. Med. Biol.* **2007**, *595*, 407–423.

72. Ovadje, P.; Roma, A.; Steckle, M.; Nicoletti, L.; Arnason, J. T.; Pandey, S. Advances in the Research and Development of Natural Health Products as Main Stream Cancer Therapeutics. *Evid. Based Complement. Altern. Med.* **2015**, *2015*, 12.

73. Palanisamy, N.; Kannappan, S.; Anuradha, C. V. Genistein Modulates NF-κB-associated Renal Inflammation, Fibrosis and Podocyte Abnormalities in Fructose-fed Rats. *Eur. J. Pharmacol.* **2011**, *667* (1–3), 355–364.

74. Peterson, J.; Dwyer, J.; Adlercreutz, H.; Scalbert, A.; Jacques, P.; McCullough, M. L., Dietary lignans: Physiology and Potential for Cardiovascular Disease Risk Reduction. *Nut. Revi.* **2010**, *68* (10), 571–603.

75. Peter, K. V. *Handbook of Herbs and Spices.* Elsevier Science: New York, **2012**; 376 pp.

76. Phillipson, J. D. Phytochemistry and medicinal plants. *Phytochemistry* **2001**, *56*(3), 237–243.

77. Pitchai, D.; Manikkam, R.; Rajendran, S. R.; Pitchai, G. Database on Pharmacophore Analysis of Active Principles, From Medicinal Plants. *Bioinformation* **2010,** *5* (2), 43–45.

78. Prakash, D.; Sharma, G.*Phytochemicals of Nutraceutical Importance.* CABI: London, **2014;** E-book; 324 pp; Available online at: DOI 10.1079/9781780643632.0000 (accessed Sept 1, 2017).

79. Ramawat, K. G. *Desert Plants: Biology and Biotechnology.* In Springer: Berlin Heidelberg, **2009;** 495 pp. Available at: http://www.springer.com/in/book/9783642025495 (accessed Sept 26, 2017).

80. Ramawat, K. G. *Herbal Drugs: Ethnomedicine to Modern Medicine.* In Springer: Berlin Heidelberg, **2008;** 376 pp. Available at: http://www.springer.com/in/book/ 9783540791157 (accessed Sept 26, 2017).

81. Ranawat, L.; Bhatt, J.; Patel, J. Hepatoprotective Activity of Ethanolic Extracts of Bark of Zanthoxylum armatum DC in CCl4 Induced Hepatic Damage in Rats. *J. Ethnopharmacol.* **2010,** *127* (3), 777 Sept 780.

82. Ramakrishna, A.; Ravishankar, G. A. Influence of Abiotic Stress Signals on Secondary metabolites in Plants. *Plant Signal. Behav.* **2011,** *6* (11), 1720–1731.

83. Roeder, E. Medicinal Plants in Europe containing Pyrrolizidine Alkaloids. *Die Pharmazie* **1995,** *50* (2), 83–98.

84. Shankar, S.; Srivastava, R. K. *Nutrition, Diet and Cancer.* Springer: Netherlands, 2012; 623 pp. Available at: http://www.springer.com/in/book/9789400729223 (accessed Sept 15, 2017).

85. Spencer, J. P. E.; Crozier, A. *Flavonoids and Related Compounds: Bioavailability and Function.* CRC Press: Boca Raton, FL, 2012; 471 pp. Available at https://www. crcpress.com/Flavonoids-and-Related-Compounds-Bioavailability-and-Function/ Spencer-Crozier/p/book/9781138199415. (accessed Sept 5, 2017).

86. Surh, Y. J. Cancer Chemoprevention With Dietary Phytochemicals. *Nat. Rev. Cancer.* **2003,** *3* (10), 768–780.

87. Scalbert, A.; Deprez, S.; Mila, I.; Albrecht, A. M.; Huneau, J. F.; Rabot, S. Proanthocyanidins and Human Health: Systemic Effects and Local Effects in the Gut. *BioFactors* **2000,** *13* (1–4), 115–120.

88. Scalbert, A.; Williamson, G. Dietary Intake and Bioavailability of Polyphenols. *J. Nutr.* **2000,** *130* (8S Suppl), 2073S–2085S.

89. Sarkar, A.; Patil, S.; Hugar, L. B.; vanLoon, G. Sustainability of Current Agriculture Practices, Community Perception, and Implications for Ecosystem Health: Indian Study. *EcoHealth.* **2011,** *8* (4), 418–431.

90. Sasidharan, S.; Chen, Y.; Saravanan, D.; Sundram, K. M.; Yoga Latha, L. Extraction, Isolation and Characterization of Bioactive Compounds from Plants' Extracts. *Afr. J. Tradit., Complement., and Altern. Med.* **2011,** *8* (1), 1–10.

91. Saxena, P. *Development of Plant-Based Medicines: Conservation, Efficacy, and Safety.* In Springer: Netherlands, **2013;** 262 pp. Available at: http://www.springer. com/in/book/9780792368717 (accessed Sept 26, 2017).

92. Sener, B. *Biodiversity: Biomolecular Aspects of Biodiversity and Innovative Utilization.* Springer: New York, 2012; 400 pp. Available at: https://link.springer. com/book/10.1007/978-1-4020-6955-0 (accessed Sept 26, 2017).

93. Shahid, M.; Shahzad, A.; Malik, A.; Sahai, A. *Recent Trends in Biotechnology and Therapeutic Applications of Medicinal Plants*. Springer: Netherlands, **2013**; 347 pp. Available at: http://www.springer.com/in/book/9789400766020 (accessed Sept 26).

94. Silva, N. C.; Barbosa, L.; Seito, L. N.; Fernandes, A. Antimicrobial Activity and Phytochemical Analysis of Crude Extracts and Essential Oils from Medicinal Plants. *Nat. Prod. Res.* **2012**, *26*(16), 1510–1514.

95. Skinner, M.; Hunter, D. *Bioactive Compounds in Fruit: Health Benefits and Functional Foods*. John Wiley: New York, **2013**; 509 pp. Available at: http://as.wiley.com/WileyCDA/WileyTitle/productCd-0470674970.html (accessed Sept 26).

96. Sunil, K. N.; Ravishankar, B.; Yashovarma, B.; Rajakrishnan, R.; Thomas, J. Development of Quality Standards of Medicinal Mistletoe - *Helicanthes elastica* (Desr.) Danser Employing Pharmacopoeial Procedures. *Saudi J. Biol. Sci.* **2016**, *23* (6), 674–686.

97. Tiwari, A. K.; Srinivas, P. V.; Kumar, S. P.; Rao, J. M. Free Radical Scavenging Active Components from *Cedrus deodara*. *J. Agric. Food Chem.* **2001**, *49* (10), 4642–4645.

98. Tringali, C. *Bioactive Compounds from Natural Sources, Second Edition: Natural Products as Lead Compounds in Drug Discovery*. CRC Press, Boca Raton; **2011**; 648 pp. Available at: https://www.crcpress.com/Bioactive-Compounds-from-Natural-Sources-Second-Edition-Natural-Products/Tringali/p/book/9781439822296 (accessed Aug, 26).

99. Tolossa, K.; Debela, E.; Athanasiadou, S.; Tolera, A.; Ganga, G.; Houdijk, J. G. M. Ethno Medicinal Study of Plants Used for Treatment of Human and Livestock Ailments by Traditional Healers in South Omo, Southern Ethiopia. *J. Ethnobiol. Ethnomed.* **2013**, *9*, 32–42.

100. van de Venter, M.; Roux, S.; Bungu, L. C.; Louw, J.; Crouch, N. R.; Grace, O. M.; Maharaj, V.; Pillay, P.; Sewnarian, P.; Bhagwandin, N.; Folb, P. Antidiabetic Screening and Scoring of 11 Plants Traditionally Used in South Africa. *J. Ethnopharmacol.* **2008**, *119* (1), 81–86.

101. Vinatoru, M.; Toma, M.; Radu, O.; Filip, P. I.; Lazurca, D.; Mason, T. J. The Use of Ultrasound for the Extraction Of Bioactive Principles from Plant Materials. *Ultrason. Sonochem.* **1997**, *4* (2), 135–139.

102. Wagner, K. H.; Elmadfa, I., Biological Relevance of Terpenoids. Overview Focusing on Mono-, Di- and Tetraterpenes. *Ann. Nutr. Metab.* **2003**, *47* (3–4), 95–106.

103. Watson, R. R. *Polyphenols in Plants: Isolation, Purification and Extract Preparation*. Elsevier Science: New York, **2014**; 475 pp. Available at: https://www.elsevier.com/books/polyphenols-in-plants/watson/978-0-12-397934-6 (accessed Sept 26, 2017).

104. Zhang, L.; Demain, A. L., *Natural Products: Drug Discovery and Therapeutic Medicine*. Humana Press: Totowa, NJ, **2007**; 384 pp. Available at: http://www.springer.com/in/book/9781588293831 (accessed Aug 26, 2017).

CHAPTER 3

ETHNOBOTANICAL SURVEY FOR MANAGING SELECTED NON-COMMUNICABLE DISEASES: CASE STUDY IN NIGERIA

GODWIN OJOCHOGU ADEJO, SUNDAY ENE-OJO ATAWODI, and KINGSLEY OKOYOMOH

ABSTRACT

It is necessary to find facts on medicinal plants that have been used traditionally by people of Idah, Kogi State in Nigeria to manage various non-communicable maladies. The survey in this chapter includes data during 2006–2011 for interviewing adult natives, who were aged between 43 and 86 years. They attested to the medicinal values of each plant presented. Findings were further corroborated with practicing and established traditional herbalists in order to confirm or refute public claims. No unsubstantiated claims were included in this chapter. 54 plants from 33 different families were widely used by most of the people, who use herbal plants to manage various non-communicable diseases. Some of the plants may perhaps form a basis for search of new compounds that possess potent therapeutic and/or prophylactic properties for treating some non-communicable diseases.

3.1 INTRODUCTION

Presently, there is an alarming increase of non-communicable diseases globally.[33] While modern medicine may be rationally argued for, in terms of utility and economics, the fact still remains that the cultural importance of traditional medicine must be upheld. Furthermore, WHO supports the identification and exploitation of aspects of traditional medicine to provide safe and effective remedies or practices that are useful in primary health care.[12] Therefore, the important role of medicinal and poisonous plants in African societies cannot be overemphasized, partly because, folkloric knowledge of medicinal plants have contributed significantly to the discovery of many important drugs that are being used in modern medicine. Herbal remedies are currently used to treat a plethora of acute and chronic disorders ranging from common acute cold to various chronic diseases like cancers.[51,57,99,107,143]

An estimated 50% of Western drugs are plant materials or are parts of the drug models.[167] Most drugs in modern medicine were used in crude form in traditional or folk healing practices, or for other purposes that suggested their efficacy. This has increased the search for natural compounds and other preparations that are plant-based and can be used to prevent or alleviate maladies. For example, some medicinal and food plants such as *Moringa oleifera*,[18] *Khaya senegalensis*,[19] *Cynara scolymus, Cinchona ledgeriana, Digitalis lanata*, and *Papaver somniferum*[99] have been known to possess useful prophylactic and thera-peutic properties. The institutionalization of herbal and traditional medi-cine has historically been identified with the Igala people.[130] Besides, many notable herbal medicinal plants have been reportedly identified and sourced from Igala land.[15,130]

Traditionally, the Igala society is largely agrarian.[157] Additionally in Idah, fishing is also practiced[27,71] due to presence of Niger River and several streams.

Idah local government is in the Eastern part of Kogi State of Nigeria on the Eastern bank of Niger River (Fig. 3.1). With land area of 36.4 km^2 and population of 79,755 in 2006,[112] it is geographically located on 7°5'0" North and 6°45'0" East.[35] The average temperature ranges 25–35°C in Idah with an annual rainfall between 1016 and 1524 mm. The vegeta-tion comprises mixed guinea woodland and forest savanna. There are two distinct rainy and dry seasons, which occur during April–October and November–March, respectively.[7]

Most ethnobotanical studies in Africa have confirmed the significant components of local plants in traditional African medicine. The ever-growing interest in tropical African medicinal plant studies emanates from the challenging realities of western medical practices in Africa, which includes limited access, low affordability, piracy, and diminishing efficacy in some cases. It is, therefore, intuitive for people to resort to what nature has already freely made available. Attempts are being made to validate and document traditional claims.

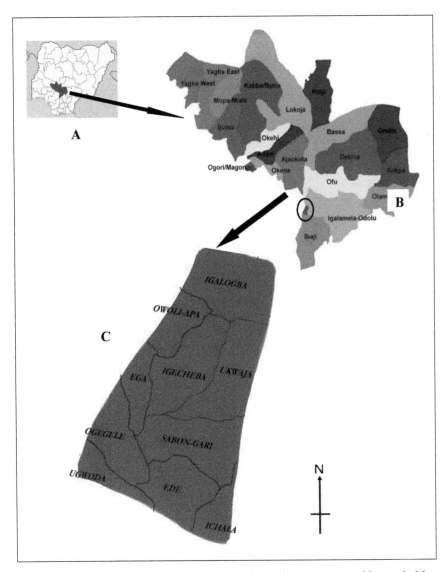

FIGURE 3.1 **(See color insert.)** Location of Idah Local Government and its wards. Map of: A—Nigeria; B—Kogi state; and C—Idah.

The main objective of the survey in this chapter was to investigate and to identify medicinal plants used by Igala people of Idah of Kogi State for managing some non-communicable diseases.

3.2 MATERIALS AND METHODS

3.2.1 SURVEY

The survey was conducted in Idah area of Kogi State in Nigeria. A total of 416 adult Idah natives irrespective of gender—who were lay people and 12 practicing or known herbalists and traditional healers—were orally interviewed. Responses were recorded in pre-planned checklist.

3.2.2 INTERVIEWS

Interviewers gathered information semi-passively and without any strict adherence to time or days during 2006–2011. They were all native speakers of Igala language, drawn from each of the 10 wards. Five interviewers drew and recorded relevant data from their respective wards and periodically submit findings to a coordinator, who proceeded to interview six practicing herbalists and traditional healers in order to corroborate recorded claims. The following five questions were asked during the interview:

- **Question 1:** Name some diseases you know? The lay Igala person may not easily differentiate between non-communicable and communicable diseases. Therefore, each respondent was allowed to name as many as she/he knew. All non-communicable diseases mentioned were noted. Then, she/he was requested to dwell on the ones on, which she/he has most knowledge or information.
- **Question 2:** Are you aware of any medicinal plant that can be used to manage the (named) disease?
- **Question 3:** Can you identify plant and do you know where to locate?
- **Question 4:** Can you supply us the plant?
- **Question 5:** Other information on the part(s) used, preparation method, dosage, and its side effects?

Guide in Table 3.1 was used to determine the rating of level of expertise of natives.

TABLE 3.1   Guide to Determine the Rating of Level of Expertise of Natives.

Response pattern	Rating
Negative response to question 1:	Respondent was thanked and politely excused.
Positive response to question 1 + Negative response to question 2:	The mentioned disease(s) was/were noted and the respondent was politely excused.
Positive response to question 1 + Positive response to question 2 only:	Low level of expertise.
Positive response to question 1 + Positive response to questions 2 and 3:	Medium level of expertise.
Positive response to question 1 + Positive response to questions 2–4 only	High level of expertise.
Positive response to question 1 + Positive response to questions 2– 5:	Very high level of expertise.

3.3   RESULTS

Figure 3.2 shows that about 60% of the people are below high expertise level in use of medicinal plants. Only 17% were at very high expertise level. These people are knowledgeable about at least non-communicable diseases, and were able to mention at least one medicinal plant used for its treatment, able to identify, supplied samples, and gave additional valid information about the plants. The number of women was 4.3% higher than the number of men and women were at very high expertise level among all ages. There was also an increase in expertise level with increase in age. Age group between 51 and 60 presented highest number of persons with increased knowledge of medicinal plants (Fig. 3.3).

From 554 respondents, 138 (24.9%) were discarded either for gross ignorance of at least one non-communicable disease or lack of knowledge of any medicinal plant used to manage the named disease. Among 416 respondents that have at least some knowledge, 111 plants were mentioned as being used for managing different non-communicable diseases. However, after expunging plants with conflicting identities and those unsubstantiated by practicing herbalists, 54 plants were finally listed (Fig. 3.4). In Figure 3.4, the following categories are mentioned:

I   Number of respondents (natives).
II   Number of respondents with any expertise.
III   Number of plants mentioned (by local names).
IV   Total number of plants identified by natives.
V   Number of plants unidentified/unsubstantiated.
VI   Number of plants identified with different names.
VII   Number of plants identified and confirmed by practicing herbalists.
VIII   Actual number of plants identified and confirmed by practicing herbalists.

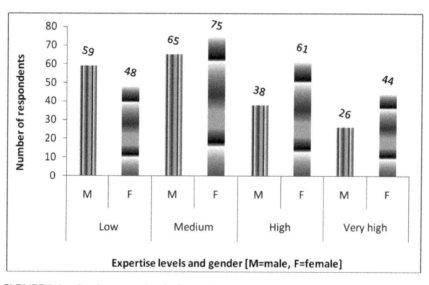

FIGURE 3.2   Gender versus level of expertise.

3.3.1   IDENTIFIED MEDICINAL PLANTS AND THEIR FAMILIES

54 medicinal plants were identified to manage some non-communicable diseases among the Igala population in Idah. The plants belonged to 32 families (Table 3.2). Out of the total number of 32 identified families, leguminaceae and euphorbiaceae families dominated eight and six representatives, respectively. Rubiaceae family was represented by three plants, while apocynaceae, bignonaceae, bombaceae, Asteracea, moraceae, sapotacea had two or one plant representative (Table 3.2).

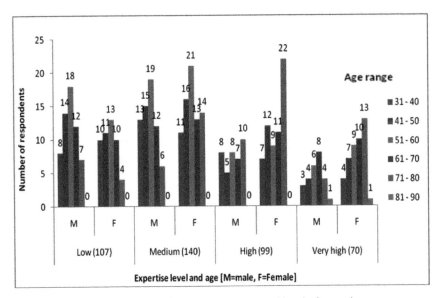

FIGURE 3.3   (See color insert.) Respondent's age and level of expertise.

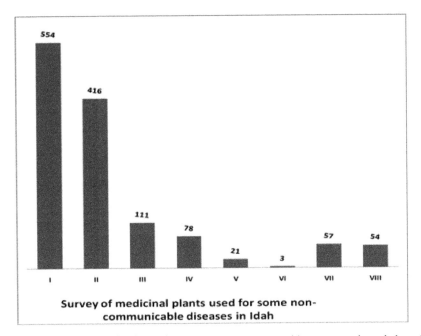

FIGURE 3.4   (See color insert.) Population's response with respect to knowledge of medicinal plants.

3.3.2 SOURCE OF MEDICINAL PLANTS IN AFRICA

Among the 54 identified medicinal plants used by Idah natives to manage non-communicable diseases, only 16 were found to be cultivated while the majority (38) are sourced from the wild (Table 3.2). The cultivated plants are mainly for other purposes, and their medicinal purposes are coincidental.

TABLE 3.2 Identified Plants to Manage Some Non-communicable Diseases in Idah, Their Families, and Occurrence.

Family	Plant	Source
Anacardiaceae	*Lannea nigritana* (Sc. Elliot) Keay	W
Annonaceae	*Annona senegalensis* var. deltiodes Robyns & Ghesq.	W
Apocynaceae	*Alstonia boonei* De Wild.	W
	Rouvolfia caffra Sond.	W
Asclepiadaceae	*Calotropis procera* Linn.	C
Asteraceae	*Vernonia amygdalina* Del.	C
Bignoniaceae	*Kigelia Africana* (Lam.) Benth.	W
	Newbouldia leavis (P. Beauv.) Seeman ex Bureau	C
Bombacaceae	*Adansonia digitata* (L.) Medic.	W
	Bombax costatum Pell. & Vuil.	W
Burseraceae	*Commiphora africana* (A. Rich.) Engl.	W
Caricaceae	*Carica papaya* L.	C
Clusiaceae-Guttiferae	*Gircinia kola* Heckel	C
Compositae	*Ageratum conyzoides* Linn.	W
- Asteraceae	*Chromolaena odorata* (L.f.) King & Robinson	W
Cucurbitaceae	*Luffa cylindrical* (L.) Roem	W
Euphorbiaceae	*Alchornia cordifolia* (Schumm. & Thonn.) Muell.Arg.	W
	Tetracarpidium conophorum (Müll. Arg.) Hutch. & Dalz.	C
	Euphorbia hirta Linn.	W
	Combretum racemosum (P. Beauv.)	W
	Phyllanthus discoides Müll.Arg.	W
	Hymenocardia acida Tul.	W
Labiatae	*Vitex doniana* Sweet	W
Lamiaceae	*Ocimum basilicum* Linn	C
Lecythidaceae	*Napoleona imperalis* P. Beauv.	W

TABLE 3.2 *(Continued)*

Family	Plant	Source
Leguminosae	*Afzelia africana* (Smith)	W
	Abrus precatorius Linn	W
	Acacia nilotica (L.) Wild. ex Del.	W
	Baphia pubescens Hook.f.	W
	Desmodium vellutinum (Wild.) D.C.	W
	Dialium guineense Wild.	W
	Cynometra vogelii Hook	W
	Erythrophleum suaveolens (Guill. & Perr.) Brenan	W
Loganiaceae	*Anthocleista nobilis* G. Don.	W
Loranthaceae	*Viscum album* L.	W
Magnoliopsida	*Elaeise guineensis* Jacq.	C, W
Meliaceae	*Azadirachta indica* A. Juss	C, W
	Ekebergia senegalensis A. Juss	W
	Khaya senegalensis (Desr.) A. Juss.	W
Moraceae	*Ficus carpensis* Thunb.	W
	Ficus thonningii Blume	C
	Treculia Africana Decne ex Trecul	W
Moringaceae	*Moringa oleifera* Lam.	C
Musaceae	*Musa paradaisica* L.	C
Myrtaceae	*Psidium guajava* Linn.	C
Poaceae	*Cymbopogon citratus* (DC) Stapf.	C
Polygalaceae	*Securidaca longipendumculata* Fres.	W
Rubiaceae	*Mitragyna inermes* O. Kze.	W
	Nuclea latifolia Sm.	W
	Morinda lucida Benth.	W
Sapotaceae	*Chrysophyllum albidum* G. Don	C
Spindaceae	*Paullinia pinnata* Linn.	W
Sterculiaceae	*Waltheria indica* Linn.	W
Verbenaceae	*Tectona grandis* Linn.	C

C, cultivated; W, wild.

Plants like *Carica papaya, Moringa oleifera, Tetracarpidium conophorum, Vernonia amygdalina, Ocimum basilicum, Psidium guajava,* and *Musa paradaisica* are cultivated primarily for food. Those cultivated as shade trees are *Ficus thonningii* and *Chrysophyllum albidium* while *Calotropis procera, Newbouldia laevis*, and *Azadirachta indica* are planted as hedge or fence plants and for ecological remediation. *Tectona granis* is cultivated purposely for timber, *Gircinia kola* for herbal remedies and stimulant, while *Cymbopogon citratus* is cultivated around houses with the belief that it wards off snakes.

3.3.3 KNOWN NON-COMMUNICABLE DISEASES IN AFRICA

The research from this survey has shown 31 non-communicable diseases that are traditionally known and are managed using herbal remedies by Igala people of Idah (Appendix A). These diseases, which are known by their Igala native names (Table 3.3), range from minor digestive problems like constipation, to complex pathological conditions like cancer and those related to reproductive, excretory, nervous, circulatory, musculo-skeletal, and respiratory systems.

TABLE 3.3 Some Non-communicable Diseases and Their Common Names in Igala.

English name	Igala name	English name	Igala name
Abnormal hiccups	Ikpeke	Aches and pains	ojibu/anwola
Allergies	Uwoli	Anaemia	oga-ebie'nya
Arthritis	oga-iko	Asthma/respiratory diseases	oga-inmi enya
Cancer	oga-orela	Constipation	efu-eji
Convulsion	oga-ailo	Diabetes	oga-ishuga
Earache	oga-eti	Epilepsy	Ajikpakpa
Food poisoning	oga-ogu ijenwu	Frigidity	Olatokutulu
Hypertension	ebie-efule	Impotence	oga-enekele
Inflammation	oga ene ola	Jaundice	iba ab'eju goo
Kidney disease	oga ef'ikpili	Liver disease	oga ef'odo
Mental disorder	oga imu	Pathological fear/ palpitation	oga edo
Pile	oga ej'of'eyodufu	Poison	oga ogu
Poor limb development	oga ekwute imoto	Spasms	oga agbiti
Stomachache	oga-efu	Spleenomegaly	Obe
Stroke	oga-ekwut'ola	Ulcer	ikete-ef'edo
Wounds/sores	ikete or agbonoko		

3.3.4   PARTS OF A PLANT USED AS A MEDICINE

The leaves of medicinal plants were most popular parts in various herbal preparations, constituting as much as 60%. Plant bark constituted 25% of medicinal plant preparations while the roots, fruits and young stem made up to 4% each. Seeds, peel, and whole plant were 2% and 1%, respectively, thus indicating their infrequent usage in medicinal preparations among the Igala natives. The relative ease or difficulty in accessing plant parts like the roots of most mature medicinal plants may have lowered their frequent inclusion in most preparations. Similarly, the sheer abundance of leaves, branches, and stems may be part of the attributing factors for their regular appearances in medicinal formulations. Results showed that aerial parts of plant were most commonly used, constituting 96% of medicinal plant preparations in Idah.

3.3.5   PREPARATION OF MEDICINAL PLANTS

Figure 3.5 indicates that most common method of preparation of herbal medicines found among the Igala people of Idah is decoction, which constitutes 39% followed by infusion (16%), pulverization (13%), and expression (of juice) from plant material (11%). The least frequently applied methods of processing were preparation into syrups or suspensions and cooking or boiling to be consumed in food as part of the diet (2.5% each). *Desmodium vellutinum* was used as aphrodisiac and the powder of *Viscum album* is added to pap and drank for stomach problems. Raw usage requiring no processing constituted only 1% of the cases.

3.3.6   COMBINATION OF MEDICINAL PLANTS FOR FORMULATION

For most of the diseases, medicinal plants are singly used. Only in 10 out of 53 (19.2%) were combined with other plants probably as adjuncts. The highest combination of four different plants was found in the treatment of bronchitis, where *Cymbopogon citratus* is macerated with the leaves of *Carica papaya*, *Gircinia cola*, and *Capsicum* spp.

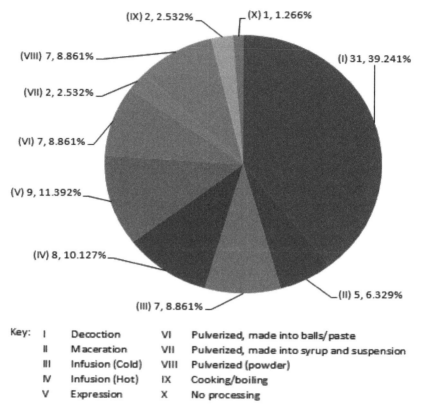

FIGURE 3.5 **(See color insert.)** Modes of extraction and preparation of medicinal plants.

3.3.7 NUMBER OF INDIVIDUAL PLANTS USED IN MANAGING A DISEASE

Several different medicinal plants managed most of the non-communicable diseases. Inflammation was managed by highest number of medicinal plants totaling up to 10, with *Calotropis procera*, *Chrysophyllum albidum*, and *Morinda lucida* being mostly used plants. For aches and pains, only seven medicinal plants were used, while for anemia and constipation, six medicinal plants each, were identified. Other conditions were managed with 2–4 different medicinal plants. However, 8% or 26% of the diseases were managed by only one medicinal plant.

3.3.8   ROUTES: SOLVENTS, DOSAGING, AND ADMINISTRATION

Water was popular medicinal plant extraction medium or vehicle known to the Igala people. Only in very few medicinal plants, palm kernel oil was used, but mainly as a topical application medium rather than for extraction. Dosage regimen commonly ranged from 150 to 250 mL for orally administered medicines while topical applications were simply liberal. Although most presented doses are for adults, yet children and teenagers are administered doses of same preparations by scaling-down prescribed adult doses. However, these are usually subject to the individuals' discretion, which are mostly inaccurate. Administration period was mostly 1–4 weeks, while some are continued until complete cure was observed. Oral route of administration was mostly used (71%), followed by topical (22%) while the least was inhalation and sub-cutaneous application.

3.3.9   SIDE EFFECTS OF MEDICINAL PLANTS

Most of the medicinal plants possess very mild or no side effects. However, stomach upset and purging were identified for the treatment of constipation and food poisoning. These could be attributable to the laxative action mechanisms either via stimulating intestinal motility, generation of myoelectric alterations in intestinal smooth muscle, and also lead to the accumulation of fluid in the intestinal lumen. Laxative effects bring about rapid transit of bowel materials.[59] Laxative-induced fluid accumulation may take place via inhibition of ion and water absorption, stimulation of fluid secretion, or both. Inhibition of cellular energy production or utilization, mucosal injury, and activation of adenylate cyclase could also be part of the mucosal action of these plants.[59]

3.4   DISCUSSION

Historically, Idah people have been trading agricultural products including fresh or partially processed medicinal plants collected from the wild. The Igala culture, including traditional healing practices and beliefs were acquired through oral traditional communication.[129] These knowledge and skills are further transmitted to their offsprings down family lines.

In the course of this survey, each plant was marked for investigation after it is severally implicated. A practicing herbalist, who identified and separately supplied the plant, substantiated the information. The harvested plant and their parts were taken to the herbarium unit of the Department of Biological Sciences, Faculty of Sciences, Ahmadu Bello University, Zaria for identification. After 2 to 3 years, same plant was requested from the herbalist, except where another herbalist was requested to make the supply due to demise of earlier supplier. They were collected and sent to the herbarium unit of the Department of Biological Sciences, National Institute of Pharmaceutical Research and Development (NIPRD), Idu, Abuja as a follow-up and revalidation. Only plants that successfully and consistently passed through these procedures are presented here.

There were probably more than a hundred medicinal plants used by the Igala people. However, most of these plants were not presented here, because they were neither known by more than just one or two natives, nor could be identified. In addition, such information was largely based on thought and hear-says rather than first- or second-hand experience. More-over, the consulted herbalists did not corroborate some of their claims. A good number of medicinal plants equally failed our list for investigation based on disagreements in the information received on such plants. Authors, however, noticed that some plants were called by different local names. For example, "Ogu-olula" and "Ejefe" are popular names by which *Mitracarpus scaber* was known by equally large and different factions of the people. *Chromdaena oderatum* was refered to 'Ile-Ameh' or 'Abilewa'. But *Napoleonia imperalis* is more popularly known as 'Otokuchi' by most of the people rather than 'Obu-anagbo' or 'Iye-anagbo', which are names known by just a few.

It was general observation that most of the medicinal plants were used for more than one malady. For instance, *Morinda lucida* was used in treating earache, stomachache, convulsion, splenomegaly, and inflammation. This was indicative of the Igala people's historic explorative attempts with individual plants in order to discover their possible alternative applications.

Additionally, there were relatively fewer number of medicinal plants used in managing some non-communicable disease like hypertension and stroke (which are examples of cardio-vascular diseases), cancer, and diabetes. The emergences of these diseases were more recent, because of the massive drift from the traditional to western lifestyles. Boutayeb and

Boutayeb[31] have predicted this tendency as the main cause of many non-communicable diseases, which would eventually account for at least 7 out of 10 deaths especially in developing countries.

Experimental findings and reported management of similar or different ailments the Igala people, in order of alphabets are discussed briefly in this section (Appendix A).

Abrus precatorius has been used as anodyne, aphrodisiac, diuretic, emetic, febrifuge, laxative, expectorant, purgative, refrigerant, sedative, abortifacient, antispasmodic, and antimicrobial. They are used in treating paralysis, stiffness, scratches, sores, and wounds caused by dogs, cats, and mice. They are ground with lime and applied for psoriasis, acne, boils and abscesses, tetanus, and to prevent rabies. The root is good in treating jaundice and gonorrhea.[41]

Acacia nilotica leaves are used in Senegal as antiscorbutic while the bark and gum were reportedly used for the treatment of cancers and tumors.[140] The plant has high levels of tannin, which contributes to its medicinal use as a powerful astringent, molluscicide, and algicide.[140] *Acacia nilotica* can be used to treat ailments, such as: cough, leprosy, and dysentery.[68]

Adansonia digitata (baobab tree) has been traditionally used as medicine and food.[36,157] Medicinally, it is used as analgesic, antipyretic, anti-inflammatory agent, and in the management of skin and intestinal problems.[174] It also possesses antibacterial, antiviral, and antitrypanosomal activities.[15,20]

Afzelia africana exhibits bioactive potentials for antimicrobial, anti-inflammatory, antimalaria, analgesic, and trypanocidal activities.[10] *Staphylococcus aureus* and *Bacillus subtilis* were investigated by Akah et al.[10] and were found to be susceptible to the plant extracts. An investigation on the trypanocidal activities of stem bark and leaves extracts of *A. africana* on *Trypanosoma brucei* showed that the protozoan was inhibited by the active component found in the extract.[21] The pulverized roots of *A. africana* combined with millet beer is used as treatment for hernia among some Cote d'Ivoire tribes.[40]

Ageratum conyzoides is being used in parts of Africa, Asia, and South America. It has been used in treating inflammation of the eye, colic, ulcers, and constipation, as an antipyretic and in wound dressing.[130] It is also used for skin diseases and wound healing in Nigeria. The decoction of the plant is taken in to treat diarrhea and to relieve pain associated with navel in children.[130] In Kenya and parts of East Africa, it is used as antiasthmatic, antispasmodic, and hemostatic medicine.[130] Tea of *A. conyzoides* is used as anti-inflammatory, analgesic, and antidiarrheal in Brazilian folk medicine, while it has found application in the treatment of gynecological diseases in Vietnam.[130] The plant contains flavonoids, alkaloids, benzofurans, chromenes, terpenoids, and many miscellaneous compounds.[94]

Alchornia cordifolia is used throughout tropical Africa in the treatment of wounds and ulcers,[122,132] toothache,[46] conjunctivitis, gum, and skin inflammation.[122] It possesses antioxidant property, which is linked with its possible use in hepatoprotection.[133]

Alstonia boonei stem bark has been on the list of antimalarials in the African Pharmacopoeia.[111] The stem bark serves traditionally as antidote for fever, insomnia, chronic diarrhea, rheumatic pains, painful micturition, as antisnake, and antiarrow poisons.[111] It contains alkaloids—particularly indoles, triterpenoids, and steroids. The plant possesses anti-inflammatory properties, while tests have shown its use in treating rheumatism and as an antinociceptive agent.[25,141]

The decoction of *Annona senegalensis* has been reported in treatment of sleeping sickness in Northern Nigeria.[80] Other reported uses include: treatment of cancers, coughs, anemia, chest pain, urinary tract infections, diarrhea, bloody stool, dysentery, arthritis, rheumatism, worms, venereal diseases, head and general body ache, leishmaniasis, trypanasomiasis, lice infestation, eyelid swelling, and snakebites.[4,14,20,23,34,70,98,103,117,156,163] The isolation of monotetrahydrofuran and bis-tetrahydrofuran acetogenins and two cytotoxic monotetrahydrofuran acetogenins from this plant is also documented.[149,150]

Anthocleista vogelii is useful in treating fever among the Igede people of Benue state in Nigeria.[79] The roots are boiled and drunk orally as a laxative. The inner part of the root bark when scraped and squeezed or macerated, salt is added and taken to cure sexually transmitted diseases particularly, syphilis, and gonorrhea.[79]

Azadirachta indica (neem) possesses antitumor,[161] hepato-protective,[66] and antimicrobial properties. The study on wound healing revealed that water extract of the plant was particularly effective on *Pseudomonas sp.* and *Staphylococcus aureus*.[175] *A. indica* has been extensively used in India as an ayurvedic medicine for the treatment of leprosy, intestinal helminthiasis, and respiratory disorders in children. The bark has been used in managing ulcer and possesses cytoprotective properties.[148]

Baphia pubescens bark has been used in folkloric treatment of arthritis and rheumatism. Oil extract from the bark is useful in managing kidney problems and as diuretic.[91] According to folklore, the sap, which is extruded from the plant, is usually collected and applied to an ailing eye.[91]

Bombax Costatum's bark, fruit, and leaves are used for the treatment of fever, epilepsy, and headaches.[12] The leaves also make excellent animal fodder.[144] The fruit is used in cooking and brewing. The fruit contains white floss called 'kapok', which makes it a traditionally useful material in filling pillows and mattresses.[144]

Calotropis procera is used for fever, rheumatism, indigestion, cough, cold, eczema, asthma, elephantiasis, nausea, vomiting, and diarrhea. The plant parts or whole, or mixed with other plant parts have been used to improve the efficacy.[51]

Carica papaya juice is used for treating warts, cancers, tumors, and skin indurations. A preparation from the roots known as 'sinapism' is used in treating uterine tumors while the green fruit is used as ecbolic. The seeds have vermifugal properties and are used to quench thirst. They also possess alexeritic, abortifacient, counter-irritant, emmenagogue, and anthelmintic properties. The roots are used in treatment of piles and yaws.[48] The latex has been locally applied as an antiseptic and has been used as ecbolic in Asia. In Africa, infusions made from the roots are used for treating syphilis. Smokes from the leaf have been to relieve asthma attack by inhalation. In Cuba, the latex is used in psoriasis, ringworm, and treatment of cancerous growths.[48]

Chromolaena odorata leave decoction is usefulness in managing cough. In combination with guava leaves and lemon grass, it is used for treating malaria. It also has antidiarrheal, antispasmodic, antihypertensive, anti-inflammatory, astringent, and diuretic properties.[85] Decoction of the flowers is used as general tonic, antipyretic, and heart tonic.[32]

Chrysophyllum albidum is used as an antimalarial medicinal plant. Odugbemi et al.[127] reported among other Nigeria medicinal plants that its bark and leaves serve as potent malaria therapy comparable to *Artemisia annua* of China. Duyilemi and Lawal[49] also reported use of the roots and leaves of *C. albidium* as cure for the treatment of yellow fever and malaria, while the leaves are used as emollients and for the treatment of skin eruptions, diarrhea, and stomachache arising from infections and inflammatory reactions.

Combretum racemosum possesses antihelmintic and antimicrobial properties.[138] It is useful for genito-urinary and gastrointestinal infections.[138]

Commiphora africana is used for the treatment of typhoid and wounds.[108] The blood-glucose lowering effect of the aqueous ethanolic extract of the bark was recently demonstrated in normoglycemic Wistar rats.[63] Phytochemical studies have shown the presence of alkaloids, tannins, flavonoids, steroids, and saponins.[63]

Cymbopogon citratus fresh or dry leaves are used in Brazil as tea. The extracts are used in medicine as a restorative, digestive and anti-tussis. It is very effective against colds and used as an antihermetic, analgesic, anticardiopatic, and antithermic agent. *C. citratus* also confers anti-inflammatory, diuretic, antispasmodic, diaphoretic, and antiallergic effects.[121] Studies have shown that the phytochemical constituents from the leaves are mainly alkaloids, saponins, âsitosterol, terpenes, alcohols, ketone, flavonoids, chlorogenic acid, caffeic acid, p-coumaric acid, and sugars.[155] However, these constituents of the essential oil and solvent extracts vary depending on the origin. The stem of *Cynometra vogelii* is used as chewing stick for dental and oral health care by the people of Ekiti State. Despite the civilization and advent of conventional tooth-paste, the people first clean with chewing sticks, which are noted for helping in maintaining strong teeth and the relative low cases of dental caries.[95] The chemical constitutes in the plant indicates the presence of tannins and saponins.[96]

Desmodium vellutinum is used in managing stomatitis[86] and diarrhoea.[43] An interesting fact that this plant was not used against any particular disease, but to diagnose if there is any disease.[2]

Dialium guineense is used as excellent chewing stick, believed to possess antimicrobial properties that help in combating tooth decay and maintaining good oral hygiene.[76]

The oils obtained from the mesocarp and from the kernel of *Elaeis guineensis* are effective in combating poison. The kernel oil can be used for convulsing children to ameliorate their body temperature. The meso-carp oil has been used for treating headaches, pains, rheumatism, arterial thrombosis, and cardiovascular diseases.[143,154]

Ekerbergia senegalensis has been a Nigerian medicinal plant, which seems to have low patronage in terms of use. However, it has been used in the treatment of dysentery.[159] A maceration of its bark used to manage ovarian cyst in Cameroon.[88]

The bark extracts *Erythrophleum suaveolens* possess potent anti-inflammatory and analgesic properties.[47,90] The constituent responsible for these observable pharmacological activities is procyanidins. The water extract of the bark can be drunk as abortifacients, ecbolics; for arthritis and rheumatism and for treating cutaneous or subcutaneous parasitic infection.[90] It is useful in managing dropsy, swellings, oedema, gout, eye problems, febrifuges, and used as laxative.[90] The bark and leaf is useful as emetics and for naso-pharyngeal infections. The bark and root is useful in treating leprosy and venereal diseases.[90]

Euphorbia hirta extracts have been used in East and West Africa in treating asthma and respiratory tract inflammations. It is also used for the treatment of cough, chronic bronchitis, and other pulmonary disorders in Malagasy.[151] The plant is also widely used in Angola against diarrhea and dysentery, especially amoebic dysentery.[151] In Nigeria, extracts of the plant are used in the treatment of boils, sore, and also in promoting wound healing.[151] The plant is used as a diuretic, an antidiarrheal, antispasmodic, and anti-inflammatory agent.[151]

Ficus capensis plant is abundant in the tropics and is widely used in Southern, Central, and West Africa as herbal remedy for various ailments such as diarrhea, rheumatism, and threatened abortion.[137] Different parts of the plant possess medicinal properties.[137] The stem bark in particular is often used in folk medicine for the treatment of threatened abortion.[28]

Garcinia kola seeds possess phytochemicals like biflavonoids, xanthone, and benzophenones,[83] which are powerful antioxidants. The phenolics found in *G. kola* have antimicrobial, anti-inflammatory, antiviral, and antidiabetic properties.[5] The seed extracts and powders of *G. kola* have been formulated into tablets, creams, and toothpaste.[82] The extract of the plant is traditionally used for ailments like larngitis, liver disease, and cough.[134] It also has anti-inflammatory, antimicrobial, antidiabetic, antiulceration, and antiviral properties.

Hymenocardia acida is usefulness as remedy for many health conditions. Decoction of the root powder has been used for treating fever, diarrhea, and dysentery. Powder of the roots is also applied in the treatment of colds, muscular pains, headaches, jaundice, hypotension, enteralgia, chest pains, and nephritis, while the root ash is used to treat mouth infections.[81] Experiments have confirmed the antifungal and antimycobacterial, antimicrobial, antisickling, antiulceration, antidiarrhoeal, antiplasmodial, and tyrpanocidal activities. *H. acida* has also been found to possess anti-HIV and anti-inflammatory activities.[3,71,112,116,118,165,169,176]

Khaya senegalensis is used as a fever remedy, as a vermifuge, taenia-cide, depurative, and for treating syphilis.[178] Extract of the bark is useful for the treatment of jaundice, dermatoses, scorpion bites, allergies, infection of the gums, hookworms, disinfection of wounds, and even constipation. In addition to the bark, both the seeds and leaves have also been used as medicine for treating fever and headache, while the preparations made from the roots have been used against syphilis, leprosy, and as an aphrodisiac.[178] In addition to its medicinal values, the bark is used for tanning of leather due to its high tannin contents.[29]

Kigelia africana leave extracts are useful for antidiarrhoeal activity,[9] antileprotic and antimalaria activities especially against drug resistant strains of *Plasmodium falciparum* superior to chloroquine and quinine.[33,105,177] The ethanolic extract of the stem bark possesses strong anti-inflammatory and analgesic activities.[142] The strong pain relieving effects painful joints, back and rheumatism of the fruit and bark extracts have been established.[77] Various parts of the plant have been identified for anticancer properties, apart from its traditional application in managing gynecological disorders. *K. africana* contains phytochemicals like terpenoid, sterols, lignans, naphthaquinones, iridoids, fatty acids, norviburtinal, and flavonoids.[135]

The bark of *Lannea nigritana* can be used in the treatment of diarrhea, dysentery, and as painkiller. The roots are used for pulmonary and for skin issues.[89] The sap from the stem has been medicine for paralysis, epilepsy, convulsion, and spasms. While the fruit serves as food, and the seeds are used as laxative.[89]

Luffa cylindrica (sponge gourd) is used for treating fever, enteritis, and swellings.[125] The extracts from vines are used as ingredients in cosmetics and medicine[106]; for bathing, detoxification, and skin regeneration. The immature fruit is used as vegetable for managing diabetes.[27]

Mitragyna inermes has the presence of sterol, triterpene, polyphenol, flavonoïd, catechic tannin, saponoside, and alkaloid. It has been used in Ivorian traditional medicine to treat diabetes.[100]

The tea from *Morinda lucida* leaves has been used for fever in treating malaria. The leaves are also used as potent laxative. *M. lucida* has been used as analgesic and general febrifuge. Weak decoction made from stem bark is used for treating severe jaundice.[136] Phytochemical constituents of *M. lucida* include distinct types of anthraquinols and alkaloids-anthra-quinones. Isolated and characterized from the stem of *M. lucida* are two compounds namely: oruwalol and oruwal, and 10 anthraquinones.[6] In vitro study, isolated anthraquinones from *M. lucida* have been demonstrated to be active against *P. falciparum*.[101] In other experiments, the antimalarial activity of *M. lucida* against *P. berghei berghei* in mice was reported.[124] Asuzu and Chineme[18] reported the trypanocidal activity of the methanol extract of *M. lucida* leaf.

Roots and bark preparations from *Moringa oleifera* have been traditionally used to treat dental caries[42]; while the gummy extrudation has been employed in the management of syphilis and typhoid,[57] and the leaves for urinary tract infection. The leaves and roots have also been found to possess cardiotonic, antihypertensive, and antitumor properties.[55–57]

Musa paradaisica (popularly called 'banana') is a perennial plant. Every part of *M. paradaisica* has been useful either orally or topically. It has been used in the treatment of diarrhea, dysentery and intestinal lesions in colitis.[160] Besides its usefulness in treatment for snakebite, it has been applied in other medicinal purposes due to its antilithic, anti-inflammatory, antinociceptive, antiulcerogenic, hypoglycemic, hypolipidemic, and antioxidant effects.[37,62,102,109,128,146] Hydroxyanigorufone is one constituent found in *M. paradisiaca*, which confers its cancer chemo preventive properties.[87]

Napoleona imperalis is used as analgesic, tonic, antitussive, antiasthmatic, and wound healing/dressing.[54,81] It is also used for treating asthma[126] according to folklore.

Newbouldia leavis has been used as antidotes for venomous stings and bites[92] and in the treatment of hemorrhoids, ear problems, and as laxatives. It is also used in managing paralysis, epilepsy, convulsions, and spasm.[92] The bark and root parts are used in the treatment of arthritis, rheumatism, and as painkillers. The leaves have been used in the treatment of eye infections. The leaves and roots are used in the treatment of dropsy, swellings, oedema, and gout. The young twigs combined with bark have been used in regulating menstrual cycle in women.[92] Usman and Osuji[170] reported that the plant is effective against elephantiasis, rheumatic swellings, syphilis, dysentery, constipation, and hemorrhoid. It is also a potent vermifuge against round worms. Akunyuli[11] investigated the medicinal uses in the treatment of chest pain, epilepsy, earache, sore foot, and convulsion in children. The leaf, stem, and fruits have been used for wound dressing, stomachache, and in alleviating fever.[84] Phytochemical investigation of the root, root bark and stem of *N. laevis* plant showed presence of alkaloids, quinoid, and phenylpropanoid, and so on.[16,60,61]

Decoction of bark and leaves of *Nuclea latifolia* has been used to treat stomach pains, malaria fever, and to expel nematodes in both man and animals.[45] It has also been combined with other plants to treat diarrhea.[166] The effectiveness of *Nuclea latifolia* in reducing pain and inflammation has been reported.[1]

Ocimum bacilicum (sweet basil) has its medicinal use as an antimicrobial and in treating muscle cramps due to its high content of potassium. It also contains high levels of magnesium, which draws water out of the body into the bowels thereby, stimulating digestion.[69] It is also used as acne treatment, antidepressant, antiseptic, antispasmodic, carminative, cephalic, digestive, expectorant, febrifuge, stomachic, and tonic.[69]

Paullinia pinnata has been used traditionally as remedy for different forms of pains.[35] The leaves are used in East African regions against mental problems, eye troubles and blindness, snakebites, and rabies. Extracts of a combination of the leaves and the roots are potent against malaria, ancylostomiasis, gonorrhea, paralysis, wounds, threatened abortion, and placenta expulsion during childbirth. The roots of *P. pinnata* have been used as a tonic and styptic medicine and to treat eczema. The processed whole plant is used to treat bad skin conditions, microbial infections and wounds.[35] The root decoction is used as antidote to nausea and vomiting.[35]

The bark of *Phyllanthus discoides* is used as purgative and gastrointestinal disorders. Phytochemical investigation on the plant has shown the presence of alkaloids.[152] Sani et al.[152] found out that the ethyl acetate extract from powdered stem bark contained β-sitosterol, euphol and kaempferol.

Psidium guajava possesses several medicinal properties. The fruit has been used as hemostatic,[26] laxative, antiemertic, treatment of sore throat, and stomach upset.[50] The leaves have antidiarrhoeal, anti-inflammatory, analgesic, antispasmodic,[164] antidiabetic,[39] and used for kidney problems like nephritis,[26] for malaria,[123] in oral care,[131] and for treating vaginal irritation and discharges.[147] A decoction of a combination of leaves and bark is used to remove placenta after childbirth.[50] A root and bark combination has been recommended for chronic infantile diarrhea in decoction of ½ oz. in 6 oz. of water, boiled down to 3 oz., and given as one teaspoon.[50]

Rouvolfia caffra roots are bitter, acrid, heating, sharp, pungent, and anthelminic.[145] Rauvolfia preparations are used as antihypertensive and as sedative. It is also used for the treatment of various central nervous system disorders associated with psychosis, schizophrenia, insanity, insomnia, and epilepsy.[145]

Securidaca longipendumculata bark and roots are consumed orally in the powdered form or as infusions for treating inflammation, constipation, headache, chest pain, abortion, tuberculosis, infertility, and venereal diseases. Chewing the roots of *S. longipendumculata* helps in relieving toothache.[170,171] Combination of the roots of *S. longipendumculata* and dwarf custard apple is used in treating gonorrhea. The powder of roots or wood scrapings has been traditionally used in treating headache by rubbing on the forehead, while the root infusions are used to wash tropical ulcers.[24]

Tectona grandis leaf extract is used in the folklore for the treatment of various kinds of wound, especially burn wound.[162] The seeds are used as diuretic and promotion of growth of hair. The flowers are used as diuretic and to treat biliousness, bronchitis, and urinary disorders. The leaves are effective against mycobacterium, tuberculosis and for preservation of meat and fish. The bark is useful against bronchitis, while the roots are used for dyeing mattings.[65]

Tetracarpidium conophorum (African walnut) possesses antimicrobial activities[8] and this must be reason for its folkloric use in treating dysentery. It is also used to improve male fertility.[8] The plant contains two isolectins: Agglutin I and II in the seed extract.[17] Other useful compounds found in the nut include: oxalates, phytates, tannins, proteins, fiber, oil, and carbohydrates.[52,53]

Treculia africana is used to treat various infectious diseases, inflammations, injuries, and other diseases.[104]

The cold water extract of *Vernonia amygdalina* using 2 to 3 crushed leaves (approximately 10–15 g fresh weight) in 300–400 mL of water has been used for parasitosis or gastrointestinal upset, to restore appetite, physical strength, urine, and stool quality.[74] In western Uganda, *V. amygdalina* has been traditionally used in aiding childbirth by hastening parturition.[119] The two most abundant and bioactive constituents, vernodalin and vernonioside B_1 are found in the plant.[75]

Viscum album alleviates the side effects of cancer therapy. It has been used to treat hypertension, epilepsy, exhaustion, anxiety, arthritis, vertigo, and degenerative inflammation of the joints.[110] The herb is known to relieve headaches due to high blood pressure. It also reduces heart rate and strengthens the capillary walls at the same time. The cardiotonic action is attributable to the presence of lignans while the hypotensive action is linked with its choline derivative related to acetylcholine, which exerts parasympathetic stimulation and vasodilation.[173]

In Nigeria, decoction of the chopped stem bark of *Vitex doniana* is prepared and consumed orally for the treatment of gastroenteritis. It is useful in treating gut infections, improving fertility, and treating eye infections.[97]

For sore throat, *Waltheria indica* buds are given to children to chew and the roots or bark is for adults. Fresh juice from the whole plant is used for treating asthma.[67,72,93] In Mexico, the decoction is taken regularly to tackle syphilis. It is used to prevent fever and as abortifacient in Ghana. It serves as a purgative in South Africa, while the leaves are used to treat excess giddiness in Nigeria. In the Caribbean, it is used in treating urinary tract infections while it is used to cure fatigue.[44,114] It is combined with *Musa. spp* and *Syzygium malaccense* for sore throat. In combination with *Musa spp.*, *Hedyotis spp.*, and *Phyllostachys nigra* shoots, it is potent for bronchial infections. When combined with *Desmodium spp.*, it is highly effective in treating asthma.[44]

The use of traditional herbal medicine seems to be appreciably developed among the Igala people, considering the dosages and quantification of administered herbal preparations. According to Sofowora,[158] the use of different parts of a plant singly in some instances or combining them in another instance, and the combination of parts of different plants go further to indicate proven evidences of good understanding based on observed efficacies.

3.5 SUMMARY

The Igala people's involvement in the use of herbal medicine and mastery over time is legendary, and Idah, which serves as the administrative headquarters of the Igala kingdom, has played significant role in this attainment. Although a vast number of herbal practices and the details of their arts of healing through medicinal plants have been lost due to neglect, yet lack of records and blackmail is enormous. Authors indicate that good number of medicinal plants presented in this chapter can be candidates for further research and development into medically prescribable drugs.

ACKNOWLEDGMENTS

Profound appreciations of the authors go to Mal. US Gallah of the Herbarium unit of the Biological Science Department, Faculty of Sciences, A.B.U.,

Zaria for the identification of plant specimens; Mrs. Jemilat A. Ibrahim of the Herbarium unit, NIPRD, Abuja for helping with plant validation.

KEYWORDS

- **ethnobotanical**
- **medicinal plants**
- **Nigeria**
- **non-communicable diseases**
- **plant bioactive compounds**

REFERENCES

1. Abbah, J.; Amos, S.; Chindo, B.; Ngazal, I.; Vongtau, H. O. Pharmacological Evidence Favoring the use of *Nauclea Latifolia* in Malaria Ethnopharmacy: Effects Against Nociception, Inflammation, and Pyrexia in Rats and Mice. *J. Ethnopharmacol.* **2010,** *127* (1), 85–90.
2. Abhijit, D.; Jitendra, N. D. Survey of Ethnomedicinal Plants Used by the Tribals of Ajoydha Hill Region, Purulia District, India. *Amer. Eur. J. Sustain. Agric.* **2010,** *4* (3), 280–290.
3. Abu, A. H.; Uchendu, C. N.; Ofukwu, R. A. In vitro Antitrypanosomal Activity of Crude Extracts of Some Selected Nigerian Medicinal Plants. *J. Appl. Biosci.* **2009,** *21*, 1277–1282.
4. Abubakar, M. S.; Musa, A. M.; Ahmed, A.; Hussaini, I. M. The Perception and Practice of Traditional Medicine in the Treatment of Cancers and Inflammations by the Hausa and Fulani Tribes of Northern Nigeria. *J. Ethnopharmacol.* **2007,** *111*, 625–629.
5. Adedeji, O. S.; Farimi, G. O.; Ameen, S. A.; Olayemi, J. B. Effects of Bitter Kola (*Garcinia Kola*) as Growth Promotes in Broiler Chicks from Day Old to Four Weeks Old. *J. Animal Vet. Adv.* **2006,** *5* (3), 191–193.
6. Adewunmi, C. O.; Adesogan, E. K. Anthraquinones and Oruwacin from *Morinda Lucida* as Possible Agents in Fasciolasis and Schistosomiasis Control. *Fitoterapia* **1984,** *55*, 259–63.
7. Adeyemi, S. O. Food and Feeding Habits of *Protopterus annectens* (Owen) (Lungfish) at Idah Area of River Niger, Nigeria. *Prod. Agric. Technol.* **2010,** *6* (2), 69–74.
8. Ajaiyeoba, E. O.; Fadare, D. A. Antimicrobial Potential of Extracts and Fractions of the African Walnut – *Tetracarpidium Conophorum. Afr. J. Biotechnol.* **2006,** *5* (22), 2322–2325.

9. Akah, P. A. Antidiarrhoeal Activity of the Aqueous Leaf Extract of *Kigelia Africana* in Experimental Animal. *J. Herbs Spices Med. Plants* **1996**, *4* (2), 31–38.

10. Akah, P. A.; Okpi, O.; Okoli, C. O. Evaluation of the Anti-inflammatory, Analgesic and Anti-Microbial Activities of *Afzelia africana* Niger. *J. Nat. Prod. Med.* **2007**, *11*, 48–52.

11. Akunyili, D. N. In *Anticonvulsant Activity of the Ethanolic Extract of Newbouldia Leavis*, Proceedings of the 2nd NAAP Scientific Conference, 2000; pp 155–158.

12. Akwaji, P. I.; Eyam, E. O.; Bassey, R. A. Ethnobotanical Survey of Commonly used Medicinal Plants in Northern Cross River State, Nigeria. *World Sci. News* **2017**, *70* (2), 140–157.

13. Aladesanmi, A. J.; Nia, R.; Nahrstedt, A. New Pyrazole Alkaloids from the Root Bark of *Newbouldia Leavis*. *Planta Medica* **1998**, *64*, 90–91.

14. Alawa, C. B. I.; Adamu, A. M.; Gefu, J. O.; Ajanusi, O. J.; Abdu, P. A.; Chiezey, N. P.; Awawa, J. N.; Bowman, D. D. In vitro Screening of Two Nigerian Medicinal Plants (*Vernonia amygdalina*) and (*Anonna senegalensis*) for Anthelmintic Activity. *Vet. Parasitol.* **2003**, *113*, 73–81.

15. Anani, K.; Hudson, J. B.; De Souza C.; Akpagana, K.; Towers, G. H. N.; Amason, J. G.; Gbeassor, M. Investigation of Medicinal Plants of Togo for Antiviral and Antimicrobial Activities. *Pharm. Biol.* **2000**, *38*, 40–45.

16. Aniama, S. O.; Usman, S. S.; Ayodele, S. M. Ethnobotanical Documentation of Some Plants Among Igala People of Kogi State. *Int. J. Eng. Sci.* **2016**, *5* (4), 33–42.

17. Animashaun, T.; Togun, R. A.; Hughes, C. R. Characterization of Isolectins in *Tetracarpidium Conophorum* Seeds (Nigerian Walnut). *Glycoconjugate J.* **1994**, *11*, 299–303.

18. Asuzu, I. U.; Chineme, C. N. Effects of *Morinda Lucida* leaf Extract on *Trypanosoma Brucei Brucei* Infection in Mice. *J. Ethnopharmacol.* **1990**, *30*, 307–313.

19. Atawodi, S. E. Antioxidant Potential of African Medicinal Plants. *Afr. J. Biotechnol.* **2005**, *4* (2), 128–133.

20. Atawodi, S. E.; Bulus, T.; Ibrahim, S.; Ameh, D. A.; Nok, A. J.; Mamman, M.; Galadima, M. In vitro Trypanocidal Effect of Methanolic Extract of Some Nigerian Savannah Plants. *Afr. J. Biotechnol.* **2003**, *2*, 317–321.

21. Atawodi, S. E.; Atawodi, J. C.; Idakwo, G. A.; Pfundstein, B.; Haubner, R.; Wurtele, G.; Spiegelhalder, B.; Bartsch, H.; Owen, R. W. Evaluation of Polyphenol Content and Antioxidant Properties of Methanol Extracts of the Leaves, Stem, And Root Barks of *Moringa Oleifera* Lam. *J. Med. Foods* **2010**, *13* (3), 710–716.

22. Atawodi, S. E.; Atawodi, J. C.; Pala, Y.; Idakwo, P. Assessment of the Polyphenol Profile and Antioxidant Properties of Leaves, Stem and Root Barks of *Khaya Senegalensis* (Desv.). *Elec. J. Biol.* **2009**, *5* (4), 80–84.

23. Audu, J. Medicinal Herbs and their Uses in Bauchi State. *Niger. Field* **1998**, *54*, 157–168.

24. Avhurengwi, P. N. *Securidaca Longepedunculata* Fresen, Walter Sisulu National Botanical Garden, 2006. http://www.plantzafrica.com/plantqrs/securidlong.htm (accessed May 15, 2018).

25. Awe, S. O.; Opeke, O. O. Effects of *Alstonia Congensis* on *Plasmodium Berghei* in Mice. *Fitoter* **1990**, *61*, 225–229.

26. Ayensu, E. S. *Medicinal Plants of West Africa*; Reference Publications: Algonac, Michigan, 1978; pp 312.

27. Bal, K. J.; Hari, B. K. C.; Radha, K. T.; Madhusudan, G.; Bhuwon, R. S.; Madhusudan, P. U. *Descriptors for Sponge Gourd (Luffa Cylindrica* (L.) Roem.); NARC, LIBIRD & IPGRI: Nigeria, 2004; pp 115.

28. Berg, C. C. Annotated Checklist of the *Ficus* Species of the African Floristic Region, with Special Reference and a Key to the Taxa of South Africa. *Kirkia* **1990**, *13*, 257.

29. Boffa, J. M. *Agroforestry Parklands in Sub-Saharan Africa*. Conservation Guide No. 34; FAO: Rome, 1999; pp 32.

30. Boston, J. Igala Political Organization. *Afr. Notes* **1968**, *4*, 2–3.

31. Boutyeb, A.; Boutayeb. S. The Burden of Non-communicable Diseases in Developing Countries. *Int. J. Equity Health* **2005**, *4* (1), 2–6.

32. Bunyapraphatsara, N.; Chokechaijaroenporn, O. *Thai Medicinal Plants*; Faculty of Pharmacy, Mahidol University and National Center for Genetic Engineering and Biotechnology: Bangkok, 2000; Vol 4, pp 622–626.

33. Carvalho, L. H.; Rocha, E. M. M.; Raslan, D. S.; Oliveira, A. B.; Krettli, A. U. In vitro Activity of Natural and Synthetic Naphthoquinones Against Erythrocytic Stages of the *Plasmodium Falciparum*. *Braz. J. Med. Biol. Res.* **1988**, *21*, 485–487.

34. Chabra, S. C.; Makunah, R. L. A.; Mshiu. E. N. Plants Used in Traditional Medicine in Eastern Tanzania: Pteridopyhtes and Angiosperms (Aquanthaceae to Canelliceae). *J. Ethnopharmacol.* **1987**, *21*, 253–277.

35. Chabra, S. C.; Makuna, R. L. A; Mshiu, E. N. Plants Used in Traditional Medicine in Eastern Tanzania. *J. Ethnopharmacol.* **1991**, *33*, 143–57.

36. Chadare, F. J.; Linnemann, A. R.; Hounhouigan, J. D.; Nout, M. J. R.; Van Boekel, M. A. J. Baobab Food Products: Review on their Composition and Nutritional Value. *Crit. Rev. Food Sci. Nutr.* **2009**, *49*, 254–274.

37. Coe, F.; Anderson, G. J. Ethnobotany of the Sumu (Ulwa) of Southeastern Nicaragua and Comparisons with Miskitu Plant Lore. *Econ. Botany* **1999**, *53*, 363–383.

38. Collins Maps. Map of Idah, Nigeria, Africa, 2011. http://www.collinsmaps. com/maps/nigeria/kogi/idah/p479594.00.aspx (accessed May 15, 2018).

39. Conway, P. *Tree Medicine–A Comprehensive Guide to the Healing Power of Over 170 Trees*; Judy Piatkus (Publishers) Ltd.: Africa, 2001; pp 113.

40. Dalziel, J. M. *The Useful Plants of West Tropical Africa*; Whitefriars Press: London, 1937; pp 164.

41. Daniel, M. *Medicinal Plants: Chemistry and Properties*, 1st Ed.; Oxford And IBH Publishing House Co. Pvt. Ltd: New Delhi, 2006; pp 118–119.

42. Das, B. R.; Kurup, P. A.; Narasimha, P. L. Antibiotic Principle from *Moringa Pterygosperma*, Part VII: Anti-Bacterial Activity and Chemical Structure of Compounds Related to Pterygospermin. *Ind. J. Med. Res.* **1957**, *45*, 191–196.

43. Dash, S. K.; Padhy, S. Review on Ethnomedicines for Diarrhoea Diseases From Orissa: Prevalence Versus Culture. *J. Human Ecol.* **2006**, *20* (1), 59–64.

44. David, B. L. *Medicine at your Feet: Healing Plants of the Hawaiian Kingdom*; Roast Duck Production Publishers: Hawaii, 2006; pp 1–9.

45. Deeni, Y. Y.; Hussain, H. S. N. Screening for Antimicrobial Activity And for Alkaloids of *Nauclea Latifolia*. *J. Ethnopharmacol.* **1991**, *35*, 91–96.

46. Delaude, C.; Delaude, J.; Breyne, H. *Plantes Medicilaes et Ingredients Magiques Du Grands Marche De Kinshasa* (Medicine Plants and Magical Ingredients of the Great Marche of Kinshasa); Tervuren: Africa, 1971; Vol 17, pp 93–103.

47. Dongmo, A. B.; Kamanyi, A.; Anchang, M. S.; Nkeh, B. C.; Njamen, D.; Nguelefack, T. B.; Nole, T.; Wagner, H. Anti-inflammatory and Analgesic Properties of the Stem Bark Extracts of *Erythrophleum Suaveolens* (Caesalpiniaceae), Guillemin and Perrottet. *J. Ethnopharmacol.* **2001**, *77* (2–3), 137–141.

48. Duke, A. J. *Handbook of Energy Crops.* Unpublished, 1983. https://www.hort.purdue. edu/newcrop/duke_energy/dukeindex.html (accessed Aug 28, 2018).

49. Duyilemi, O. P.; Lawal, I. O. Antibacterial Activity and Phytochemical Screening of *Chrysophyllum Albidum* Leaves. *Asian J. Food. Agro-Indus.* **2009**, Special Issue, S75–S79.

50. Dweck C. A. *A Review of Guava (Psidium Guajava)*; FLS FRSC FRSH: Dweck Data, 2004; pp 94.

51. Dwivedi, A.; Chaturvedi, M.; Gupta, A; Argal, A. Medicinal Utility of *Calotropis Procera* (Ait.) R. Br. as Used by Natives of Village Sanwer of Indore District, Madhya Pradesh. *Int. J. Pharm. Life Sci.* **2004**, *1* (3), 188–190.

52. Enujiugha, V. N. Chemical and Functional Characteristics of Conophor Nut. *Pak. J. Nutr.* **2003**, *2*, 335–338.

53. Enujiugha, V. N.; Ayodele-Oni, O. Evaluation of Nutrient and Anti-nutrients in Lesser Known Under-utilized Oil Seeds. *Int. J. Food Sci. Technol.* **2003**, *38*, 525–528.

54. Esimone, C. O., Ibezim, E. C.; Chah, K. F. The Wound Healing Effect of Herbal Ointments Formulated with *Napoleona Imperialis*. *J. Pharm. Allied Sci.* **2005**, *3* (1), 294–299.

55. Faizi, S. Bioactive Compounds from the Leaves and Pods of *Moringa Oleifera*. *New Trends Nat. Prod. Chem.* **1998**, *10*, 175–183.

56. Faizi, S. Novel Hypotensive Agents, Niazimin A, Niazimin B, Niazicin A and Niazicin B from *Moringa Oleifera*: Isolation of First Naturally Occurring Carbamates. *J. Chem. Soc. Perkin Trans. I*, **1994**, *12*, 3035–3040.

57. Fuglie, L. J. *The Miracle Tree: Moringa Oleifera: Natural Nutrition for the Tropics*; Church World Service: Dakar, 1999; pp 68.

58. François, M. N.; Amadou, D.; Rachid, S. Chemical Composition and Biological Activities of *Ficus Capensis* Leaf Extracts. *J. Nat. Prod.* **2010**, *3*, 147–160.

59. Gaginella, T. S.; Bass, P. Laxatives: An Update on Mechanism of Action. *Life Sci.* **1978**, *23* (10), 1001–1009.

60. Gafner, S.; Wolfender, J. L.; Nianga, M.; Hostettmann, K. Phenylpropanoid Glycosides from *Newbouldia Laevis* Roots. *Phytochemistry* **1997**, *44* (4), 687–690.

61. Germann, K., Kaloga, M.; Ferreira, D.; Marais, J. P.; Kolodziej, H. Newbouldioside A-C Phenyl-Ethananoid Glycosides from the Stem Bark of *Newbouldia Leavis*. *Phytochemistry* **2006**, *67* (8), 805–811.

62. Goel, R. K.; Sairam, K.; Rao, C. V. Role of Gastric Antioxidant and Anti-*helicobacter Pylori* Activities in Anti-ulcerogenic Activity of Plantain Banana (*Musa Sapientum* Var. Paradisiaca). *J. Exp. Biol.* **2001**, *39*, 719–722.

63. Goji, A. D. T.; Dikko, A. A. U.; Bakari, A. G.; Mohammed, A.; Ezekiel, I.; Tanko, Y. Effect of Aqueous-Ethanolic Stem Bark Extract of *Commiphora Africana* on Blood

Glucose Levels on Normoglycemic Wistar Rats. *Int. J. Animal Vet. Adv.* **2009**, *1* (1), 22–24.

64. Gonzalez-Avila, M.; Arriaga-Alba, M.; De La Garza, M.; Del Carmen Hernandez-Pretelin, M.; Dominguez-Ortiz, M. A.; Fattel-Fazenda, S.; Villa-Trevino, S. Antigenotoxic, Antimutagenic and ROS Scavenging Activities of a *Rhoeo Discolor* Ethanolic Crude Extract. *Toxicol.* in vitro **2003**, *17*, 77–83.

65. Gurmartine, T.; Goudzwaard, L. *Tree Factsheet: Tectona Grandis L*; Forest Ecology and Forest Management: Wageningen University, Wageningen, 2011; pp 38.

66. Haque, E.; Mandal, I.; Pal, S.; Baral, R. Prophylactic Dose of Neem (*Azadirachta Indica*) Leaf Preparation Restricting Murine Tumor Growth is Nontoxic, Hematostimulatory and Immune-Stimulatory. *Immunopharmacol. Immunotoxicol.* **2006**, *28* (1), 33–50.

67. Haselwood, E. L.; Motter, G. G., Eds. *Handbook of Hawaiian Weeds*, 2nd ed.; University of Hawaii Press: Honolulu, 1983; pp 99.

68. Herbal Ayurveda Medicine, 2011. http://www.herbtreatment. com/herbal-ayurveda-medicine/page1.html (accessed May 15, 2018).

69. Herbal-How-To, 2011. *Ocimum Bacilicum*. http://www.herbal-howto-guide.com/ basil-ocimum-basilicum.html (accessed May 15, 2018).

70. Hirschmann, G. S.; Rojas De Arias, A. Survey of Medicinal Plants of *Minas Gerais*, Brazil. *Braz. J. Ethnopharmacol.* **1990**, *29*, 237–260.

71. Hoet, S.; Opperdoes, F.; Brun, R.; Adjakidje, V.; Quetinleclercq, J. In vitro Antitry-panosomal Activity of Ethnopharmacology of Selected Beninese Plants. *J. Ethnopharmacol.* **2004**, *91*, 37–42.

72. Honychurch, P. N. *Carribbean Wild Plants and their Uses*; Macmillan Education Limited: London, 1986; pp 134.

73. Hudson, J. B.; Anani, K.; Lee, M. K.; De Souza, C.; Arnason, J. T.; Gbeassor, M. Further Investigations on the Antiviral Activities of Medicinal Plants of Togo. *Pharm. Biol.* **2000**, *38*, 46–50.

74. Huffman, M. A.; Seifu, M. Observations on the Illness and Consumption of A Possibly Medicinal Plant *Vernonia Amygdalina* (Del.) by a Wild Chimpanzee. In: *Primates*; The Mahale Mountains National Park: Tanzania, 1989; Vol 30, pp 51–63.

75. Huffman, A. M.; Gotoh, S.; Izutsu, D.; Koshimizu, K.; Kalunde, S. M. Further Observations on the Use of the Medicinal Plant, *Vernonia Amygdalina* (Del), by a Wild Chimpanzee, its Possible Effect on Parasite Load, and its Photochemistry. *Afr. Study Monographs* **1993**, *14* (4), 227–240.

76. Hugues, A. A.; Akpona, J. D. T.; Awokou, S. K.; Yemoa, A.; Dossa, L. O. S. N. Inventory, Folk Classification and Pharmacological Properties of Plant Species Used as Chewing Stick in Benin Republic. *J. Med. Plants Res.* **2009**, *3* (5), 382–389.

77. Hutching, A.; Scott, A. H.; Lewis, G.; Cunningham, A. B. *Zulu Medicinal Plants: An Inventory*; University of Natal Press: Pietermaritzburg, 1996; pp 53–54.

78. Idah Local Government. History of the Igala-People, 2013. http://idahgov.org/ history-of-the-igala-people (accessed Jan 9, 2013).

79. Igoli, J. O.; Ogaji, O. G.; Tor-Anyiin, T. A.; Igoli, N. P. Traditional Medicine Practice Amongst the Igede People of Nigeria, Part II. *Afr. J. Trad. Complemen. Altern. Med.* **2005**, *2* (2), 134–152.

80. Igwe, A. C.; Onabanjo, A. O. Chemotherapeutic Effects of *Annona Senegalensis* in Trypanosoma. *Ann. Trop. Med. Parasitol.* **1989**, *83* (5), 527–533.
81. Irvine, F. R. *Woody Plants of Ghana*; Oxford University Press: London, 1961; p 78.
82. Iwu, M. M. Antihepatotic Constituents of *Garcinia Kola* Seeds. *Experimental* **1985**, *4*, 699–700.
83. Iwu, M. M. *Handbook of African Medicinal Plants*; CRC Press: London, 1993; p 244.
84. Iwu, M. M. *Handbook of African Medicinal Plants*; CRC Press: London, 2000; p 19.
85. Iwu, M. M.; Duncan, A. R.; Okunji, C. O. *New Antimicrobials of Plant Origin: Perspectives on New Crops*; ASHS Press: Alexandria, VA, 1999; pp 457–462.
86. Jagtap, S. D.; Deokule, S. S.; Pawar, P. K.; Harsulkar, A. M. Traditional Ethnomedicinal Knowledge Confined to the Pawra Tribe of Satpura Hills, Maharashtra, India. *Ethnobotan. Leaflets* **2009**, *13*, 98–115.
87. Jang, D. S.; Park, E. J.; Hawthorne, M. E.; Vigo, J. S. Constituents of *Musa Paradisiaca* Cultivar with the Potential to Induce the Phase II Enzyme, Quinone Reductase. *J. Agric. Food Chem.* **2002**, *50*, 6330–6334.
88. Jiofack, T.; Fokunang, C.; Guedje, N.; Kemeuze, V.; Fongnzossie, E.; Nkongmeneck, B. A.; Mapongmetsem, P. M.; Tsabang, N. Ethnobotanical Uses of Some Plants of Two Ethnoecological Regions of Cameroon. *Afr. J. Pharm. Pharmacol.* **2009**, *3* (13), 664–684.
89. JSTOR. *Lannea Nigiritana*, 2011. http://plants.jstor.org/upwta/1_189 (accessed May 15, 2018).
90. JSTOR. *Erythropleum Suaveolens*, 2011. http://plants.jstor.org/upwta/3_185 (accessed May 15, 2018).
91. JSTOR. *Baphia Pubescens*, 2011. http://plants. jstor.org/upwta/ (accessed May 15, 2018).
92. JSTOR. *Newbouldia Laevis*, 2011. http://plants.jstor.org/upwta/ (accessed May 15, 2018).
93. Kaaiakamanu, D. M. *Hawaiian Herbs of Medicinal Value*. Trans.; Akaiko Akana, Ed.; Tuttle Company Maui Community College Database: Maui, 1982; p 32.
94. Kamboj, A.; Saluja, A. K. *Ageratum Conyzoides* L.: Review on its Phytochemical and Pharmacological Profile. *Int. J. Green Pharm.* **2008**, *Apr–Jun*, 59–68.
95. Kayode, J.; Omotoyinbo, M. A. Conservation of Botanicals Used for Dental and Oral Healthcare in Ekiti State, Nigeria. *Ethnobotan. Leaflets*. **2008**, *12*, 7–18.
96. Kayode, J.; Omotoyinbo, M. A. Cultural Erosion and Biodiversity: Conserving Chewing Stick Knowledge in Ekiti State, Nigeria. *Afr. Sci.* **2008**, *9* (1), 41–51.
97. Kilani, A. M. Antibacterial Assessment of Whole Stem Bark of *Vitex Doniana* Against Some Enterobactriaceae. *Afr. J. Biotechnol.* **2006**, *5* (10), 958–959.
98. Klaus, V.; Adala, H. S. Traditional Herbal Eye Medicine in Kenya. *World Health Forum* **1994**, *15*, 138–143.
99. Konaté, K.; Alain, S.; Roland, M.; Ahmed, Y. C.; Martin, K.; Aline. L.; Maroufath, L.; Jeanne, M.; Odile, G. N. Polyphenol Contents, Antioxidant and Anti-inflammatory Activities of Six Malvaceae Species Traditionally Used to Treat Hepatitis B in Burkina Faso. *Eur. J. Sci. Res.* **2010**, *44* (4), 570–580.
100. Konkon, N. G.; Adjoungoua, A. L.; Manda, P.; Simaga, D.; N'Guessan, K. E.; Kone, B. D. Toxicological and Phytochemical Screening Study of *Mitragyna Inermis* (Willd.) O Ktze (Rubiaceae) Antidiabetic Plant. *J. Med. Plants Res.* **2008**, *2* (10), 279–284.

101. Koumaglo, K.; Gbeassor, M.; Nikabu, O.; De Souza, C.; Werner, W. Effects of Three Compounds Extracted from *Morinda Lucida* on *Plasmodium Falciparum*. *Planta Medica* **1992**, *58*, 533–538.

102. Krishnan, K.; Vijayalakshmi, N. R.; Alterations in Lipids & Lipid Peroxidation in Rats Fed with Flavonoid Rich Fraction of Banana (*Musa Paradisiaca*) From High Background Radiation Area. *Ind. J. Med. Res.* **2005**, *122*, 540–546.

103. Kudi, A. C.; Myint, S. H. Antiviral Activity of Some Nigerian Medicinal Plant Extracts. *J. Ethnopharmacol.* **1999**, *68*, 289–294.

104. Kuete, V.; Metuno, R.; Ngameni, B.; Mbaveng, A. T.; Ngandeu, F.; Bezabih, M.; Etoa, F. X.; Ngadjui, B. T.; Abegaz, B. M.; Beng, V. P. Antimicrobial Activity of the Methanolic Extracts and Compounds from *Treculia Africana* and *Treculia Acuminata* (Moraceae). *South Afr. J. Botany* **2008**, *74*, 111–115.

105. Lal, S. D.; Yadar, B. K. Folk Medicines of Kurkeshetra District (Haryana), India. *Econ. Botany* **1983**, *37*, 299–305.

106. Lee, S.; Yoo, J. G. Method for Preparing Transformed *Luffa Cylindrica* Roem, 2006. http://www.wipo.int/pctdb/en/wo.jsp?ia=kr20-04002745&display=status (accessed May 15, 2018).

107. Leslie, T. Plant Based Drugs and Medicines. In: *The Healing Power of Rainforest Herbs*; Raintree Nutrition Inc.: Carson City, 2000; pp 113–117.

108. Lewis, W. H.; Elvin-Lewis, M. P. E. *Medicinal Botany: Plants Affecting Man's Health*; John Wiley and Sons: New York, USA, 1977; pp 261–340.

109. Lewis, D. A.; Fields, W. N.; Shaw, G. P. Natural Flavonoid Present in Unripe Plantain Banana Pulp (*Musa Sapientum* L. Var. Paradisiaca) Protects the Gastric Mucosa From Aspirin-Induced Erosions. *J. Ethnopharmacol.* **1999**, *65*, 283–288.

110. Loeper, M. E. Mistletoe (*Viscum Album* L.). Longwood Herbal Task Force, 1999. http://www.mcp.edu/herbal/default.htm (accessed May 15, 2018).

111. Majekodunmi, S. O.; Adegoke, O. A.; Odeku, O. A. Formulation of the Extract of the Stem Bark of *Alstonia Boonei* as Tablet Dosage Form. *Trop. J. Pharm. Res.* **2008**, *7* (2), 987–994.

112. Mann, A.; Amupitan, J. O., Oyewale, A. O.; Okogun, J. I.; Ibrahim, K.; Oladosu, P.; Lawson, L.; Olajide, J.; Nnamdi, A. Evaluation of in vitro Antimycobacterial Activity of Nigerian Plants Used for Treatment of Respiratory Diseases. *Afr. J. Biotechnol.* **2008**, *7* (11), 1630–1636.

113. Mbarek, L. A.; Mouse, H. A.; Elabbadi, N.; Bensalah, M.; Gamouh, A.; Aboufatima, R.; Benharref, A.; Chait, A.; Kamal, M.; Dalal, A.; Zyad, A. Anti-tumor Properties of Blackseed (*Nigella Sativa* L.) Extracts. *Braz. J. Med. Biol. Res.* **2007**, *40*, 839–847.

114. Mcbride, L. R. *Practical Folk Medicine of Hawaii*; Petroglyph Press: Honolulu, Hawaii, 1975; pp 45.

115. Mohammad, S. Anticancer Agents from Medicinal Plants. *Bang. J. Pharmacol.* **2006**, *1*, 35–41.

116. Mpiana, P. T.; Tshibanga, D. S. T.; Shetonde, O. M.; Ngbolua, K. N. In vitro Antidrepanocytary Activity of Some Congolese Plants. *Phytomedicine* **2007**, *14*, 192–195.

117. Muanza, D. N.; Kim, B. W.; Euler, K. L.; Williams, L. Antibacterial and Antifungal Activities of 9 Medicinal Plants from Zaire. *Int. J. Pharmacog.* **1994**, *32*, 337–345.

118. Muanza, D. N.; Euler, K. L.; Williams, L.; Newman, D. S. Screening for Antitumor and Anti-HIV Activities of Nine Medicinal Plants from Zaire. *Int. J. Pharmacol.* **1995**, *33* (2), 98–106.

119. Mugisha-Maud, K.; Oryem-Origa, H.; Odyek, O.; Makawiti, D. W. In *Ethno Pharmacological Screening of Vernonia Amygdalina and Cleome Gynandra Traditionally Used in Childbirth in Western Uganda.* Proceedings of the 11th NAPRECA Symposium Book; Antananarivo: Madagascar, 2006; pp 110–122.

120. NPC (National Population Commission). *Federal Republic of Nigeria. Population and Housing Census, Priority Table (Volume III): Population Distribution by Sex, State, Local Government Area and Senatorial District*; National Population Commission: Abuja-Nigeria, 2010; pp 37.

121. Negrelle, R. R. B.; Gomes, E. C. *Cymbopogon Citratus* (D.C) Stapf: Chemical Composition and Biological Activities. *Revista Brasileira De Plantas Medicinais* **2007**, *9*, 80–92.

122. Neuwinger, H. D. *African Traditional Medicine–A Dictionary Of Plant Use And Applications*; Medical Pharmacology: Stuttgart, Germany, 2000; pp 29–30.

123. Nundkumar, N.; Ojewole, J. A. Studies on the Antiplasmodial Properties of Some South African Medicinal Plants Used as Antimalarial Remedies in Zulu Folk Medicine. *Methods Find. Exp. Clin. Pharmacol.* **2002**, *24* (7), 397–401.

124. Obih, P. O.; Makinde, J. M.; Laoye, J. O. Investigations of Various Extracts of *Morinda Lucida* for Anti-malaraial Actions on *Plasmodium Berghei Berghei* in Mice. *Afr. J. Med. Med. Sci.* **1985**, *14*, 45–49.

125. Oboh, I. O.; Aluyor, E. O. *Luffa Cylindrica*—An Emerging Cash Crop. *Afr. J. Agric. Res.* **2009**, *4* (8), 684–688.

126. Odo, G. N. Preliminary Pharmacological Investigation into the Antispasmodic Properties of the Aqueous Root Extract of *Napoleona Imperialis*. A B. Pharm. Research Project Submitted to the University of Nigeria, Nsukka, 1986.

127. Odugbemi O. T.; Akinsulire, O. R.; Aibinu, I. E.; Fabeku, P. O. *Afr. J. Trad. Complem. Altern. Med.* **2007**, *4* (2), 191–198.

128. Ojewole, J. A.; Adewunmi, C. O. Hypoglycemic Effect of Methanolic Extract of *Musa Paradisiaca* (Musaceae) Green Fruits in Normal and Diabetic Mice. *Methods Find. Exp. Clin. Pharmacol.* **2003**, *25*, 453–456.

129. Okpaku, J. O.; Opubor, A. E.; Oloruntimehin, B. O. The Arts and Civilization of Black and African People: Black Civilization and African Government. *Centre Black Afr. Arts Civil.* **1986**, *9*, 126–130.

130. Okunade L. A. *Ageratum Conyzoides* L. (Asteraceae). *Fitoterapia* **2002**, *73*, 1–16.

131. Okwu, D. E.; Ekeke, O. Phytochemical Screening and Mineral Composition of Chewing Sticks in South Eastern Nigeria. *Global J. Pure Appl. Sci.* **2003**, *9* (2), 235–238.

132. Okwu, D. E.; Ukanwa, N. Isolation, Characterization and Antibacterial Activity Screening of Anthocyanidine Glycosides from *Alchornea Cordifolia* (Schumach. and Thonn.) Mull. Arg. Leaves. *E-J. Chem.* **2010**, *7* (1), 41–48.

133. Olaleye, M. T.; Adegboye, O. O.; Akindahunsi, A. A. *Alchornea Cordifolia* Extract Protects Wistar Albino Rats Against Acetaminophen-Induced Liver Damage. *Afr. J. Biotechnol.* **2006**, *5* (24), 2439–2445.

134. Olaleye, M. T.; Farombi, E. O.; Adewoye, E. A.; Owoyele, B. V.; Onasanwo, S. A.; Elegbe, R. A. Analgesic and Anti-inflammatory Effects of Kolaviron (*Garcinia Kola* Seed Extract). *Afr. J. Biomed. Res.* **2000**, *3*, 171–174.

135. Olatunji, A. G.; Atolani, O. Comprehensive Scientific Demystification of *Kigelia Africana*: A Review. *Afr. J. Pure Appl. Chem.* **2009**, *3* (9), 158–164.

136. Oliver-Bever, B. *Medicinal Plants in Tropical West Africa*; Cambridge University Press: Cambridge, 1986; pp 89–90.

137. Omonkhelin J. O.; Zulekhai A. N.; Abiodun, F.; Buhiyaminde, A. A.; Nwako, C. N. Evaluation of Tocolytic Activity of Ethanol Extract of the Stem Bark of *Ficus Capensis* Thunb. (Moraceae). *Acta Poloniae Pharm. Drug Res.* **2009**, *66* (3), 293–296.

138. Onocha, P. A.; Audu, E. O.; Ekundayo, O.; Dosumu, O. O. Phytochemical and Antimicrobial Properties of Extracts of Combretum Racemosum. *Acta Horti.* **2005**, *1*, 97–101.

139. Opata, C. C.; Agu, S. C. Traditional Medicine and the Promotion of Inter-Group Relations: The Igbo and Igala (Igara) Experiences in Nigeria. *Histor. Res. Lett.* **2012**, *4*, 1–7.

140. Oruwa, C.; Mutua, A.; Kindt, R.; Jamnadass, R.; Anthony, S. *Agroforestree Database: A Tree Reference and Selection Guide* Version 4.0, 2009. http://www.world agroforestry.org/sites/treedbs/treedatabases.asp (accessed May 15, 2018).

141. Osadebe, P. O. Antiinflmmatory Properties of the Root Bark of *A. Boonei*. *Niger. J. Nat. Prod. Med.* **2002**, *6*, 39–41.

142. Owolabi, O. J.; Omogbai, E. K. Analgesic and Anti-inflammatory Activities of Ethanolic Stem Bark Extract of *Kigelia Africana* (Bignoniacea). *Afr. J. Biotechnol.* **2007**, *6* (5), 582–585.

143. Owoyele B. V.; Owolabi, G. O. Traditional Oil Palm *Elaeis Guineensis* Jacq. and its Medicinal Uses: A Review. *TANG* **2014**, *4* (3), 1–8.

144. Oyen, L. P. A. *Bombax Costatum*, 2011. http://www.prota4u.org/search.asp (accessed May 19, 2018).

145. Pankaj, O. *Rauvolfia Serpentine*, 2002. http://www.hort.purdue.edu/newcrop/crop factsheets/rauvolfia.html (accessed May 15, 2018).

146. Prasad, K. V.; Bharathi, K.; Srinivasan, K. K. Evaluation of *Musa Paradisiaca* Linn. Cultivar) – Puttubale Stem Juice for Antilithiatic Activity in Albino Rats. *Ind. J. Physiol. Pharmacol.* **1993**, *37*, 337–341.

147. Raintree. *Carica Papaya*, 2011. http://www.rain-tree.com/papaya.htm (accessed May 15, 2018).

148. Rupesh, M. K.; Mohamed, K. N.; Mani, T. T.; Fasalu, O. M. R.; Satya, K. Review on Medicinal Plants for Peptic Ulcer. *Der Pharm. Lett.* **2011**, *3* (2), 180–186.

149. Sahpaz, S.; Bories, C. H.; Loiseau, P. M.; Cartes, D.; Hocquemiller, R.; Laurens, A. Cave, A. Cytotoxic and Antiparasitic Activity from *Annona Senegalensis* Seeds. *Planta Medica* **1994**, *60*, 538–540.

150. Sahpaz, S., Gonzalez, M. C.; Hocquemiller, R.; Zafra-Polo, M. C.; Cortes, D. Annosenegalin and Annogalene: Two Cytotoxic Monotetrahydrofuran Acetogenins From *Annona Senegalensis* and *Annona Cherimolia*. *Phytochemistry* **1996**, *42*, 103–107.

151. Sandeep, B. P.; Naikwade, N. S.; Magdum, C. S. Phytochemical Screening and Evaluation of the Diuretic Activity of Aqueous Methanol Extract From Aerial Parts

of *Mentha Viridis Linn* (Labiatae) in Albino Rats. *Trop. J. Pharm. Res.* **2014**, *37* (4), 1307–1314.

152. Sani A. A.; Ilyas, N.; Abdulraheem, O. R.; Sule, I. M.; Haruna, A. K.; Ilyas, M.; Abdulkareem, S. S.; Alemika, T. E.; Ekhator, O. Chemical Investigation on the Stem Bark of *Phyllanthus Discoides* (Euphorbiaceae). *Res. J. Pharm. Biol. Chem. Sci.* **2011**, *2* (2), 612–614.

153. Santanu S.; Sambit, P.; Patro, V. J.; Mishra, U. S.; Ashish, P. Antioxidant and Anti-Inflammatory Potential of *Pterospermum Acerifolium*. *Int. J. Pharm. Sci. Rev. Res.* **2010**, *2* (1), Article 001.

154. Sasidharan, S.; Selvarasoo, L.; Latha, L. Y. Wound Healing Activity Of *Elaeis Guineensis* Leaf Extract Ointment. *Int. J. Mol. Sci.* **2012**, *13*, 336–347.

155. Saulo, M. S., Silva, P. S.; Viccini, L. F. Cytogenotoxicity of *Cymbopogon Citratus* (DC) Stapf (Lemon Grass) Aqueous Extracts in Vegetal Test Systems. *Ann. Braz. Acad. Sci.* **2010**, *82* (2), 305–311.

156. Selvanayahgam, Z. E.; Gnanevendhan, S. G.; Balakrishna, K. Antisnake Venom Botanicals from Ethnomedicine. *J. Herbs Spices Med. Plants* **1994**, *2*, 45–100.

157. Sidibe, M.; Wiliams, J. T. Baobab: *Adansonia Digitata*; International Centre for Underutilized Crops, University of Southampton: Southampton, UK, 2002; pp 23.

158. Sofowora, A. *Medicinal Plants and Traditional Medicine in Africa*; Spectrum Books Limited: Ibadan, Nigeria, 1993; pp 9–10.

159. Sofidiya, M. O.; Odukoya, O. A.; Familoni, O. B.; Inya-Agha. S. I. Free Radical Scavenging Activity of Some Nigerian Medicinal Plant Extracts. *Pak. J. Biol. Sci.* **2006**, *9* (8), 1438–1441.

160. Stover, R. H.; Simmonds, N. W. *Bananas*. Tropical Agriculture Series; Longman Scientific and Technical: Essex, Harlow, 1987; pp 86–101.

161. Subhash, C. S. C.; Sandhir, R.; Rai, D. V.; Kaul, A. Preventive Effects of *Azadirachta Indica* on Benzo(A)Pyrene-DNA Adduct Formation in Murine Fore-Stomach and Hepatic Tissues. *Phytother. Res.* **2006**, *2* (10), 889–895.

162. Sumthong, P.; Damveld, R. A.; Choi, Y. H.; Arentshorst, M.; Ram, A. F.; Van Den Hondel, C. A.; Verpoort, R. Activity of Quinines Frim Teak (*Tectona Grandis*) on Fungal Wall Stess. *Planta Medica* **2006**, *72* (10), 943–944.

163. Tabuti, J. R. S.; Lye, K. A.; Dhillion, S. S. Traditional Herbal Drugs of Bulamogi, Uganda: Plant's use and Administration. *J. Ethnopharmacol.* **2003**, *88*, 19–44.

164. Ticzon, R. *Ticzon Herbal Medicine Encyclopaedia*. Ticzon Publishing: Philippines, 1997; pp 48.

165. Tona, L.; Kambu, K.; Masia, K.; Cimanga, R.; Aspers, S.; Rebruyne, T.; Piatens, L.; Totten, J. Biological Screening of Traditional Preparations from Some Medicinal Plants Used as Antidiarrhoeal in Kinshasha, Congo. *Phytomedicine* **1999**, *6* (1), 59–66.

166. Tona, L.; Kambu, K.; Ngimbi, N. Antiamoebic and Spasmolytic Activities of Extracts from Some Anti-diarrhoeal Traditional Preparations Used in Kinshasa, Congo. *Phytomedicine* **2000**, *7* (1), 31–38.

167. Trager, W.; Jensen, J. B. Human Malaria Parasites in Continuous Culture. *Science* **1976**, *193*, 673–675.

168. Udo, K. R. *Geographical Regions of Nigeria*. University of California Press: Davies, USA, 1970; pp 49–51.

169. Ukwe, C. V. Evaluation of the Anti-ulcer Activity of Aqueous Stem Bark Extract of *Hymenocardia Acida*. *Niger. J. Pharm. Res.* **2004,** *3* (1), 86–89.

170. Usman, H.; Osuji, J. C. Phytochemical and in vitro Antibacterial Assay of the Leaf of *Newbouldia Leavis*. *Afr. J. Trad. Complement. Altern. Med.* **2007,** *4* (4), 476–480.

171. Van Wyk, B. E.; Van Oudtshoorn, B.; Gericke, N. *Medicinal Plants of South Africa*; Briza Publications: Pretoria, 1997; pp 98.

172. Van Wyk, B. E.; Gericke, N. *Guide to Useful Plants of Southern Africa*; Briza Publications: Pretoria, 2000; pp 243.

173. Viable Herbal Solutions. *Mistletoe*, 2011. www.viable-herbal.com/singles/herbs/s861.htm (accessed May 15, 2018).

174. Vimalanathan, S.; Hudson, J. B. Multiple Inflammatory and Antiviral Activities in *Adansonia Digitata* (Baobab) Leaves, Fruit and Seeds. *J. Med. Plants Res.* **2009,** *3* (8), 576–582.

175. Vinoth, R. R.; Ramanathan, T.; Savitha, S. Studies on Wound Healing Property of Coastal Medicinal Plants. *J. Biosci. Technol.* **2009,** *1* (1), 39–44.

176. Vonthron-Senecheau, C.; Weniger, B.; Quattara, M.; Tra, B. F.; Kamenan, A.; Lobstein, A. In vitro Antiplasmodial Activity and Cytotoxicity of Ethno Botanically Selected Ivorian Plants. *J. Ethnopharmacol.* **2003,** *87*, 221–225.

177. Weenen, H.; Nkunya, M. H. H.; Bray, D. H.; Mwasumbi, L. B.; Kinabo, L. S.; Kilimali, V. A. E. B. Antimalaria Activity of Tanzanian Medicinal Plants. *Planta Medica* **1990,** *56*, 368–370.

178. World Agroforestry Centre. *Khaya Senegalensis*. Agroforestry Database, 2004. http://www.worldagroforestry.org/sites/treedbs/aft/species (accessed May 15, 2018).

APPENDIX A. List of Some Medicinal Plants for Managing Some Non-communicable Diseases in Idah, Kogi State, Nigeria.

Disease—Plant used (botanical name) [family]	Native name (Igala)	Voucher No.	Native implicating plant	Part used	Preparation method	Dosage/Mode of application	Known side effects
Abnormal hiccups							
Tetracarpidium conophorum (Müll. Arg.) Hutch. & Dalz [Euphorbiaceae]	Ukwa [a]	2144	6	Leaves	Leave decoction.	200 mL drink 6 to 7 times/day.	None
Aches/pains							
Adansonia digitata (L.) Medic. [Bombacaceae]	Obobo [j]	1226	7	Leaves and bark	Cold water infusion.	100–200 mL, 3×/day/1 wk.	None
Afzelia africana (Smith) [Fabaceae]	Anwa [i]	900054	7	Leaves and Bark	Squeezing juice from leaves. Cold water infusion.	Apply liberally to incisions made on the head/headache. 100–250 mL, 3×/day/1–4 wks.	None
Alchornia cordifolia (Schumm. & Thonn.) Muell. Arg. [Euphorbiaceae]	Olufiafia/ Olufofo [a]	1868	6	Leaves and soft stem.	Decoction	150–250 mL, drink 1ce/ day until pain stops.	None
Combretum racemosum (P. Beauv.) [Euphorbiaceae]	Itado [k]	14838	6	Leaves and young stem	Hot water infusion.	100–250 mL, 3×/day/3–7 days.	None
Ficus thonningii (Blume) [Moraceae]	Oda [i]	1885	10	Leaves	Warm water decoction.	As much as can be drunk/ tolerated.	None
Mitragyna inermes O.Kze. [Rubiaceae]	Ohiapele [i]	587	5	Leaves and bark	Decoction.	150–250 mL, 3× daily/1 wk.	None
Phyllanthus discoides Müll.Arg. [Euphorbiaceae]	Ode [l]	221	7	Bark	Bark + black plum (bark) + velvet tamarind (bark) pounded, molded into small bolls and sun-dried.	1 ground boll + water or pap, taken 3×/day/1 wk.	None

APPENDIX A. *(Continued)*

Disease—Plant used (botanical name) [family]	Native name (Igala)	Voucher No.	Native implicating plant	Part used	Preparation method	Dosage/Mode of application	Known side effects
Alergies							
Moringa oleifera Lam. [Moringaceae]	Geli-gedi [h]	571	10	Leaves	Decoction Powder	150–250 mL, drink 1ce/day/as necessary. 1–2 table spoonfuls in food.	None
Anaemia							
Carica papaya L. [Caricaceae]	Echibakpa [a]	195	8	Leaves and roots	Juice squeezed from ripe leaves + water.	150–250 mL, 2×/day/1–2 wks.	None
Ficus carpensis Thunb. [Moraceae]	Ogbaikolo [a]	2368	6	Leaves, Barks	Decoction of leaves combined with leaves of teak, fig, and guava.	50–200 mL, 2× daily depending on age.	None
Kigelia Africana (Lam.) Benth. [Bignoniaceae]	Ebie [f]	2379	5	Leaves and bark	Decoction	100–250 mL, 3×/day/1 wk.	None
Moringa oleifera Lam. [Moringaceae]	Geli-gedi [h]	571	10	Leaves	Decoction	150–250 mL, drink 1ce/day/as long as necessary.	None
Tectona grandis Linn. [Verbenaceae]	Oli-are, Iloba [f]	900056	7	Leaves	Decoction	150–250 mL drink cold or warm, 2× daily	None
Vitex doniana Sweet [Labiatae]	Ejiji [b]	3544	7	Leaves	Decoction of leaves mixed with *Ficus capensis* leaves.	100–250 mL, 2×/day/3 to 4 wks.	None
Arthritis							
Baphia pubescens Hook.f. [Leguminosae-Papilionoideae]	Akpoti [g]	718	5	Leaves	Decoction	100–250 mL, 2×/day/1–4 wks.	None

APPENDIX A. *(Continued)*

Disease—Plant used (botanical name) [family]	Native name (Igala)	Voucher No.	Native implicating plant	Part used	Preparation method	Dosage/Mode of application	Known side effects
Viscum album L. [Loranthaceae]	Oche-oli [h]	90073	5	Leaves	Maceration	Affected part is massaged with juice, using cloth.	None
Carica papaya L. [Caricaceae]	Echibakpa[h]	195	7	Leaves	Decoction + lime orange	100–250 mL, 2–3×/day/3–4 wks.	None
Asthma/respiratory diseases							
Erythrina senegalensis A.Juss. [Leguminaceae (Papilion-aceae)]	Acheche [i]	1675	7(A)	Bark	Powder, made into syrup with palm oil.	2–4 spoonfuls of syrup 3×/day/1 wk.	None
Hymenocardia acida Tul. [Euphorbiaceae]	Enache [i]	318	6(A), 5(T)	Leaves and bark	Decoction with leaves of *Combretum racemosum*.	150–250 mL, 3×/day/1 wk.	None
Euphorbia hirta Linn [Euphorbiaceae]	Omiaku-ikede [f]	583	10(A)	Whole plant	Cold water infusion	150–250 mL, 3cc/day until cured.	None
Cymbopogon citratus (DC) Stapf [Poaceae]	Egbe- Oyibo [b]	1882	7(B)	Leaves	Maceration + leaves of *Carica papaya* + *Gircinia cola* + pinch of *Capsicum* spp.	100–200 mL, 2×/day/2–4 wks Steam inhaled/5–10 min every evening/4 wks.	None
Erythrina senegalensis A.Juss. [Leguminaceae (Papilion-aceae)]	Acheche [i]	1675	7(A)	Bark	Powder, made into syrup with palm oil.	2–4 spoonfuls of syrup 3×/day/1 wk.	None
Cancer							
Lannea nigritana (Sc. Elliot) Keay [Anacardiaceae]	Echikala [f]	220	5	Leaves and bark	Decoction.	Hot liquid used to press breast 2× daily/14 days.	None (avoid poisonous relative)

APPENDIX A. *(Continued)*

Disease—Plant used (botanical name) [family]	Native name (Igala)	Voucher No.	Native implicating plant	Part used	Preparation method	Dosage/Mode of application	Known side effects
Constipation							
Tetracarpidium conophorum (Müll. Arg.) Hutch. & Dalz. [Euphorbiaceae]	Ukwa [a]	2144	6	Fruits	Fruit, cooked, and chewed.	About 10–20 fruits.	None
Abrus precatorius Linn [Leguminaceae]	Epu [g]	476	5	Seeds	Pulverization of mature seeds.	½ to 1 teaspoon full of powder in food 1ce/day until cured.	Stomach upset
Moringa oleifera Lam. [Moringaceae]	Geli-gedi [h]	571	10	Leaves	Par-boiled	Mixed with groundnut cake, pepper, and salt.	Stomach upset
Napoleona imperalis P.Beauv [Lecythidaceae]	Otokuchi [b]	2464	7	Leaves	Decoction	100–250 mL, every evening/3 days	Mild stomach upset
Anthocleista nobilis G.Don. [Loganiaceae]	Odogwu, Okpocha [b]	900202	9	Roots and Leaves	Decoction of combination of leaves.	1 glass 3×/day/4 wks	Stomach upset in overdose
Elaeis guineensis Jacq. [Magnoliopsida]	Ekpe	—		Kernel	Kernel oil.	Adults: 2–4 table spoons 3×/day. Children: 1 teaspoon 3×/day for 3 days and applied into anus.	None
Convulsion							
Morinda lucida Benth. [Rubiaceae]	Ogele [c]	1862	9	Leaves and bark	Hot water infusion of leaves + *Parkia biglobosa* leaves.	In infants, 50–100 mL, 3×/day/1–2 wks. Liquid is also used/bathing.	Nausea when taken orally
Carica papaya L. [Caricaceae]	Echibakpa [e]	195	8	Leaves and roots	Dried powder leave + whole *Phyllantus amarus* palm kernel oil.	Rubbed all over baby's body 4 to 5× daily.	None

APPENDIX A. *(Continued)*

Disease—Plant used (botanical name) [family]	Native name (Igala)	Voucher No.	Native impli- cating plant	Part used	Preparation method	Dosage/Mode of application	Known side effects
Diabetes							
Treculia Africana Decne ex Trecul [Moraceae]	Ehio [a]	9108	6	Fruit and bark	Decoction	150–250 mL, 2× daily until cured.	None
Vernonia amygdalina Del. [Asteraceae]	Ilo [e]	675	6	Leaves	Juice squeezed from fresh leaves and diluted with cold water.	100–250 mL, 2–4×/day, depending on age and severity of ailment.	Purging at higher doses.
Moringa oleifera Lam. [Moringaceae]	Geli-gedi [h]	571	10	Leaves	Decoction	150–250 mL, drink 1ce/ day/as long as necessary.	None
Vernonia amygdalina Del. [Asteraceae]	Ilo [e]	675	6	Leaves	Juice squeezed from fresh leaves and diluted with cold water.	100–250 mL, 2–4×/day, depending on age and severity of ailment.	Purging at higher doses.
Moringa oleifera Lam. [Moringaceae]	Geli-gedi [h]	571	10	Leaves	Decoction	150–250 mL, drink 1ce/ day/as long as necessary.	None
Ear ache							
Morinda lucida Benth. [Rubiaceae]	Ogele [f]	1862	9	Leaves and bark	Decoction of Bark + leaves.	1 drop/ear, 3×/day/1 wk.	None
Epilepsy							
Alstonia boonei De Wild. [Apocynaceae]	Anno [a]	9121	10	Bark and leaves	Hot water infusion.	150–250 mL, 2×/day until cured.	None
Ekebergia senegalensis A. Juss [Meliaceae]	Orachi [f]	766	5	Leaves and Bark	Decoction of leaves and bark	200 mL of warm infusion 3×/day/4 wks.	Not sure
Food poisoning							
Ocimum Basilicum Linn. [Lamiaceae]	Anyeba [e]	90021	8	Leaves	Cold pressing/ squeezing in water + salt.	150–250 mL, 3×/day.	Purging.

APPENDIX A. *(Continued)*

Disease—Plant used (botanical name) [family]	Native name (Igala)	Voucher No.	Native implicating plant	Part used	Preparation method	Dosage/Mode of application	Known side effects
Frigidity/aphrodisiac							
Desmodium vellutinum (Wild.) D.C. [Leguminosae]	Umoga-chi [g]	1553	6	Seeds and leaves	Dried seeds and leave powder applied to food.	1 tablespoon per meal when desired.	None
Paullinia pinnata Linn. [Spindaceae]	Ijili [f]	579	5	Leaves and young stem	Cold water infusion.	200 mL, when needed as aphrodisiac.	None
Hypertension							
Newbouldia leavis (P.Beauv.) Seeman ex Bureau [Bignoniaceae]	Ogichi [j]	2881	9	Leaves	Decoction of dried or young leaves.	150–250 mL, 2×/day until blood pressure normalizes.	Hypotension in prolonged usage.
Vernonia amygdalina Del. [Asteraceae]	Ilo [e]	675	6	Leaves	Juice squeezed from fresh leaves and diluted with cold water.	100–250 mL, 2–4×/day, until normalcy is attained.	Purging at higher doses.
Impotence							
Viscum album L. [Loranthaceae]	Oche-oli [a]	90073	9	Leaves	Dried and powdered leaves.	1 teaspoonful taken daily in food or pap.	None
Cynometra vogelii Hook [Leguminosae-Caesalpinioideae]	Uli [i]	1836	6	Leaves	Leaves ground moist into paste with overripe pawpaw leaves + water.	150 mL, 4× daily and undiluted paste applied round the waist (for impotence or body generally (for measles).	None
Inflammation							
Acacia nilotica (L.) Willd. ex Del. [Leguminoceae (Mimosoideae)]	Anaka [g]	698	5	Leaves and soft stem	Maceration	100–250 mL, 3×/day/1 wk.	None

APPENDIX A. *(Continued)*

Disease—Plant used (botanical name) [family]	Native name (Igala)	Voucher No.	Native implicating plant	Part used	Preparation method	Dosage/Mode of application	Known side effects
Annona senegalensis var. deltiodes Robyns & Ghesq. [Annonaceae]	Abo [g]	900167	7	Leaves, roots, and bark	Decoction of equal parts combined.	100–250 mL, 2×/day/1 to 2 wks.	None
Calotropis procera Linn. [Asclepiada-ceae]	Ebo [a]	900219	10	Leaves	Juice squeezed from pre-heated leaves.	Applied liberally to base of umbilical cord.	None
Chrysophyllum albidum G.Don [Sapotaceae]	Ute, Ehia [f]	2680	9	Leaves/bark.	Decoction of equal portion of leaves and bark.	100 mL, 3× daily/1 wk.	None
Erythrophleum suaveolens (Guill. & Perr.) Brenan [Leguminosae Caesal-pinioideae]	Igbegbe [i]	242	6	Leaves	Ground with local chalk into paste.	Topically applied lavishly over the stomach daily/2 to 3 wks.	None
Khaya senegalensis (Desr.) A.Juss. [Meliaceae]	Ago [f]	900181	5	Leaves and bark	Ground to paste	Liquid wash/ application to affected part.	None
Kigelia africana (Lam.) Benth. [Bignoniaceae]	Ebie [f]	2379	5	Leaves and bark	Ground to paste	Liquid wash/application of paste on affected part.	None
Luffa cylindrical (L.) Roem [Cucurbitaceae]	Ugboche [f]	2552	5	Leaves and unripe fruit	Powdered leaves and fruit made into paste.	Applied to affected part.	None
Morinda lucida Benth. [Rubiaceae]	Ogele [c]	1862	9	Leaves and bark	Hot water infusion of leaves + *Parkia biglobosa* leaves.	150–250 mL, 3×/day/1 wk.	Nausea
Treculia Africana Decne ex Trecul [Moraceae]	Ehio [b]	9108	6	Whole fruit and leaves	Cold or warm water infusion.	150–250 mL, 2×/day/ to curb inflammatory conditions.	None
Acacia nilotica (L.) Willd. ex Del. [Leguminoceae (Mimosoideae)]	Anaka [g]	698	5	Leaves and soft stem	Maceration	100–250 mL, 3×/day/1 wk.	None

APPENDIX A. *(Continued)*

Disease—Plant used (botanical name) [family]	Native name (Igala)	Voucher No.	Native implicating plant	Part used	Preparation method	Dosage/Mode of application	Known side effects
				Jaundice			
Afzelia africana (Smith) [Fabaceae]	Anwa [i]	900054	7	Leaves and Bark	Cold water infusion.	100–250 mL, 3×/day/1–4 wks.	None
Ekebergia senegalensis A.Juss [Meliaceae]	Orachi [f]	766	5	Leaves and Bark	Decoction of leaves + guava leaves + sugar cane leaves.	200 mL of warm infusion 3×/day/2 wks.	Not sure
				Kidney disease			
Baphia pubescens Hook.f. [Leguminosae-Papilionoideae]	Akpoti [g]	718	5	Leaves	Decoction	100–250 mL, 2×/day/1–4 wks.	None
				Liver diseases			
Securidaca longipendumcula Fres. [Polygalaceae]	Ichoko [g]	221	7	Bark, roots and leaves.	Warm water infusion	150–300 mL/day/3 to 4 wks.	None
Gircinia kola Heckel [Clusiaceae (guttiferae)]	Igoligo [f]	239	9	Fruits	Mature fruits chewed alone or ground wet and mixed with honey.	4–10 fruits/day, 2–4 months.	None
Securidaca longipendumcula Fres. [Polygalaceae]	Ichoko [g]	221	7	Bark, roots and leaves.	Warm water infusion	150–300 mL/day/3–4 wks.	None
				Mental disorder			
Rouvolfia caffra Sond. [Apocynaceae]	Okata [i]	900130	5	Leaves	Dried leave powder applied to food	As often as food is demanded.	None
				Pathological fear/palpitations			
Waltheria indica Linn. [Sterculiaceae]	Achifufu [g]	600	5	Leaves, stem, and roots.	Hot water infusion of leaves and stem. Decoction of roots only.	100–250 mL, 1 cc/day.	Insensitivity in pro-longed use

APPENDIX A. *(Continued)*

Disease—Plant used (botanical name) [family]	Native name (Igala)	Voucher No.	Native implicating plant	Part used	Preparation method	Dosage/Mode of application	Known side effects
Pile							
Carica papaya L. [Caricaceae]	Echibakpa [e]	195	8	Leaves, roots	Decoction of fresh roots.	150–250 mL, 2×/day/1–2 wks.	None
Azadirachta indica A. Juss [Meliaceae]	Inimu or Oli-iba [b]	90015	7	Leaves	Cold-water infusion.	150–250 mL, 2×/day until cured.	Nausea
Poison							
Mitragyna inermes O.Kze. [Rubiaceae]	Ohiapele [i]	587	5	Leaves and bark	Decoction.	150–250 mL, 3x daily/1 wk.	None
Commiphora africana (A. Rich.) Engl. [Burseraceae]	Ochimichi [g]	2848	7	Leaves	Warm water infusion.	150–250 mL, daily/1 wk.	None
Ocimum basilicum Linn [Lamiaceae]	Anyeba [e]	90021	8	Leaves	Cold pressing/ squeezing in water + salt.	150–250 mL, 3×/day.	Purging.
Poor limb development in infants							
Securidaca longipedunculata Fres. [Polygalaceae]	Ichoko [f]	221	7	Bark, roots, and leaves.	Warm water infusion	50–200 mL/day and massage of limbs until cured.	None
Spasms							
Ocimum basilicum Linn [Lamiaceae]	Anyeba [e]	90021	8	Leaves	Cold pressing/ squeezing in water + salt.	150–250 mL, 3×/day.	Purging.
Stomach ache							
Ageratum conyzoides Linn. [Compositae]	Iloji-anagbo [k]	261	8	Leaves and young stem	Decoction	150–250 mL, 3×/day /1 wk.	None

APPENDIX A. *(Continued)*

Disease—Plant used (botanical name) [family]	Native name (Igala)	Voucher No.	Native implicating plant	Part used	Preparation method	Dosage/Mode of application	Known side effects
Morinda lucida Benth. [Rubiaceae]	Ogele [c]	1862	9	Leaves and bark	Expressing leaves to obtain juice. Maceration.	In infants, 5–10 mL drops, 3×/day/1 to 2 wks. Adults, 200–300 mL of extract 3×/day until ache subsides.	Nausea when taken orally
Newbouldia leavis (P.Beauv.) Seeman ex Bureau [Bignoniaceae]	Ogichi [a]	2881	9	Leaves and bark	Decoction or maceration.	About 50–100 mL taken with pap.	None
Psidium guajava Linn. [Myrtaceae]	Igwaba [e]	8881	7	Leaves	Decoction	150–250 mL, 2×/day/2–5 days.	None
Spleenomegaly							
Morinda lucida [Rubiaceae]	Ogele [c]	1862	6	Leaves	Decoction	100–250 mL, 2×/day/2–3 wks.	Nausea
Nuclea latifolia Sm. [Rubiaceae]	Ogbahi [h]	005	10	Leaves & Roots	Decoction of parts combined	100 – 250 ml, 2×/day/2 wks.	None
Stroke							
Dialium guineense Willd. [Caesalpiniaceae]	Ayigele [f]	243	5	Bark	Bark is ground to paste. A portion is diluted with warm water.	Paste is applied to affected part. 150–250 mL is taken 3×/day until cured.	None
Ulcer							
Bombax costatum Pell. & Vuil. [Bombacaceae]	Agwugwu-ikeke [i]	90010	6	Leaves	Ground to powder	1 table spoon eaten in soup/day until healed.	None
Musa paradaisica. [Musaceae]	Ogede-agboh b]	1626	6	Peel	Dried and powdered unripe peel.	1 to 2 table spoon in half cup of pap, 2× daily/ 4–6 month.	None

APPENDIX A. *(Continued)*

Disease—Plant used (botanical name) [family]	Native name (Igala)	Voucher No.	Native implicating plant	Part used	Preparation method	Dosage/Mode of application	Known side effects
				Wounds/sores			
Chromolaena odorata (L.f.) King & Robinson [Asteraceae (Compositae)]	Abilewa, IleAmeh [b]	1128	9	Leaves	Raw drops of juice squeezed from fresh leaves applied to site.	Applied lavishly to injury site.	Burning (but harm-less) sensation
Ekebergia senegalensis A.Juss [Meliaceae]	Orachi]	766	5	Leaves and Bark	Leaves and bark ground to paste.	Applied 2×/day after hot water wash/2–3 wks.	Not sure

A, Asthma; T, Tuberculosis; B, Bronchitis.

Herbalists/addresses in Column 2 of Appendix A:

[a] Ikani Dansoho Aduku/Igalogba,

[b] Simon Shaibu/Ubi-Egbe,

[c] Mal. Garba/Ede-Alaba,

[d] Anomo'ne/Ayija,

[e] Baba Ebiloma/Edeh,

[f] Atai-Omoba/Iyegu,

[g] Ajeletu Momoh/Igalogba,

[h] Iye-Inachalo/Inachalo,

[i] Tahiru Osu-ma/Okenya,

[j] Umama Ayija/Ayija,

[k] Udale/Igalogba,

[l] wenya Gabrelu/Ayija.

PART II
Bioactive Compounds and Health Potentials

CHAPTER 4

BIOFLAVONOIDS: SOURCES, TYPES, AND NUTRACEUTICAL MANEUVERS

MUHAMMAD SAJID ARSHAD, UROOJ KHAN, ALI IMRAN, and HAFIZ ANSAR RASUL SULERIA

ABSTRACT

Bioflavonoids have gained attention in the field of research because of their enormous applications in food sciences. Chief dietary sources of bioflavonoids are composed of fruits and vegetables providing a variety of health benefits to the consumer, as well as promoting the growth of plants, by acting as growth promoters and help in combating oxidative stress. They are being used in the prevention of cancer, gastrointestinal ailments, diabetes, and have notable radical scavenging, antiulcerogenic, hepatoprotective, anti-inflammatory, and antimicrobial activity. Bioflavonoids provide helpful role against cardiovascular diseases by working as a vasorelaxant, antiatherosclerotic, antithrombogenic, cardioprotective, and antineoplastic agent. This chapter summarizes the data of past few decades based on potential source and nutraceutical health benefits of bioflavonoids with latest studies.

4.1 AN OVERVIEW OF BIOFLAVONOIDS

Bioflavonoids are phenolic compounds belonging to a major class of secondary metabolites, displaying a vast array of structures and are accountable for organoleptic properties of food and beverage derived from plants, color, and flavor along with nutritional characters. Flavonoids consist of one of the most important groups of plant phenolic and now more than 8000 verities of flavonoids are present.[35,75] The concept is known for centuries that the compounds originated from plants have a vast array of biological significance. Szent-Gyorgyi in 1930 discovered a novel substance from the family of orange and named it as vitamin P. Later it appeared that the substance was infected with flavonoid.[93] The letter "P" was given to this group for the permeability factor, as they enhance the integrity and permeability of capillary lining.

Bioflavonoids originate the term from a Latin word "flavus" which means yellow. The water-soluble substances own the molecular weight range of 300–700, having a basic skeleton comprising of 15 carbon atoms and are generally called as "vitamin-like" substances.[75] Flavonoids gained more interest at the time of French Paradox wherein Mediterranean population an obvious decrease was observed in the cardiovascular disease with the consumption of red wine and saturated fats.

This chapter summarizes the data of past decades on potential source and nutraceutical health benefits of bioflavonoids with latest studies.

4.2 NATURAL SOURCES OF BIOFLAVONOIDS

Strawberries and red grapes are naturally packed with an abundant number of bioflavonoids.[56] There are different types of bioflavonoids and their sources (Table 4.1). Recent research studies reported that the apple poly-phenols aid in the prevention of spikes in blood sugar by means of various mechanisms. Flavonoids present in apple include quercetin majorly and other traces as well, which inhibit alpha-glucosidase and alpha-amylase as these enzymes are responsible for the complex carbohydrates breakdown into simple sugars. Ultimately, this can reduce the absorption of sugar into the body.[11] Peaches contain abundant amount of bioflavonoids poly phenolic antioxidants, including *β-cryptoxanthin, lutein,* and *zea-xanthin.* These compounds work as defensive scavengers against reactive oxygen species (ROS) and oxygen derived free radicals playing an important role in various disease processes.[110]

The tomato family is provided with plenty of useful components for a healthy life, bioflavonoids being one of their chief components.[76] They possess major flavonols, flavones, and polymethoxylated flavones, like kaempferol, quercetin, isorhamnetin, myricetin, apigenin, luteolin, nobi-letin, tangeretin, and their derivatives.[79,82,83] Moreover, in several studies it is evidenced that quercetin is found to be chief flavonoids in garlic, onions, and leeks. As quercetin is known for its sparking potential to boost up the immune system, it is a potent antioxidant.[12] All affiliates of tea family are sufficiently enriched with bioflavonoids, including catechin, epigallocat-echin, epicatechin, and epicatechin gallate. In diabetes prevention, they play their imperative role and directly work on replicating DNA sequences for the preclusion of mutations that lead to cancer[21] Bioflavonoids in our daily diet are also found in cereals and herbs, presenting many benefits.[120]

TABLE 4.1   Bioflavonoids and Their Food Sources.

Food source	Bioflavonoids	References
Apple, berries, cherries, onion, broccoli, kale	Quercetin, kaempherl, rutin, myricetin	[105]
Soybean	Daidzein and genistein	[8]
Strawberry, cherry, and raspberry	Apigenidin, cyaniding	[14]
Tea and its products	Epicatechin and catechin	[51]
Tomato peel, celery, parsley, and thyme	Rutin, apigenin, luteolin	[109]

4.3   FLAVONOIDS PRODUCTION FROM MICROBES

In the reaction of the low production potency from chemical and plants synthesis, scientists have directed their concentration to the assembly of flavonoids in microorganisms by employing metabolic engineering and biology.[115] Chemical synthesis of flavonoids needs intense reaction conditions and virulent chemicals.[88] Attributing with rapid development in biological science and also the flooding of genome information from a range of organisms, combinatorial biosynthesis proposes a bonus for production of uncommon and pricy natural products.[114] Many of the prokaryotes and eukaryotes, including *Saccharomyces cerevisiae, Phelli-nusigniarius, E. coli, Streptomyces venezuelae,* and medicinal mushrooms, are used for flavonoids production.

4.4   PHENYLPROPANOID PATHWAY

Naringeninchalcone, in plants, is the precursor for an outsized range of flavonoids made from phenylpropanoid (PP) artificial pathway. Produc-tion from fermentation through *E. coli,* carrying artificially arranged PP pathway, is the primary example to point out an almost complete synthesis of plant biosynthetic pathway in the heterologous micro-organism for bioflavonoid production from the amino acid precursors, tyrosine, and phenylalanine. As the primary step in the plant PP pathway, cinnamic acid is produced by the activity of phenylalanine ammonialyase after deamination of phenylalanine. Cinnamate-4 hydroxylasehydroxylate the cinnamic acid to *p*-coumaric acid and through the action of 4-coumarate: Coenzyme A (CoA) ligase it is activated ultimately to the *p*-coumaroyl-CoA. Chalcone synthase after catalyzing the malonyl-CoA acetate units along with *p*-coumaroyl-CoA produces naringeninchalcone, which is then converted to naringenin.[10]

4.5 NUTRACEUTICAL HEALTH BENEFITS OF BIOFLAVONOIDS

4.5.1 FLAVONOIDS AND THEIR MEDICINAL CHATTELS

Bioflavonoids account for the chief active medicinal ingredient found in plants. Like other phenolic compounds, they typically show a persuasive antioxidant action. They also show antiallergenic, antiviral, anti-inflammatory, hepatoprotective, anticarcinogenic, and antithrombotic activities as well as have long been renowned as potent metal chelators.

4.5.2 ANTIOXIDANT PROPERTY OF BIOFLAVONOIDS

Nearly each group of bioflavonoids has the ability to perform as an antioxidant among which catechins and flavones are considered to be at the top of the list against ROS. The ROS and free radicals, which are persuaded by any exogenous damage or by usual oxygen metabolism, incessantly intimate the body tissues and cells by damage.[87] ROS and free radicals have incremented large numerous human diseases. Some bioflavonoids, such as rutin, morin, quercetin, myricetin, and kaempferol act as a potent antioxidant, hence, imparting beneficial activities including anticancer, antiallergic, anti-inflammatory, and antiviral. Moreover, they are known to play a significant role in the care of cataracts, cardiovascular and liver diseases. Silybin and quercetin have been suggested to provide a protective effect by performing free radical scavenger role in liver reperfusion ischemic tissue damage.[87] The scavenging activity of some flavonoids follows myricetin on the top, having highest scavenging activity, following the quercetin having slightly low scavenging activity than myricetin, subsequently followed by rhamnetin, morin, diosmetin, naringenin, apigenin, 5-7-dihydroxy-3,4,5-trimethoxy flavones ribinin, kaempferol, and flavone show least scavenging activity, respectively.[92]

4.5.3 ANTIMICROBIAL POTENTIAL OF BIOFLAVONOIDS

The investigated review of biophenols and phenolic acid esters has been discussed in terms of antifungal, antibacterial, and antiviral properties. Flavonoids extracted from orange peel were considered to be fungicidally active toward deutro-phomatracheiphila. Langeritin demonstrated weak activity compared to nobiletin with strong activity, and slightly stimulated fungal growth was observed in the case of hesperidin. Chlorflavonin is flavonoid type antifungal having antibiotic action.[87,104]

Plant phenolics have great potential to against wide array of microorganisms. Hesperidin, quercetin, catechin, dihydroquercetin, morin, apigenin, and rutin are recognized as antiviral against 11 types of viruses. Antiviral activity depends upon nonglycosidic compounds. It is documented that flavonols are more effective against Herpes simplex virus (Type 1) as compared to flavones, in which galangin gain importance and it is followed by kaempferol. The least important is quercetin, a natural bioflavonoid polymer, having a molecular weight of 2100 Daltons, has been considered active against Type 1 and Type 2 Herpes simplex virus's strains.[13,65] Research study is in progress for the application of bioflavonoids against human immunodeficiency virus (HIV). Bioflavonoids were also found to have anti-acquired immunodeficiency syndrome (AIDS) activity. It has been tested that out of 28 flavonoids, flavans were considered to be more effective in discriminatory inhibition of HIV-1 and HIV-2 than flavonones and flavones.[20]

4.6 EFFECT ON GASTROINTESTINAL SYSTEM

4.6.1 ANTIULCEROGENIC POTENTIAL

There are numerous health benefits of bioflavonoids as depicted in Table 4.2. Synthesis of flavonoid plants is an adaptive response to stress conditions (cold stress, wounding, infection, water stress). Flavonoids

TABLE 4.2 Bioflavonoids with Their Health Benefits.

Bioactivity	Bioflavonoids	Functions	References
Antiatherosclerotic agent	Isoflavone, genistein	These flavonoids show several antioxidant activities associated with the prevention of atherosclerosis.	[54]
Anti diabetic effects	Quercetin	Quercetin enhances uptake of Ca^{2+} as well as motivates the release of insulin especially in the case of noninsulin-dependent diabetes.	[111]
Anti-inflammatory potential	Apigenin, hesperidin, quercetin, and luteolin	These have anti-inflammatory and analgesic effects. Hesperidin imparts analgesic and anti-inflammatory effect.	[47,84]
Anti microbial potential	Chlorflavonin, hesperidin, quercetin, catechin, dihydroquercetin, morin, apigenin, and rutin	Antifungal and antiviral activity was found. Antiviral activity was due to nonglycosidic compounds.	[87]
Anti neoplastic impact	Coumestrol, genistein, kaempferol, taxifolin, catechin, and fisetin	Reduce the risk of prostate cancer and suppressed the cell growth.	[102, 56]
Anti oxidant property	Catechins, flavones, rutin, morin, quercetin, myricetin, and kaempferol	Potent antioxidant, hence imparting beneficial possessions including anticancer, antiallergic, anti-inflammatory, and antiviral activity.	[87]
Antithrombogenic agent	kaempferol, myricetin, and quercetin	Bioflavones properly maintain the concentration of nitric oxide and prostacyclin by directly scavenging free radicals, thus show the antithrombotic effect.	[36, 85]
Antiulcerogenic potential	Kaempferol, quercetin, and rutinintraperitonial	Help in reducing the gastric injury and ulcer.	[20]
Cardio protective impact	7-Monohydroxyethylrutoside and 7',3',4'- trihydroxyethylrutoside	Cardio-toxicity of doxorubicin can be subdued by the Bioflavonoids.	[4]
Vaso-relaxant effect	Kaempferol, anthocyanin, delphinidin	Help in the enhancement of vasorelaxation process, preventing endothelial dysfunction	[5, 34]

help to protect the body from free radicals,[50] because of strong antioxidant property. They constitute an essential part of our daily diet (the diet given on daily basis should provide flavonoids of1 gram per day). It has been shown by several epidemiological studies that the consumption of phenolic-rich compounds plants has a direct link with the risk of cancer and other degenerative diseases[40] including immune dysfunction,[52] allergic,[53] and cardiovascular[81] issues. The free radicals play a significant role in the gastrointestinal ulcers and erosive lesions formation.

It is reported that flavonoids such as quercetin, naringin, anthocyanoside, sophoradin, and silymarin derivatives have antiulcer properties.[61] Flavonoid glycosides of Labiatae have shown to reduce ulcer index with the inhibition of pepsin and gastric acid in aspirin-persuaded ulcer among rats.[1] The administration (25–100 mg/kg) of kaempferol, quercetin, and rutinintraperitonial significantly prevented the dose-dependent gastric injury in rats caused by acidified ethanol.[20,48]

Acute or lethal damage to the liver has been observed through various substances, such as galactosamine, phalloidin, ethanol, CCl_4, and other compounds; and this injury can be controlled by the bioflavonoids having a hepatoprotective response. A study investigated the reputed remedial action of apigenin, naringenin, silymarin, and quercetin, against hepatotoxicity, is induced by microcystin LR, among which the most effective one was silymarin.[19] In investigational cirrhosis, the bioflavonoid, venoruton, and rutin have shown hepatoprotective and regenerative effects.[20,67]

4.6.2 ANTI-INFLAMMATORY POTENTIAL OF BIOFLAVONOIDS

For tissue damage, microbial pathogen infection and chemical stimulation reactions are normal biological process through the immune cells from the vascular migration and release of the media, at the site of injury to start. Subsequently, ROS, reactive nitrogen species (RNS), and proinflammatory cytokines are raised and released in inflammatory cells to eliminate foreign pathogens and repair damaged tissue. In general, normal inflammation is fast and self-limiting, but abnormal solutions and prolonged inflammation can lead to various chronic diseases.[86]

The anti-inflammatory potential of flavone/flavonol glycosides and flavonoid aglycons have been reported to impart anti-inflammatory action in case animals of both chronic and acute inflammation when applied topically or given orally.[63,70] It is reported that some flavonoids, such as apigenin, hesperidin, quercetin, and luteolin have anti-inflammatory and analgesic effects. Flavonoids may specifically relate to the function of enzymes involved in severe inflammatory processes, particularly tyrosine, and serine–threonine protein kinases.[47,84] Citrus flavonoid, known as hesperidin, imparts analgesic and anti-inflammatory effect significantly. Moreover, according to recent research, quercetin, apigenin, and luteolin are reported to show potent anti-inflammatory role.[29] It is documented thatarachidonic acid metabolism can be modulated by the inhibition of lipoxygenase (LO) activity and cyclo-oxygenase (COX). It is also hypothesized that the inhibitory action of flavonoids on the metabolism of arachidonic acid results in their antiallergic and anti-inflammatory properties.[26] Among all the flavones, myricetin, kaempferol, fisetin, and quercetin possess LO activity and COX inhibitory action.

4.6.3   ANTIDIABETIC EFFECTS

American adults' intake of flavonoids is mainly flavan-3-ol, followed by flavonols, flavanones, flavones, isoflavones, and anthocyanidins.[24] Epidemiological studies and meta-analysis showed that dietary consumption of flavonoids was inversely proportional to the development of many aging-related diseases, including cardiovascular disease, cancer, neurodegenerative diseases, osteoporosis, and diabetes.[9,33] Bioflavonoids acquire antidiabetic action, especially quercetin. Quercetin is known as one of the many extensively used flavonoids, in our dietary nutrition. It is widely distributed in different types of fruits, tea, pepper, fennel, coriander, dill, radish, onions, wine, berries, and apples.[3,91] It follows the mechanism of pancreatic islets regeneration and increases the release of insulin studied in diabetic rats by streptozotocin induction.[111] In another study, quercetin enhanced the uptake of Ca^{2+} and motivated the release of insulin in the case of noninsulin-dependent diabetes.[43,70] Another study evaluated the quercetin effect on Caco-2E cells of the intestine.[61] Studies have shown that quercetin strongly inhibits the transport of fructose and glucose by GLUT2. Blocking tyrosine kinases is a mechanism by means of which quercetin is reported to have a diabetic effect. Similarly, quercetin diet resulted in cell proliferation in diabetic mice.[57]

4.7   IMPACT OF BIOFLAVONOIDS FOR CARDIOVASCULAR SYSTEMS

4.7.1   VASO-RELAXANT EFFECT

The flavonoid, kaempferol 3-O-(6"-trans-p-coumaroyl)-β-D-glucopyranoside (also known as tiliroside), has significant biological properties to reduce risk of the cardiovascular diseases (CVDs), including lowering low-density lipoprotein human oxidation.[95] Flavonoid consumption leads to enhancement of vasorelaxation process, preventing endothelial dysfunction, which ultimately leads to the decrease in arterial pressure.[49] Studies have shown that vasodilators can be used to treat cerebral vasospasm and hypertension, improving peripheral circulation. A number of endothelium-dependent vasodilators, such as histamine, acetylcholine, and bradykinin have been reported to increase Ca^{2+} levels in endothelial cells and activate NO release, leading to vasodilatation.[73,74] On the other hand, the contraction reaction in the smooth muscle is induced by the Ca^{2+} inflow through the receptor-operated Ca^{2+} channel and/or the voltage-dependent Ca^{2+} channel (VDC).[55] Independent endothelial vasodilators (such as nifedipine, nicardipine, verapamil, and diltiazem) inhibited VDC and resulted in a decrease in intracellular Ca^{2+} contraction in the smooth muscle causing vasorelaxation.[55] Another study reported that the consumption of bioflavonoids prevents CVD with a reduction in atherosclerosis and hypertension. Many other reports support the fact that the phenolic compounds have the potential to induce a vasorelaxant effect in rats and to decrease the arterial pressure. Bioflavonoids (especially *Anthocyanin delphinidin*) persuade endothelium-dependent vasorelaxation effect in subjected organisms.[5,34]

The long-term consumption of flavanols-containing foods can lead to a sustained increase in endothelial function or prevention of future cardiovascular diseases.[17] The antioxidant activity possessed by flavanones relies on the phenolic OH group number and their spatial location. The flavanones present in the hydrophilic environment show higher antioxidant activity.[18] It is believed that flavanones have the potential to resist atherosclerosis. Studies have shown that due to the diminution of atherosclerosis in mice on a diet that was high in fat and cholesterol with the supplements of naringenin at nutrition-related levels. The results can be used to improve dyslipidemia and the dysfunction of endothelial biomarkers, with changes in the expression of the gene. The flavanones can thus prevent the risk of CVDs.[23] Pentamethoxyflavone (PMF) was isolated from the rhizomes of *Kaempferia parviflora*, which has antihypertensive effects. This study has clearly shown that PMF has an endothelium-dependent and independent relaxation activity with isolated thoracic aortic rings. A research study was conducted to explore the possible mechanisms for relaxation in order to determine whether PMF acts (1) as a nitric oxide (NO) stimulant, guanylatecyclase, adenylatecyclase, or H_2S stimulant and/or through an open K^+ channel, (2) off the voltage-dependent calcium channel, (3) by inhibiting intracellular calcium mobilization, (4) as a calcium-channel inhibitor for storage operations, or (5) as a Rho-kinase inhibitor.[89]

The research shows that Quercetin and luteolin are vasorelaxant that have the capacity to enhance production of precontracted rings of aorta to 100% with the relaxation of pulmonary embolism (PE).[22] As mentioned earlier, the presence of a sugar substitute reduces the vasorelaxation of flavonoids.[39] Thus, relatively less activity of quercetagetin-7-Ob-D-glucopyranoside (glycosylation on C-3) has been shown rather than the lack of these substituted quercetin and luteolin, respectively. The less vasorelaxant activity of the flavanols and chalcone further indicates that the presence of the conjugate structure (i.e., the C ring containing the double bonds) is essential for the vasorelaxant activity of these compounds.

Vascular tension is regulated by a number of endothelium-dependent and endothelium-independent factors, such as endothelium-derived factors, autonomic nervous system, and local mediators, of which calcium $(Ca_2þ)$ is a key element in control of vascular contractility.[117] In general, a contractile response is being generated by an increase in $Ca_2þ$ found in smooth muscle cells.[66] As elicited by KCl, the contraction mainly results from the influx of extracellular $Ca_2þ$ that was induced by the depolarization of cell membrane following the opening of the voltage-dependent $Ca_2þ$ channels.[82] PE in comparison to this is an a-adrenergic agonist that induces contractions in the smooth muscle cells via a $Ca_2þ$ influx by means of receptor-operated $Ca_2þ$ channels and by a $Ca_2þ$ release from the sarcoplasmic reticulum.[69]

4.7.2 ANTIATHEROSCLEROTIC EFFECTS

The endothelium plays a significant in the intravascular homeostasis. Oxidative stress involves many vascular diseases. Cardiovascular risk factors, such as diabetes, smoking, and hypertension, change the redox state within the blood vessels and can lead to endothelial dysfunction leading to atherosclerosis. Endothelial dysfunction and loss of NO are early manifestations and characteristics of vascular disease. As a result of the endothelial dysfunction and decrease in the bioavailability of NO, increased oxidative stress changes a number of physiological functions, such as the aggregation of platelets, adhesion of leukocytes, and the flow of blood in the endothelium.[107] The plants like *Mangifera indica*,[6] *Garciniacambogia*,[60] *Asparagus racemosus*,[112] and *Hypericum perforatum L*[112] have been confirmed to appreciably lower the threat of cardiovascular diseases and atherosclerosis. In the initiation of atherosclerosis by the reaction of free radicals, the induction of the oxidation of low-density lipo-protein (LDL) results in their modification and this serves as an initial step in the commencement of the disease. These oxidation-induced-modified LDLs are rapidly uprooted through the scavenger receptors, which lead to foam cell formation. Here, the bioflavonoids act as a chain-breaking antioxidant and scavenge some of the radical species directly.[15]

Flavonoids have shown to prevent atherosclerosis in animal models.[42] Various flavonoids show several antioxidant activities associated with atherosclerosis prevention. Isoflavone genistein in the cell-free and the cell-mediated systems has revealed antioxidant activity. These flavo-noids also include LDL particles that prevent the oxidation of LDL and protect the vascular cells from oxidation.[54] Nobiletin has a wide range of activities similar to that of antioxidants. It inhibits the development of oxidants by means of three systems produced by the RAW 264.7 cell line, such as xanthine oxidase system, oxidative stress induced by TPA and also NO.[77,78]

The use of quercetin and the derivatives of quercetin can inhibit the impairment caused due to oxidation in numerous systems, together with those systems that are significantly associated with atherosclerosis. The flavonoids, such as myricetin and quercetin, prevent the preoxidation of lipids and they can possibly be active in the alpha-tocopherol regen-erations.[32,72] The synergistic effect of rutin (derivative of quercetin) with ascorbic acid and c-terpinene antioxidants played a role in inhibiting the oxidation of LDL.[171] Quercetin effectively binds with LDL particles and by the transfer of intramolecular electrons repairing phenoxyl radicals.[27,28] Therefore, quercetin and the derivatives of quercetin, in a number of major components that are involved in atherosclerosis, can inhibit the oxidation. Kaempferol can increase the production of nitric acid, helps in asymmetric dimethylarginine level reduction that prevents the endothelium injuries and in cells, boosts the endothelium-dependent vasorelaxation and the damage caused by oxidation.[119]

The capability of kaempferol to reduce the oxidative stress can possibly be a beneficial positive effect on cardiovascular diseases.[97] Kaempferol also prevents arteriosclerosis by inhibiting LDL oxidation and platelet formation. Kong evaluated the kaempferol effect on rabbit models encouraged by atherosclerosis and, after treatment of kaempferol for a period of 10 weeks, with a high-level cholesterol regimen, the intercellular adhesion molecule-1 expression, the vascular adhesion molecule-1, and the monocyte chemoattractant protein-1 (MCP-1) expression in the aorta of the rabbits was considerably reduced. This suggests that kaempferol can relieve vasoconstriction and can prevent atherosclerosis.[58] *Ampelopsis cantoniensis* is mainly composed of myricetin, which inhibits the oxidation of -LDL induced by the (Cu^{2+}) metal ion and (AAPH) free radical.[38] The *A. cantoniensis* extract can hence be used similar to a natural remediation agent to inhibit the oxidized low-density lipoprotein, which contributes to the formation of atherosclerotic lesions. To prevent the oxidation of low-density lipoprotein, myricetin blocks the uptake of macrophage oxidized low-density lipoprotein and plays an important role in the prevention of atherosclerosis.[64]

4.7.3 ANTITHROMBOGENIC EFFECTS

Platelet aggregation plays a polar role in case of thrombotic diseases. Oxygen free radicals and lipid peroxides were generated by the activated platelets, which adhere to the vascular epithelium. These peroxides and free radicals restrain the endothelial formation of nitrous oxide and prostacyclin. It was reported that the pigments of tea can enhance fibrinolysis, decrease the coagulability of blood, and prevent platelet aggregation and adhesion.[68] The platelets play a central role in the pathogenesis of the coronary heart disease (CHD). These cellular components activation releases a diverse range of the pro-atherogenic factors and this activation is an important feature of the thrombosis development through acute coronary events. Aspirin reduces the activity of platelets and for the general population, aspirin has been suggested as a prophylactic agent, for several years. The exposure of flavonoids may also reduce the activity of platelets and lower the chances of CHD.

Further studies have shown to enhance the platelet aggregation inhibition by enhancing the release of NO (derived from platelets).[30] Daily consumption of juice of purple grapes (7 mL/kg) for a period of 14 days resulted in the reduction of platelet aggregation in vitro, increased NO release and reduced the formation of superoxide.[30]

In another study, some selected bioflavonoids (such as kaempferol, myricetin, and quercetin) significantly inhibited platelet aggregation in monkeys and dogs.[85] Bioflavones properly maintain the concentration of nitric oxide and prostacyclin by directly scavenging free radicals, thus show antithrombotic effect.[36] Moreover, they inhibit the activity of lipoxygenase, and cyclo-oxygenase pathway, thus acting as in vivo and in vitro antithrombotic agents.[2,116]

Catechin also has a role in preventing CVD due to its involvement in the process of atherosclerosis.[113] As an antioxidant, catechins are able to regulate cell signaling pathways leading to increased vascular responsiveness, platelet aggregation, and reduced inflammation.[96,100,101]

4.7.4 CARDIOPROTECTIVE EFFECTS

As a result of the discovery of the "French paradox," which led to the study of flavonoids, the red wine consumption was high in cardiovascular mortality rate. In addition, epidemiological studies have shown that dietary flavonoids have a protective effect on CHD. The relationship between the intake of flavonoid and its long-lasting consequences on the mortality were studied afterward and it was then suggested that the intake of flavonoid with CHD mortality was negatively correlated.

The high-fat solubility of flavonoids makes them easy to enter into cells, where free radicals cause the greatest harm. In addition, they have a wide range of applications in cardiovascular studies (antioxidant, anti-inflammatory, free radical scavenging, and antiplatelet aggregation effects). These two factors prompted to assess their cardio-protective effects in ischemia-reperfusion (I/R) injury.[7] Owing to these properties, bioflavonoids are thought to be the potential protectors of heart against chronic cardiotoxicity rooted by the cytostatic drug doxorubicin. This doxorubicin could be an efficient antineoplastic agent. However, its clinical use is proscribed by the incidence of accumulative dose-related cardiotoxicity, leading to the congestive heart failure. A recent study documented that the cardiotoxicity of doxorubicin on the left atrium of a mouse has been subdued by the bioflavonoids (e.g., 7-monohydroxyethylrutoside and 7',3',4'-trihydroxyethylrutoside.[4,62]

Dexrazoxane is a cardioprotective agent approved by the FDA for anthracycline-induced cardiotoxicity. Previously, it was thought that dexrazoxane provides cardioprotective effects from anthracyclines primarily through the metal-chelating activity of its intracellular hydrolysis products in the myocardium. This activity involves chelation of free iron and iron bound in anthracycline complexes, thereby preventing the formation of cardiotoxic reactive oxygen radicals.[25] It could also act as a catalytic inhibitor of DNA topoisomerase II.[41] It has recently been suggested that the explanation is the interaction with the topoisomerase IIb isoform[108]. HIF activation is another mechanism that contributes to the protective effect of dexrazoxane against anthracycline cardiotoxicity.[99]

4.7.5   ANTINEOPLASTIC EFFECTS

Numerous recent reviews have illuminated antineoplastic activity. Flavonoids potential as antitumor agents is based on mechanisms that include cell cycle arrest, induction of apoptosis, and modulation of protein kinase activities on cancer cells. Several of these compounds and their derivatives, such as catechins, quercetin, and flavopiridol have been studied in clinical trials.[52,106] Epidemiological studies have shown that flavonone, vitamin C is effective in reducing the risk of prostate cancer.[94] It has also been shown that green tea[80] reduces the risk of cancer, lung, liver, and colon, delaying the progression of cancer and a lower risk of breast cancer. In the same way, coumestrol, genistein, and daidzein-rich foods reduce the risk of prostate cancer.[102]

Elaborated studies have discovered that quercetin applies a dose-reliant inhibition of growth as well as colony formation.[59] Quercetin, with the scientific name of 3,3',4',5,7-pentahydroxyflavone, is a natural bioflavonoid derived from a variety of edible fruits, tea leaves, vegetables, and seeds.[118] Recent studies have shown that quercetin has a variety of pharmacological activities, and its anticancer activity has been extensively studied in many cancer cell lines, which can induce tumor cell apoptosis, regulate tumor cell proliferation, and tumor angiogenesis, inhibit signal transitions leading many important target pathways.[37,31] However, the therapeutic effect of quercetin is hampered by its low solubility of water, and the effect of quercetin on the cure of colon cancer is rarely documented.

Therefore, the antitumor effect and mechanism of quercetin on colonic neoplasms can be studied in vitro and in vivo by using an efficient delivery system for the effective solubilization of quercetin in aqueous solutions. Some bioflavonoids (such as kaempferol, taxifolin, catechin, and fisetin) suppressed cell growth.[56] On screening the antileukaemic efficacy of 28 naturally present and artificial flavonoids on human promyelocytic-leukaemic HL-60 cells, genistein (the isoflavone) was found to possess sturdy effect.[44,103]

Recently, apigenin has been considered a beneficial health promoter because of its low internal toxicity and a different control result of cancer cells relative to other structural flavonoids. In addition, it has been shown by several studies that apigenin supports the chelating of metals, scavenging the free radicals and stimulating cell culture plus detoxification enzymes at stage II in the in vivo tumor models. As a severe inhibitor of ornithine decarboxylase, the anticarcinogenic effects of apigenin have been demonstrated in the carcinogenesis model of skin.[46]

4.8 SUMMARY

Cure and prevention of illness through bioflavonoids are well known. Naturally, these are present in various fruits and vegetables, including apples, peach, strawberry, even among cereals as well as imparting color and flavor along with nutritional benefits. Like other phenolic compounds, bioflavonoids typically show a persuasive antioxidant, antiallergenic, antiviral, anti-inflammatory, hepatoprotective, anti-carcinogenic, and antithrombotic activities as well as have long been renowned as potent metal chelators. Diversity of bioflavonoids instituted in nature acquires their own chemical, physical, and physiological properties. Medicinal efficiency of various bioflavonoids is well recognized. In developing countries, they are used commonly, whereas their therapeutic potential and the use of newly derived compounds must be validated by employing some specific biochemical tests. On the other hand, it is easier to produce large number of bioflavonoids through microbial means. Supplementary research is needed to provide innovative insights, which will surely escort to a novel era of bioflavonoid based pharmaceuticals for the cure and management of various degenerative and infectious ailments.

KEYWORDS

- antiatherosclerotic
- anti-inflammatory
- antithrombogenic
- bioflavonoids
- cardiovascular disorders
- nutraceutical health benefits

REFERENCES

1. Alarcon, D.; Martin, M.; Locasa, C.; Motilva, V. Antiulcerogenic Activity of Flavonoids and Gastric Protection. *Ethnopharmacol.* **1994,** *42,* 161–170.
2. Alcaraz, M.; Ferrandiz, M. Modification of Arachidonic Metabolism by Flavonoids. *J. Ethnopharmacol.* **1987,** *21* (3), 209–229.
3. Alinezhad, H.; Azimi, R.; Zare, M.; Ebrahimzadeh, M. A.; Eslami, S.; Nabavi, S. F.; Nabavi, S. M. Antioxidant and Antihemolytic Activities of Ethanolic Extract of Flowers, Leaves, and Stems of *Hyssopusofficinalis* L. Var. angustifolius. *Int. J. Food Prop.* **2013,** *16* (5), 1169–1178.
4. Almeida-Rezende, B.; Carvalho-Pereira, A.; Cortes, S.; Soares Lemos, V. Vascular Effects of Flavonoids. *Curr. Med. Chem.* **2016,** *23* (1), 87–102.
5. Andriambeloson, E.; Kleschyov, A. L.; Muller, B.; Beretz, A.; Stoclet, J. C.; Andriantsitohaina, R. Nitric Oxide Production and Endothelium-Dependent Vasorelaxation Induced by Wine Polyphenols in Rat Aorta. *Br. J. Pharmacol.* **1997,** *120* (6), 1053–1058.
6. Anila, L.; Vijayalakshmi, N. R. Antioxidant Action of Flavonoids from *Mangiferaindica* and *Emblicaofficinalis* in Hypercholesterolemic Rats. *Food Chem.* **2003,** *83* (4), 569–574.
7. Annapurna, A.; Reddy, C. S.; Akondi, R. B.; Rao, S. R. Cardioprotective Actions of Two Bioflavonoids, Quercetin and Rutin, in Experimental Myocardial Infarction in

Both Normal and Streptozotocin-induced Type I Diabetic Rats. *J. Pharm. Pharmacol.* **2009**, *61* (10), 1365–1374.

8. Antunes, P. M.; De Varennes, A.; Rajcan, I.; Goss, M. J. Accumulation of Specific Flavonoids in Soybean (*Glycine max* (L.) Merr.) as a Function of the Early Tripartite Symbiosis with Arbuscularmycorrhizal Fungi and Bradyrhizobiumjaponicum (Kirchner) Jordan. *Soil Biol. Biochem.* **2006**, *38* (6), 1234–1242.

9. Arts, I. C.; Hollman, P. C. Polyphenols and Disease Risk in Epidemiologic Studies. *Am. J. Clin. Nutr.* **2005**, *81* (1), 317S–325S.

10. Austin, M. B.; Noel, J. P. The Chalcone Synthase Superfamily of Type III Polyketide Synthases. *Nat. Prod. Rep.* **2003**, *20* (1), 79–110.

11. Awad, M. A.; de Jager, A.; van Westing, L. M. Flavonoid and Chlorogenic Acid Levels in Apple Fruit: Characterisation of variation. *Scientia Horticulturae* **2000**, *83*(3), 249–263.

12. Ayaz, E.; Alpsoy, H. Garlic (*Allium sativum*) and Traditional Medicine. *Turkiye Parazitolojii Dergisi/Turkiye Parazitoloji Dernegi (Acta parazitologica Turcica/ Turkish Society for Parasitology)* **2006**, *31* (2), 145–149.

13. Babii, C.; Bahrin, L. G.; Neagu, A. N.; Gostin, I.; Mihasan, M.; Birsa, L. M.; Stefan, M. Antibacterial Activity and Proposed Action Mechanism of a New Class of Synthetic Tricyclic Flavonoids. *J. App. Microbiol.* **2016**, *120* (3), 630–637.

14. Banjarnahor, S. D.; Artanti, N. Antioxidant Properties of Flavonoids. *Med. J. Indones.* **2014**, *23* (4), 239–245.

15. Basu, A.; Das, A. S.; Majumder, M.; Mukhopadhyay, R. Anti-atherogenic Roles of Dietary Flavonoids Chrysin, Quercetin, and Luteolin. *J. Cardiovasc. Pharmacol.* **2016**, *68* (1), 89–96.

16. Borrelli, F.; Izzo, A. A. The Plant Kingdom as a Source of Anti-ulcer Remedies. *Phytother. Res.* **2000**, *14* (8), 581–591.

17. Brodowska, K. M. Natural Flavonoids: Classification, Potential Role, and Application of Flavonoid Analogues. *Eur. J. Biol. Res.* **2017**, *7* (2), 108–123.

18. Cai, Y. Z.; Sun, M.; Xing, J.; Luo, Q.; Corke, H. Structure–Radical Scavenging Activity Relationships of Phenolic Compounds from Traditional Chinese Medicinal Plants. *Life Sci.* **2006**, *78* (25), 2872–2888.

19. Carlo, G.; Izzo, A.; Maiolino, P.; Mascolo, N.; Viola, P.; Diurno, M.; Capasso, F. Inhibition of Intestinal Motility and Secretion by Flavonoids in Mice and Rats: Structure-activity Relationships. *J. Pharm. Pharmacol.* **1993**, *45* (12), 1054–1059.

20. Castro-Vazquez, L.; Alañón, M. E.; Rodríguez-Robledo, V.; Pérez-Coello, M. S.; Hermosín-Gutierrez, I. Bioactive Flavonoids, Antioxidant Behavior, and Cytoprotective Effects of Dried Grapefruit Peels (*Citrus paradisi* Macf.). *Oxid. Med. Cell. Longev.* **2016**, *3*, 1–12.

21. Chakraborty, K.; Bhattacharjee, S.; Pal, T. K.; Bhattacharyya, S. Evaluation of In Vitro Antioxidant Potential of Tea (*Camelia sinensis*) Leaves Obtained from Different Heights of Darjeeling Hill, West Bengal. *J. App. Pharm. Sci.* **2015**, *5* (1), 063–068.

22. Chan, E. C.; Pannangpetch, P.; Woodman, O. L. Relaxation to Flavones and Flavonols in Rat Isolated Thoracic Aorta: Mechanism of Action and Structure-activity Relationships. *J. Cardiovasc. Pharmacol.* **2000**, *35* (2), 326–333.

23. Chanet, A.; Milenkovic, D.; Manach, C.; Mazur, A.; Morand, C. Citrus Flavanones: What is their Role in Cardiovascular Protection? *J. Agricul. Food Chem.* **2012,** *60* (36), 8809–8822.

24. Chun, O. K.; Chung, S. J.; Song, W. O. Estimated Dietary Flavonoid Intake and Major Food Sources of US Adults. *J. Nutr.* **2007,** *137* (5), 1244–1252.

25. Cvetković, R. S.; Scott, L. J. Dexrazoxane. *Drugs* **2005,** *65*(7), 1005–1024.

26. Ferrandiz, M.; Alcaraz, M. Anti-inflammatory Activity and Inhibition of Arachidonic Acid Metabolism by Flavonoids. *Agents and Actions* **1991,** *32* (3–4), 283–288.

27. Filipe, P.; Morlière, P.; Patterson, L. K.; Hug, G. L.; Mazière, J. C.; Mazière, C.; Santus, R. Mechanisms of Flavonoid Repair Reactions with Amino Acid Radicals in Models of Biological Systems: A Pulse Radiolysis Study in Micelles and Human Serum Albumin. *Biochimicaet. Biophysica. Acta.-General Subjects* **2002,** *1571* (2), 102–114.

28. Filipe, P.; Morlière, P.; Patterson, L. K.; Hug, G. L. Repair of Amino Acid Radicals of Apolipoprotein B100 of Low-density Lipoproteins by Flavonoids. A Pulse Radiolysis Study with Quercetin and Rutin. *Biochemistry* **2002,** *41* (36), 11057–11064.

29. Formica, J.; & Regelson, W. Review of the Biology of Quercetin and Related Bioflavonoids. *Food Chem. Toxicol.* **1995,** *33* (12), 1061–1080.

30. Freedman, J. E.; Parker, C.; Li, L.; Perlman, J. A.; Frei, B. Select Flavonoids and Whole Juice from Purple Grapes Inhibit Platelet Function and Enhance Nitric Oxide Release. *Circulation* **2001,** *103* (23), 2792–2798.

31. Gao, X.; Wang, B.; Wei, X.; Men, K.; Zheng, F. Anticancer Effect and Mechanism of Polymer Micelle-encapsulated Quercetin on Ovarian Cancer. *Nanoscale* **2012,** *4* (22), 7021–7030.

32. Gordon, M. H.; Roedig-Penman, A. Antioxidant Activity of Quercetin and Myricetin in Liposomes. *Chem. Phys. Lipids* **1998,** *97* (1), 79–85.

33. Graf, B. A.; Milbury, P. E.; Blumberg, J. B. Flavonols, Flavones, Flavanones, and Human Health: Epidemiological Evidence. *J. Med. Foods* **2005,** *8* (3), 281–290.

34. Grassi, D.; Ferri, C.; Desideri, G. Brain Protection and Cognitive Function: Cocoa Flavonoids as Nutraceuticals. *Curr. Pharm. Des.* **2016,** *22* (2), 145–151.

35. Groot, H. d.; Rauen, U. Tissue injury by Reactive Oxygen Species and the Protective Effects of Flavonoids. *Fund. Clin. Pharmacol.* **1998,** *12* (3), 249–255.

36. Gryglewski, R. J.; Korbut, R.; Robak, J.; Święs, J. On the Mechanism of Antithrombotic Action of Flavonoids. *Biochem. Pharmacol.* **1987,** *36* (3), 317–322.

37. Gupta, S. C.; Kim, J. H.; Prasad, S.; Aggarwal, B. B. Regulation of Survival, Proliferation, Invasion, Angiogenesis, and Metastasis of Tumor Cells Through Modulation of Inflammatory Pathways by Nutraceuticals. *Cancer Metastasis Rev.* **2010,** *29* (3), 405–434.

38. Ha, D. T.; Thuong, P. T.; Thuan, N. D. Protective Action of *Ampelopsis cantoniensis* and its Major Constituent Myricetin Against LDL Oxidation. *Vietnam J. Chem.* **2014,** *45* (6), 768.

39. Hammad, H. M.; Abdalla, S. S. Pharmacological Effects of Selected Flavonoids on Rat Isolated Ileum: Structure-activity Relationship. *Gen. Pharmacol.* **1997,** *28* (5), 767-771.

40. Harris, C. S.; Mo, F.; Migahed, L.; Chepelev, L.; Haddad, P. S.; Wright, J. S.; Bennett, S. A. Plant Phenolics Regulate Neoplastic Cell Growth and Survival: A Quantitative

Structure–activity and Biochemical Analysis. *Can. J. Physiol. Pharmacol.* **2007,** *85* (11), 1124–1138.

41. Hasinoff, B. B. The use of dexrazoxane for the Prevention of Anthracycline Extravasation Injury. *Expert Opin. Investig. Drugs* **2008,** *17* (2), 217–223.

42. Hayek, T.; Fuhrman, B.; Vaya, J. Reduced Progression of Atherosclerosis in Apolipoprotein E–deficient Mice Following Consumption of Red Wine, or its Polyphenols Quercetin or Catechin, is Associated with Reduced Susceptibility of LDL to Oxidation and Aggregation. *Arter., Thromb., Vasc. Biol.* **1997,** *17* (11), 2744–2752.

43. Hii, C.; Howell, S. Effects of Flavonoids on Insulin Secretion and 45Ca^{2+} Handling in Rat Islets of Langerhans. *J. Endocrinol.* **1985,** *107* (1), 1–8.

44. Hirano, T.; Gotoh, M.; Oka, K. Natural Flavonoids and Lignans are Potent Cytostatic Agents Against Human Leukemic HL-60 Cells. *Life Sci.* **1994,** *55* (13), 1061–1069.

45. Hollman, P. C.; de Vries, J. H.; van Leeuwen, S. D.; Mengelers, M. J.; Katan, M. B. Absorption of Dietary Quercetin Glycosides and Quercetin in Healthy Ileostomy Volunteers. *Am. J. Clin. Nutr.* **1995,** *62* (6), 1276–1282.

46. Horinaka, M.; Yoshida, T.; Shiraishi, T. The Dietary Flavonoid Apigenin Sensitizes Malignant Tumor Cells to Tumor Necrosis Factor–related Apoptosis-inducing Ligand. *Mol. Cancer Ther.* **2006,** *5* (4), 945–951.

47. Hunter, T. Protein Kinases and Phosphatases: The Yin and Yang of Protein Phosphorylation and Signaling. *Cells* **1995,** *80* (2), 225–236.

48. Izzo, A.; Carlo, G. D.; Mascolo, N.; Capasso, F. Antiulcer effect of flavonoids. Role of endogenous PAF. *Phytotherapy Research*, **1994,** *8* (3), 179-181.

49. Jayakody, R.; Senaratne, M.; Thomson, A.; Kappagoda, C. Cholesterol Feeding Impairs Endothelium-Dependent Relaxation of Rabbit Aorta. *Can. J. Physiol. Pharmacol.* **1985,** *63* (9), 1206–1209.

50. Jeong, J. M.; Choi, C. H.; Kang, S. K.; Lee, I. H.; Lee, J. Y.; Jung, H. Antioxidant and Chemosensitizing Effects of Flavonoids with Hydroxy and/or Methoxy Groups and Structure-activity Relationship. *J. Pharm. Pharmac. Sci.* **2007,** *10* (4), 537–546.

51. Joshi, R.; Gulati, A. Fractionation and Identification of Minor and Aroma-active Constituents in Kangra Orthodox Black Tea. *Food Chem.* **2015,** *167*, 290–298.

52. Kale, A.; Gawande, S.; Kotwal, S. Cancer Phytotherapeutics: Role for Flavonoids at the Cellular Level. *Phytother. Res.* **2008,** *22* (5), 567–577.

53. Kale, A.; Gawande, S.; Kotwal, S.; Netke, S.; Roomi, W.; Ivanov, V.; Rath, M. Studies on the Effects of Oral Administration of Nutrient Mixture, Quercetin and Red Onions on the Bioavailability of Epigallocatechingallate from Green Tea Extract. *Phytoth. Res.* **2010,** *24* (S1), 23–30.

54. Kapiotis, S.; Hermann, M.; Held, I.; Seelos, C.; Ehringer, H.; Gmeiner, B. M. K. Genistein, the Dietary-Derived Angiogenesis Inhibitor, Prevents LDL Oxidation and Protects Endothelial Cells from Damage by Atherogenic LDL. *Arter., Thromb., and Vasc. Biol.* **1997,** *17* (11), 2868–2874.

55. Karaki, H.; Ozaki, H.; Hori, M.; Mitsui-Saito, M.; Amano, K. I.; Harada, K. I.; Sato, K. Calcium Movements, Distribution, and Functions in Smooth Muscle. *Pharmacol. Rev.* **1997,** *49* (2), 157–230.

56. Kim, Y.J.; Shin, Y. Antioxidant Profile, Antioxidant Activity, and Physicochemical Characteristics of Strawberries from Different Cultivars and Harvest Locations. *J. Korean Soc. Appl. Biol. Chem.* **2015,** *58* (4), 587–595.

57. Kobori, M.; Masumoto, S.; Akimoto, Y.; Takahashi, Y. Dietary Quercetin Alleviates Diabetic Symptoms and Reduces Streptozotocin-induced Disturbance of Hepatic Gene Expression in Mice. *Mol. Nutr. Food Res.* **2009,** *53* (7), 859–868.

58. Kong, L.; Luo, C.; Li, X.; Zhou, Y.; He, H. The Anti-inflammatory Effect of Kaempferol on Early Atherosclerosis in High Cholesterol Fed Rabbits. *Lipids Health Dis.* **2013,** *12* (1), 115.

59. Kontruck, S.; Radecki, T.; Brozozowski, T. Antiulcer and Gastroprotective Effects of Solon, a Synthetic Flavonoid Derivative of Sophorandin. Role of endogenous Prostaglandins. *Bur. J. Pharmac.* **1986,** *125,* 185–192.

60. Koshy, A. S.; Anila, L.; Vijayalakshmi, N. R. Flavonoids from Garciniacambogia Lower Lipid Levels in Hypercholesterolemic Rats. *Food Chem.* **2001,** *72* (3), 289–294.

61. Kwon, O.; Eck, P.; Chen, S.; Corpe, C. P. Inhibition of the Intestinal Glucose Transporter GLUT2 by Flavonoids. *FASEB J.* **2007,** *21* (2), 366–377.

62. Lackeman, G.; Claeys, M.; Rwangabo, P.; Herman, A.; Vlietinck, A. Chronotropic Effect of Quercetin on Guinea Pig Right Atrium. *J. Planta Medica* **1986,** *52,* 433–439.

63. Lee, S. J.; Son, K. H.; Chang, H. W.; Do, J. C.; Jung, K. Y.; Kang, S. S.; Kim, H. P. Antiinflammatory Activity of Naturally Occurring Flavone and Flavonol Glycosides. *Arch. Pharmacal Res.* **1993,** *16* (1), 25–28.

64. Lian, T. W.; Wang, L.; Lo, Y. H.; Huang, I. J.; Wu, M. J. Fisetin, Morin and Myricetin Attenuate CD36 Expression and oxLDL uptake in U937-derived macrophages. *Acta (BBA)-Mol. Cell Biol. Lipids* **2008,** *1781* (10), 601–609.

65. Loewenstein, W. R. Junctional Intercellular Communication and the Control of Growth. *Acta (BBA)-Rev. Cancer* **1979,** *560* (1), 1–65.

66. Löhn, M.; Fürstenau, M.; Sagach, V. Ignition of Calcium Sparks in Arterial and Cardiac Muscle Through Caveolae. *Circ Res.* **2000,** *87* (11), 1034–1039.

67. Lorenz, W.; Kusche, J.; Barth, H.; Mathias, C. Action of Several Flavonoids on Enzyme of Histidine Metabolism In Vivo. In *Histamine*; Cz Maslinski, Ed.; Hutchinson and Ross Publishers: Pennsylvania, **1994**; pp 265–269.

68. Lou, F.; Zhang, M.; Zhang, X.; Liu, J.; Yuan, W. A Study on Tea-pigment in Prevention of Atherosclerosis. *Chinese Med. J.* **1989,** *102* (8), 579–583.

69. McFadzean, I.; Gibson, A. The Developing Relationship Between Receptor-operated and Store-operated Calcium Channels in Smooth Muscle. *Br. J. Pharmacol.* **2002,** *135* (1), 1–13.

70. Michel, T. K.; Ottoh, A. A. Bio-flavonoids and Garcinoic Acid from Garcinia Kola seeds with Promising Anti-inflammatory Potentials. *Pharmacogn. J.* **2016,** *8* (1), 12–23.

71. Milde, J.; Elstner, E. F.; Grassmann, J. Synergistic Inhibition of Low-density Lipoprotein Oxidation by Rutin, γ-Terpinene, and Ascorbic Acid. *Phytomedicine* **2004,** *11* (2-3), 105–113.

72. Morel, I.; Abaléa, V.; Sergent, O.; Cillard, P.; Cillard, J. Involvement of Phenoxyl radical Intermediates in Lipid Antioxidant Action of Myricetin in Iron-treated Rat Hepatocyte Culture. *Biochem. Pharmacol.* **1998,** *55* (9), 1399–1404.

73. Muller, B.; Kleschyov, A. L.; Gyorgy, K.; Stoclet, J. C. Inducible NO Synthase Activity in Blood Vessels and Heart: New Insight into Cell Origin and Consequences. *Physiol. Res.* **2000,** *49* (1), 19–26.

74. Muller, J. M.; Davis, M. J.; Kuo, L.; Chilian, W. M. Changes in Coronary Endothelial Cell Ca^{++} Concentration During Shear Stress and Agonist-induced Vasodilation. *Am. J. Physiol.-Heart Circ. Physiol.* **1999,** *276* (5), H1706–H1714.

75. Mulvihill, E. E.; Huff, M. W. Citrus Flavonoids as Regulators of Lipoprotein Metabolism and Atherosclerosis. *Annu. Rev. Nutr.* **2016,** *36* (1), 11–21.

76. Muniz, F. W. M. G.; Nogueira, S. B.; Mendes, F. L. V. The Impact of Antioxidant Agents Complimentary to Periodontal Therapy on Oxidative Stress and Periodontal Outcomes: A Systematic Review. *Arch. Oral Biol.* **2015,** *60* (9), 1203–1214.

77. Murakami, A.; Nakamura, Y.; Ohto, Y. Suppressive Effects of Citrus Fruits on Free Radical Generation and Nobiletin, an Anti-inflammatory Polymethoxy Flavonoid. *Biofactors* **2000,** *12* (1–4), 187–192.

78. Murakami, A.; Nakamura, Y.; Torikai, K. Inhibitory Effect of Citrus Nobiletin on Phorbol Ester-Induced Skin Inflammation, Oxidative Stress, and Tumor Promotion in Mice. *Cancer Res.* **2000,** *60* (18), 5059–5066.

79. Murphy K.J, Chronopoulos A.K. Dietary Flavanols and Procyanidin Oligomers from Cocoa (*Theobroma cacao*) Inhibit Platelet Function. *Am. J. Clin. Nutr.* **2003,** *77* (6), 1466–1473

80. Nakachi, K.; Suemasu, K.; Suga, K.; Takeo, T.; Imai, K.; Higashi, Y. Influence of Drinking Green Tea on Breast Cancer Malignancy Among Japanese Patients. *Cancer Sci.* **1998,** *89* (3), 254–261.

81. Naruszewicz, M.; Łaniewska, I.; Millo, B. Combination Therapy of Statin with Flavonoids Rich Extract from Chokeberry Fruits Enhanced Reduction in Cardiovascular Risk Markers in Patients After Myocardial Infraction. *Atherosclerosis* **2007,** *194* (2), e179–e184.

82. Nelson, M. T.; Quayle, J. M. Physiological Roles and Properties of Potassium Channels in Arterial Smooth Muscle. *Am. J. Physiol.-Cell Physiol.* **1995,** *268* (4), C799–C822.

83. Nishiumi, S.; Miyamoto, S.; Kawabata, K.; Ohnishi, K.; Mukai, R.; Murakami, A.; Terao, J. Dietary Flavonoids as Cancer-preventive and Therapeutic Biofactors. *Front. Biosci.* **2011,** *3*, 1332–1362.

84. Nishizuka, Y. The Molecular Heterogeneity of Protein Kinase C and its Implications for Cellular Regulation. *Nature* **1988,** *334* (6184), 661–665.

85. Osman, H. E.; Maalej, N.; Shanmuganayagam, D.; Folts, J. D. Grape Juice but Not Orange or Grapefruit Juice Inhibits Platelet Activity in Dogs and Monkeys (*Macaca fasciularis*). *J. Nutr.* **1998,** *128* (12), 2307–2312.

86. Pan, M. H.; Lai, C. S.; Ho, C. T. Anti-inflammatory Activity of Natural Dietary Flavonoids. *Food Funct.* **2010,** *1* (1), 15–31.

87. Pandey, R. P.; Parajuli, P.; Koffas, M. A.; Sohng, J. K. Microbial Production of Natural and Non-natural Flavonoids: Pathway Engineering, Directed Evolution and Systems/Synthetic Biology. *Biotechnol. Adv.* **2016,** *3*, 112–115.

88. Park, S. R.; Yoon, J. A.; Paik, J. H. Engineering of Plant-specific Phenyl Propanoids Biosynthesis in *Streptomyces venezuelae*. *J. Biotechnol.* **2009,** *141* (3), 181–188.

89. Perez-Vizcaino, F.; Duarte, J.; Jimenez, R.; Santos-Buelga, C.; Osuna, A. Antihypertensive Effects of the Flavonoid Quercetin. *Pharmacol. Rep.* **2009,** *61* (1), 67–75.

90. Peterson, J. J.; Beecher, G. R. Flavanones in Grapefruit, Lemons, and Limes: A Compilation and Review of the Data from the Analytical Literature. *J. Food Comp. Anal.* **2006**, *19*, S74–S80.

91. Petersen, B.; Egert, S.; Bosy-Westphal, A. Bioavailability of Quercetin in Humans and the Influence of Food Matrix Comparing Quercetin Capsules and Different Apple Sources. *Food Res. Int.* **2016**, *88*, 159–165.

92. Ratty, A.; Das, N. Effects of Flavonoids on Nonenzymatic Lipid Peroxidation: Structure-activity Relationship. *Biochem. Med. Metabol. Biol.* **1988**, *39* (1), 69–79.

93. Renaud, S. d.; de Lorgeril, M. Wine, Alcohol, Platelets, and the French Paradox for Coronary Heart Disease. *The Lancet* **1992**, *339* (8808), 1523–1526.

94. Rossi, M.; Garavello, W.; Talamini, R.; Negri, E. Flavonoids and the Risk of Oral and Pharyngeal Cancer: A Case-control Study from Italy. *Cancer Epidemiol. Prev. Biomar.* **2007**, *16* (8), 1621–1625.

95. Schinella, G. R.; Tournier, H. A.; Máñez, S.; de Buschiazzo, P. M.; del Carmen Recio, M.; Ríos, J. L. Tiliroside and gnaphaliin Inhibit Human Low-density Lipoprotein Oxidation. *Fitoterapia* **2007**, *78* (1), 1–6.

96. Shenouda, S. M.; Vita, J. A. Effects of Flavonoid-containing Beverages and EGCG on Endothelial Function. *J. Am. College Nutr.* **2007**, *26* (4), 366S–372S.

97. Singh, R.; Singh, B.; Singh, S.; Kumar, N.; Kumar, S.; Arora, S. Anti-free Radical Activities of Kaempferol Isolated from *Acacia nilotica* (L.) Wild. Ex. Del. *Toxicol. In Vitro* **2008**, *22* (8), 1965–1970.

98. Sökmen, M.; Serkedjieva, J.; Daferera, D. In Vitro Antioxidant, Antimicrobial, and Antiviral Activities of the Essential Oil and Various Extracts from Herbal Parts and Callus Cultures of Origanumacutidens. *J. Agricul. Food Chem.* **2004**, *52* (11), 3309–3312.

99. Spagnuolo, R. D.; Recalcati, S.; Tacchini, L.; Cairo, G. Role of Hypoxia-inducible Factors in the Dexrazoxane-Mediated Protection of Cardiomyocytes from Doxorubicin-induced Toxicity. *Br. J. Pharmacol.* **2011**, *163* (2), 299–312.

100. Stangl, V.; Dreger, H.; Stangl, K.; Lorenz, M. Molecular Targets of Tea Polyphenols in the Cardiovascular System. *Cardiovasc. Res.* **2007**, *73* (2), 348–358.

101. Stangl, V.; Lorenz, M.; Stangl, K. The Role of Tea and Tea Flavonoids in Cardiovascular Health. *Mol. Nutr. Food Res.* **2006**, *50* (2), 218–228.

102. Strom, S. S.; Yamamura, Y.; Duphorne, C. M. Phytoestrogen Intake and Prostate Cancer: A Case-control Study Using a New Database. *Nutr. Cancer* **1999**, *33* (1), 20–25.

103. Swanson, H. Flavonoids and Cancers of the Gastrointestinal Tract. In *Flavonoids, Inflammation, and Cancer*; World Scientific: New York, 2016; Vol. 3, pp 105–142.

104. Ten Cate, J.; Van Haeringen, N.; Gerritsen, J.; Glasius, E. Biological Activity of a Semisynthetic Flavonoid, O-(β-hydroxyethyl) Rutoside: Light Scattering and Metabolic Studies of Human Red Cells and Platelets. *Clin. Chem.* **1973**, *19* (1), 31–35.

105. Thilakarathna, S. H.; Rupasinghe, H. P. Flavonoid Bioavailability and Attempts for Bioavailability Enhancement. *Nutrients* **2013**, *5* (9), 3367–3387.

106. Thomas, C. M.; Wood III, R. C.; Wyatt, J. E. Anti-neoplastic Activity of Two Flavone Isomers Derived from Gnaphaliumelegans and Achyroclinebogotensis. *PloS One* **2012**, *7* (6), e39806.

107. Thomas, S. R.; Witting, P. K.; Drummond, G. R. Redox Control of Endothelial Function and Dysfunction: Molecular Mechanisms and Therapeutic Opportunities. *Antioxid. Redox Signal.* **2008,** *10* (10), 1713–1766.

108. Thougaard, A. V.; Langer, S. W.; Hainau, B.; Grauslund, M.; Juhl, B. R.; Jensen, P. B.; Sehested, M. A Murine Experimental Anthracycline Extravasation Model: Pathology and Study of the Involvement of Topoisomerase II Alpha and Iron in the Mechanism of Tissue Damage. *Toxicology* **2010,** *269* (1), 67–72.

109. Tuladhar, S. M.; Bajracharya, G. B.; Pokharel, D. R.; Giri, R. The Potentially Anticarcinogenic Flavonoids in Vegetables, Fruits, and Spices. *Nepal J. Sci Technol.* **2015,** *2* (1), 178–187.

110. Upadhyay, A.; Agrahari, P.; Singh, D. A Review on Salient Pharmacological Features of *Momordica charantia.* *Int. J. Pharmacol.* **2015,** *11* (5), 405–413.

111. Vessal, M.; Hemmati, M.; Vasei, M. Antidiabetic Effects of Quercetin in Streptozocin-induced Diabetic Rats. *Compar. Biochem. Physiol. Part C: Toxicol. Pharmacol.* **2003,** *135* (3), 357–364.

112. Visavadiya, N. P.; Narasimhacharya, A. V. R. L. Asparagus Root Regulates Cholesterol Metabolism and Improves Antioxidant Status in Hypercholesteremic Rats. *Evid.-Based Complementary and Altern. Med.* **2009,** *6* (2), 219–226.

113. Vita, J. A. Tea Consumption and Cardiovascular Disease: Effects on Endothelial Function. *J. Nutr.* **2003,** *133* (10), 3293S–3297S.

114. Wang, A.; Zhang, F.; Huang, L.; Yin, X.; Li, H.; Wang, Q.; Xie, T. New Progress in Biocatalysis and Biotransformation of Flavonoids. *J. Med. Plants Res.* **2010,** *4* (10), 847–856.

115. Wang, Y.; Chen, S.; Yu, O. Metabolic Engineering of Flavonoids in Plants and Microorganisms. *Appl. Microbiol. Biotechnol.* **2011,** *91* (4), 949–956.

116. Warner, E. F.; Zhang, Q.; Raheem, K. S.; O'Hagan, D.; O'Connell, M. A.; Kay, C. D. Common Phenolic Metabolites of Flavonoids, but not their Unmetabolized Precursors, Reduce the Secretion of Vascular Cellular Adhesion Molecules by Human Endothelial Cells. *J. Nut.* **2016,** *146* (3), 465–473.

117. Wellman, G. C.; Nelson, M. T. Signaling Between SR and Plasmalemma in Smooth Muscle: Sparks and the Activation of Ca^{++}-Sensitive Ion Channels. *Cell Calcium* **2003,** *34* (3), 211–229.

118. Wu, Q.; Deng, S.; Li, L.; Sun, L.; Yang, X.; Liu, X.; Gong, C. Biodegradable Polymeric Micelle-encapsulated Quercetin Suppresses Tumor Growth and Metastasis in both Transgenic Zebrafish and Mouse Models. *Nanoscale* **2013,** *5* (24), 12480–12493.

119. Xiao, H. B.; Lu, X. Y.; Chen, X. J.; Sun, Z. L. Protective Effects of Kaempferol Against Endothelial Damage by an Improvement in Nitric Oxide Production and a Decrease in Asymmetric Dimethylarginine Level. *Eur. J. Pharmacol.* **2009,** *616* (1), 213–222.

120. Yakubu, O.; Nwodo, O.; Joshua, P.; Ugwu, M.; Odu, A.; Okwo, F. Fractionation and Determination of Total Antioxidant Capacity, Total Phenolic and Total Flavonoids Contents of Aqueous, Ethanol and n-Hexane Extracts of *Vitex doniana* Leaves. *Afr. J. Biotechnol.* **2015,** *13* (5), 223–228.

CHAPTER 5

ANTIDIABETIC POTENTIAL OF MEDICINAL MUSHROOMS

VIVEK K. CHATURVEDI, SUSHIL K. DUBEY, and M. P. SINGH

ABSTRACT

Diabetes mellitus is a chronic metabolic disease caused by autoimmune devastation of insulin-producing β cell of the pancreas. Although various types of antidiabetic agents are available, including insulin, yet, they have many side-effects. Antidiabetic properties of medicinal plants are on increasing demand. Medicinal mushrooms are an imitable source of numerous bioactive compounds with antidiabetic potential and antihyperglycemic agents. The bioactive compound like polysaccharide, protein, lipids, and fiber and some low molecular weight compounds like alkaloids, terpenoids, lactones, lectines, and phenolic substances are involved in many diseases as a therapeutic mediator and also have shown to be a major milestone for the cure of diabetes. This chapter discusses antidiabetic properties of mushrooms. The chapter explores medicinal properties of bioactive compounds and metabolites of mushroom in control of diabetes. The previously published literature shows that mushrooms can be used as a safe and effective treatment for diabetes.

5.1 INTRODUCTION

Diabetes mellitus is now a common problem worldwide.[2] It is a chronic metabolic disease caused by the devastation of β cells of the pancreas by which there is less amount secretion of insulin and/or insulin sensitivity.[79] Pancreas β cells are the only cells responsible for the insulin secretion, which regulates the glucose concentration in blood.[10] Diabetes mellitus gives rise to many health-related complications like high mortality rate with increased risk of cardiovascular diseases, cancer, and so on.[14,22,34,52,68,72,87,99,116] Moreover, it causes retinopathy, neuropathy, renal dysfunction, and so on.[1,86,93,98,104,105]

Type 1 diabetes is aroused by autoimmune devastation of β cells. This type of diabetes accounts for only 5–10% among all types of diabetes. Type 2 diabetes is aroused by the resistance of β-producing cells and has comparatively insulin deficiencies. This diabetes accounts 90–95% among all cases.[6] The cause of diabetes is not yet well known. The defects in the genome of β cells also cause diabetes mellitus. Insulin-dependent diabetes mellitus (IDDM), called as type 1 diabetes, is and also called juvenile diabetes because the production of insulin is less or negligible. In type 2 diabetes, body requires a daily injection of insulin and called noninsulin-dependent diabetes mellitus (NIDDM). It requires balanced diet, exercise, or oral antidiabetic drug for the control of blood sugar level.[101]

Currently, many therapeutic agents for diabetes are available, such as insulin, sulfonylureas, biguanides, β-glucosidase inhibitors and glinides, and so on. Many of these antidiabetic agents have some adverse effects leading to search for safer and more effective antidiabetic agents.[50] Medic-inal mushrooms have served to cure diabetes since ancient time. These mushrooms are used for the cure of many other diseases in traditional medicine, especially in India, China, Korea, and Japan.[12,37–40]

Mushrooms are edible fungi, which contain a high amount of protein, vitamins (A, B, C, D, and K) and have a low calorific value.[26] They produce secondary metabolites, such as sterols, lectin, terpenoids, alkaloids, lactones, and phenolic compounds,[5,17,54,106,115] which provide numerous health benefits. These edible fungi contain minerals also like calcium, phosphorous, potassium, iron, and copper, which are instrumental for the cure of heart disease, hypertension, cancer, constipation, blood pressure, and hyperglycemia.[95] Owing to their medicinal properties, mushrooms are used in cosmetic products also as a good anti-aging factor.[45]

The presence of β glucans and dietary fibers in mushrooms can restore the function of β cell of pancreas, causing an increase in the output of insulin so that blood glucose level can be controlled. Mushroom is a perfect diet for diabetic patients because of low fats, cholesterol, and low in carbo-hydrates.[28,82,84,85,100] The study shows that the consumption of mushroom reduces the level of cholesterol, triglyceride, and high-density lipoprotein level (HDL) and low-density lipoprotein (LDL) augments.[3,47,56,103]

Mushroom contains some enzymes, which are insulin-like compounds to break down sugar and complex carbohydrates in foods to enhance the insulin resistance.[58,59,63] They are also helpful for proper functioning of liver, pancreas, endocrine glands and ensure the proper metabolic functioning by the formation of related hormones and insulin.

This chapter focuses on bioactive compounds with their antidiabetic properties, which are present in various species of edible mushrooms.

5.2 WHAT IS DIABETES?

Diabetes is a major chronic metabolic disorder caused by either insufficient insulin secretion or cellular mediated autoimmune devastation of pancreatic β cells. In simple word, the inability to control the blood glucose level is called diabetes. It is classified into two clinical syndromes: one can be diagnosed as insulin dependent which is called type 1 (IDDM). This type of diabetes is caused in early age with ketonuria and weight loss, and the second type is characterized by insensitivity to IDDM or NIDDM, in later age. Insulin is a hormone that converts sugar, complex carbohydrates, and other food into energy used for body metabolic activity. Insulin also regulates the blood glucose level. If the insulin level is not optimum, then it prevents the metamorphosis of glucose into glycogen in the liver.[15]

Hyperglycemia is a condition when blood sugar is in excessive level. Under normal conditions, blood sugar level is maintained between 4.0 and 6.0 mmol/L (72–108 mg/DL); but in case of fasting, the sugar level is increased up to 7.8 mmol/L (140 mg/DL) within 2 h after meal.[19] The blood sugar level with about 10 mmol/L (180 mg/DL) do not show any symptoms, but when it exceeds 15–20 mmol/L (270–360 mg/DL), then most of the organs of the body are affected particularly the cardiovascular and nervous system.[8,11]

5.2.1 TYPE 1

Type 1 diabetes mellitus can be classified by the cellular-mediated autoimmune devastation of the β cell in the pancreas, which leads to insulin deficiency; thus, it is called IDDM.[65] In most of the cases, children are affected by type 1 diabetes, but adult persons are also affected. In this catabolic disorder, insulin and glucagon are absent in plasma. Pancreatic β cells are incapable to respond to all stimuli of insulin through signaling pathway.[83] It is thought that the type 1 results from an infection or by some toxic substance. The immune systems of such people are predisposed for development of an autoimmune response against β-cell antigens of the pancreas.

There are some extrinsic factors, which generally affect the β-cell functioning, such as viruses like mumps virus, coxsackie virus B4, destructive cytotoxins, and antibodies. HLA genes can raise susceptibility to diabetogenic virus that can trigger the destructive autoimmune response against β cells of the pancreas. The immunosuppressive drugs such as cyclosporine or azathioprine are given.[83]

5.2.2 TYPE 2

Type 2 diabetes occurs in the young persons and is called as IDDM. As the name itself reveals, the patients with NIDDM do not require external insulin for the prevention of ketonuria and they are not enduring from ketosis. If ketosis and hyperglycemic conditions cannot be controlled by oral agents and diet, then the injection of insulin is necessary.[36]

In type 2 diabetes, the peripheral tissue becomes insulin resistant, where insulin receptors or other intermediates do not respond according to insulin-signaling pathways within the body cells and become insensitive to insulin, and glucose does not go into the tissue, resulting in rising of blood glucose concentration.[4] Type 2 diabetes can be divided into two subtypes: obese and nonobese. Type 2 has no correlation with human leucocytes antigen (HLA) and has a strong genetic influence. Individuals suffering from type 2 diabetes have a 50% probability to transmit the disorder to their children.

5.2.3 CAUSES OF DIABETES

Patients generally do not inherit type 1 diabetes, but they inherit some serious genetic changes by which type 1 diabetes can develop. Certain HLA has been associated with persons having the tendency of inheriting genetic predisposition. HLA is a group of genes to play a vital role in the transplantation of antigens and takes part in many immunological processes. HLA-DQA1, HLA-DQB1, and HLA-DRB1 genes give instructions for making proteins, which play a crucial role in immunity. Type 1 diabetes patient has an autoimmune response, which is an abnormal response. In autoimmune disorders, the antibodies target normal tissues of the body. In patients with diabetes type 1, the antibodies have been developed against islet cells of the pancreas and insulin produced by the β cells. The genetic mechanism of type 2 diabetes development is not known, but there are some risk factors like family history, age, obesity, stress, and depression, which are allied to the type 2 diabetes.

It is hypothesized as the protective effect of breast milk having cytokines, growth factors, and many necessary substances required for the maturation of intestinal mucosa. Breast milk also defends against enteric infection and also immune regulation of gut can clarify the epidemiologic associated between viral infection and type 1 diabetes.[66] Insulin resistance and impaired insulin secretion are the two problems that arise in type 2 diabetes. The sensitivity of tissue gets decreased for the insulin. Generally, insulin binds to the insulin receptors present on the cell surfaces and triggers a series of reaction, which direct to the metabolism of glucose within the cell, but in type 2 diabetes, cells become less sensitive toward the insulin stimulus. Following mushroom species have been shown to provide antidiabetic benefits (Table 5.1):

- *Agaricus brasiliensis*
- *Agaricus bisporus*
- *Agaricus blazei*
- *Agaricus campestris*
- *Agaricus subrufescens*
- *Agaricus sylvaticus*
- *Auricularia auricula-judae*
- *Coprinus comatus*
- *Cordyceps militaris*
- *Cordyceps takaomantana*
- *Ganoderma lucidum*
- *Grifola frondosa*
- *Hericium erinaceus*
- *Inonotus obliquus*
- *Laetiporus sulphureus*
- *Lentinus laedodes*
- *Lentinus strigosus*
- *Ophiocordyceps sinensis*
- *Phellinus baumii*
- *Phellinus linteus*
- *Phellinus merrillii*
- *Phellinus ribis*
- *Pleurotus eryngii*
- *Pleurotus pulmonarius*
- *Sparassis crispa*
- *Tremella aurantia*
- *Tremella fuciformis*
- *Tremella mesenterica*
- *Wolfiporia extensa*

TABLE 5.1 Bioactive Compounds from Some Medicinal Mushrooms with Antidiabetic Property.

Mushroom species	Family	Bioactive compound	Antidiabetic effect	References
Agaricus bisporus	Agaricaceae	Dehydrated fruiting body extracts	Reduces blood glucose and cholesterol levels	[47]
Auricularia auricula-judae	Auriculariaceae	Dried mycelia powder	Decreases the level of Blood glucose, total cholesterol, and triglyceride levels	[62]
Coprinus comatus	Agaricaceae	4,5-Dihydroxy-2 methoxybenzaldehyde (comatin)	Decreases fructosamine, triglycerides, and cholesterol level; maintains blood glucose level	[21]
Cordyceps sinensis	Ophiocordycipitaceae	Polysaccharide fraction CSP-1	Augmented insulin levels in blood, discharge insulin from the pancreatic cells	[73]
Ganoderma lucidum	Ganodermataceae	(3β,24E)-Lanosta 7,9(11),24-trien-3,26-diol (ganoderol B)	Strong α-glucosidase inhibition Nonenzymic and enzymic antioxidants are dose dependent	[23,48]
Hericium erinaceus	Hericiaceae	Methanol extract of the mushroom	The hypoglycemic effects lower the elevation rates of blood glucose levels	[102]
Phellinus linteus	Hymenochaetaceae	Polysaccharides	Slows down the autoimmune diabetes by regulating cytokine expression	[58]
Pleurotus pulmonarius	Pleurotaceae	Aqueous extract of the mushroom	Decreases the serum glucose level	[9,92]

5.3 *Ganoderma lucidum* (LINGZHI)

G. lucidum (Reishi or Lingzhi) species having medicinal properties are being used by the Chinese for better health and longevity. The word Ganoderma is composed of two words: "Gano" which means shiny or bright and "Derma" means skin; thus shiny skin fruiting body is the physical property of mushroom species. Many bioactive compounds have been identified in *G. lucidum*, such as polysaccharides (β glucans), terpenoids, mannitol, alkaloids, coumarin, and triterpenes (ganoderic acid, chemically similar to steroid hormones).

Sterols are also isolated from *Ganoderma*, that is, ganoderol, ganoderenic acid, ganodermanontriol, lucidadiol, ganoderiol, and ganodermadiol. These bioactive components and many other derivatives help to get better health. The *G. lucidum* has been used as a therapeutic agent for the cure of various diseases like cancer, hypertension, arthritis, hepatopathy, chronic hepatitis, immunomodulating effects, and some other diseases.[109] *G. lucidum* has tremendous antihyperglycemic and antihypocholesterolemic effects and has likelihood therapeutic activities.[77,90] Antihyperglycemic effects of *G. lucidum* are due to the polysaccharide, which facilitates the influx of Ca^{2+} into the pancreatic β cells.[112]

The placebo-controlled study was carried on 71 patients diagnosed with type 2 diabetes mellitus, who received polysaccharide fraction of extract from *G. lucidum*.[24] Isolated polysaccharides from *G. lucidum* significantly increased antioxidant activity, increased the insulin level in blood serum, also helped to reduce blood glucose level and reduced the lipid peroxidation in streptozotocin (STZ)-treated rats.[48,89]

β-Glucosidase enzyme secreted by the epithelium of small intestine converts disaccharides and oligosaccharides into glucose. A triterpenoid lanostane, called as ganoderol B [(3β,24*E*)-lanosta-7,9(11),24-trien-3,26-diol] isolated from *G. lucidum*, is a potent inhibitor of β-glucosidase enzyme.[23] Since lanostane inhibits the activity of β glucosidase, it hinders the carbohydrate digestion and prevents absorption of glucose. The in-vitro study revealed that ganoderol B has high β-glucosidase inhibition rate with an IC50 of 48.5 μg/mL (119.8 μM).

Researchers have identified protein tyrosine phosphatase 1B activity inhibitor from the fruiting body of *G. lucidum*, named as Fudan-Yueyang *G. lucidum* (FYGL). Proteoglycan extract, with FYGL with IC 5005.12 ± 0.05 μg/mL, was administrated to STZ-induced rats, showing a reduction in blood glucose level.[97]

It has been experimentally proven that the polysaccharide extract of *G. lucidum* reduced the serum glucose level in high-fat and STZ-diet-induced type 2 diabetes in mice.[107] A protein extracted from *G. lucidum* named as Ling Zhi-8 shows activation of the immune system with reduced the plasma glucose level.[64]

5.4   *Agaricus bisporus* (WHITE BUTTON MUSHROOM)

A. bisporusis is an edible medicinal mushroom having therapeutic relevance. The fruiting bodies of *A. bisporus* have 39.9% carbohydrate, 17.5% protein, 2.9% fats, and minerals. It is a prosperous source of vitamin C, B2, B12, D, niacin, folates dietary fiber, antioxidants, ergothioneine, and polyphenol.[20,81] Oral administration of fruiting bodies of *A. bisporus* mushrooms has the capability to increase hepatic glycogen composition and activate glucokinase (GK) activity in STZ-diabetic mice. Hence, button mushrooms have shown hypoglycemic effect, which lowers the blood glucose level in STZ-induced diabetic mice.[110]

Fruiting body of *A. bisporus* was fed to the STZ-induced diabetic mice, showing significant hypoglycemic and hypocholesterolemic activities.[47] Hypercholesterolemia can be a reason for the progression of type 2 diabetes/hyperglycemia in human. β-Cell metabolism can be affected directly by cholesterol, which leads to dysfunctioning of β cells and causes diabetes.[35] Dietary fibers have a defensive activity to the β cells and have shown potent antidiabetic effect. At high intake, it shows effective hypocholesterolemic effect.[7]

5.5   *Pleurotus* Spp. (OYSTER MUSHROOMS)

Various bioactive components are present in oyster mushroom, which shows medicinal properties, such as antidiabetic, anti-aging, anticancerous, antimicrobial, antiviral, immune response-stimulating, hepatoprotective, antihypertensive, hypocholesterolemic, and antioxidant activities. From ancient times, mushrooms have been perceived as essential nutritional foods because of taste, flavor, high dietary esteems, and few restorative properties.

P. pulmonarius shows hypoglycemic activity in diabetic mice induced by alloxan. The results demonstrated that no mortality in the ordinary mice up to 5000 mg/kg, while oral administration decreases the blood glucose level in alloxan-treated diabetic mice.[9]

When the diabetic rats fed with polysaccharide extract, the blood glucose levels were lowered by 44% than the negative control with the destruction of islets of Langerhans. Another study demonstrated that when the 4% extracts were fed to STZ rats with type 1 diabetes for 2 months, these play a major role in decreasing more than 40% cholesterol level.[16] A comparative clinical study was done with 120 diabetic patients to show the efficiency of *Pleurotus* spp. on the blood glucose level. The outcomes demonstrated a huge continuous diminution in hyperglycemia in type 2 diabetic and exhibited the prospective utilization for improvement of glycemic control.[2] *P. eryngii* (the king oyster mushroom) shows antihyper-glycemic and antihyperlipidemic impacts in mice.

The secondary metabolites containing dietary polysaccharides from the mushroom essentially decreased the level of cholesterol, triglyceride, and expanded with HDL cholesterol levels with enhanced insulin affect-ability.[59] Fruiting bodies of *P. ostreatus* was given to diabetic rats bring down the aggregate blood glucose level.[42] Similar effects were shown when powder of *P. citrinopileatus* was orally administered in hyperlipidemic rats; there was lowering of total cholesterol, while there was an increase the HDLs.[43] *P. ostreatus* ethanolic extract showed antidiabetic potential in alloxan-induced diabetic mice balb/c mice (25–30 μg), reduction in serum glucose level, cholesterol, triglyceride, and LDL-cholesterol level with increasing level of HDL cholesterol.[25,88]

5.6 *Grifola frondosa* (MAITAKE)

Maitake is an edible and therapeutic mushroom, with a Japanese name. The name of the mushroom was taken from the Griffin, which means leaf-like. *G. frondosa* is being perceived as a prospective source of many bioactive compounds and numerous other compounds with remarkable health-promoting potential. They are being used in the cure of cancer, diabetes, HIV, hyperlipidemia, hypertension, and hepatitis. Researcher Kubo *et al.*[67] did a research study on genetically diabetic mice and bioactive compounds of Maitake having therapeutic applications for lowering blood glucose level with an increase in insulin sensitivity.[67] The fruiting body of Maitake mushrooms having β glucan demonstrated an antidiabetic impact on mice. This may be identified with its impact on insulin receptors that can expand insulin sensitivity for improving insulin protection of peripheral-target tissues.[41] Many investigations have discovered numerous useful activities in which Maitake brings down glucose because of the mushroom content, having a β-glucosidase inhibitor. These inhibitors are available in liquid methanol concentrates of the seeds of *Momordica charantia* and the fruiting bodies of *Grifola*.[80]

5.7   *Phellinus* Spp.

Phellinus belongs to basidomycetes genera and is the largest species having medicinal properties. Derivatives of polyphenols and some other metabolites like polysaccharides, proteoglycans are reported to have anti-tumor[18,32,33,69] and antioxidant activities and are used in diabetes treatment and its complications. They reduce the plasma glucose and total triglyceride level in STZ-induced diabetic rates.[57,61] The polysaccharides found extracellular in mycelia showed a hypoglycemic effect in which there is a decrease of 49% in plasma glucose, 32% of total cholesterol, and 28% of triacylglycerol.[57] The polysaccharides of these species regulate the expression of cytokines and prevent the development of autoimmune diabetes.[57,61]

Phellibumins, derivative of hispidin, is isolated from *P. linteus* showing quenching effects in a dose-dependent regulations. Hispidin increased insulin production in hydrogen peroxide-treated cells and inhibited the apoptosis induced by hydrogen peroxide. Results show that hispidin can function as antidiabetic agent and can protect the β cells from free radicals.[46] Hispidin also shows aldose reductase inhibition to avoid diabetic complications.[69]

Crude exopolysaccharide of *P. baumii* reduces the plasma glucose level (52%) when fed to rats. *P. baumii* exoploysaccharides show significant decrease in alanine aminotransferase and aspartate aminotransferase activities, thereby showing a remedial property in liver function.[44] *P. merrillii*'s ethanolic extract showed β-glucosidase and aldose reeducates activities. Hispidin, hispolon, and inotilone showed an inhibition effect for β-glucosidase and aldose reductase. Methanolic extract of *P. ribis* contains different polychlorinated compounds, such as chlorophellins A-C. Chlorophellin C exhibits the most potent PPAR-γ agonistic outcome of the type 2 diabetes therapy.[70]

5.8 *Cordyceps sinensis* (CATERPILLAR FUNGUS)

O. sinensis have many therapeutic activities, such as antitumor, antiinflammatory, antioxidant, antidepressant, and antiosteoporosis. In China, *Cordyceps sinensis* is called as "soft gold" because of the elevated price of the mushroom, which is approximately USD $20,000–40,000 per kg. Capsule extract of *C. sinensis*, randomized trial on 95% of patients given mushroom dose at the rate of 3 g/day, lowers the blood glucose levels. However, control group showed 54% change when treated with alternate methods.[29] Fruiting bodies of the mushroom improved the level of hyperglycemia, weight loss, and polydipsia, and these data demonstrated that *Cordyceps* is a rich source for food for diabetes patients.[74]

Extracted polysaccharides from *Cordyceps* enhanced the blood glucose digestion and incremented insulin affectability in normal animals to bring down glucose levels in hereditarily diabetic animals.[55,114] A current report explored the potential hypoglycemic impacts of fermented mycelium concentrate on *C. sinensis*. They activated the β-cell survival, enhanced activity of renal NKA, and declined collagen accumulation; and meningeal matrix collection recommends that this mushroom may be a potential medication for enhancing β-cell work, which bears a hopeful treatment for diabetes mellitus.[51] Researcher's isolated polysaccharide cysteine string protein 1 (CSP-1) showed strong hypoglycemic activity. CSP-1, an oral administration in the diabetic model for 7 days, delivered significant dropping in blood glucose levels and expanded serum insulin levels. This strongly recommends that polysaccharides from *Cordyceps* (CSP-1) may invigorate pancreatic arrival of insulin and decrease insulin metabolism.[73]

5.9 *Hericium erinaceus* (LION'S MANE MUSHROOM)

H. erinaceus mushroom belongs to the Basidiomycetes class of fungi and is known as Houtou (Chinese), Yamabu shitake (Japanese), Lion's mane, Monkey's mushroom, and PomPom. *H. erinaceus* is exceptional and useful as some food-containing-effective constituents to endorse synthesis of nerve growth factor, which prevent from dementia and improves mild cognitive impairment with no antagonistic impacts.[49,53] Lion's mane mushroom has potential therapeutic and anticancerous properties.[60] Leaf extract of *Ginkgo biloba* biotransformation by *H. erinaceus* increases the serum superoxide dismutase activity by reducing of serum blood glucose level.[113] An aqueous extract of the *H. erinaceus* may thrive to bring down blood glucose levels in diabetic rats. The oral administration of water-soluble extract of Lion's mane mushroom decreased the glucose levels in rats by 19–26% along with the decrease of 20% serum lipids as compared with the control diabetic rats.[102]

5.10   *Coprinus comatus*

C. comatus has bioactivites, such as antidiabetic, antioxidant, and antimutagenic. The absorbing ability of trace elements is a great advantage for *C. comatus*. *C. comatus* shows hypoglycemic effect with vanadium. Most of the studies have suggested that vanadium has the ability to imitate insulin and showed good antidiabetic effects.[27,78,91] When *C. cosmatus* are treated with vanadium, it showed significant positive results on diabetic mouse model and is confirmed as a hypoglycemic medicine.[30,31,76,108]

When *C. comatus* rich in vanadium is fermented, it can lower the blood glucose level and inhibit the gluconeogenesis in diabetic mice. Researchers also showed that *C. comatus* has comatin (4,5-dihydroxy-2-methoxy-benzaldehyde) compound, which shows antidiabetic activity. Comatin inhibits the nonenzymatic glycosylation reaction. In an experiment on rats treated with comatin at the rate of 80 mg/kg of body weight, blood glucose level was lowered from 5.14 to 4.28 mM in 3 h. Not only the blood glucose level but also fructosamine, triglycerides, and total cholesterol concentration were also reduced. These results confirm that comatin can balance the blood glucose level and can improve the glucose tolerance.[21]

The exopolysaccharides produced by submerged culture of *C. comatus* shows more inhibitory effects on nonenzymatic glycosylation. Extracts of ethanol extracts, protein, polysaccharide, crude fiber, and alkali-soluble polysaccharide were prepared with *C. comatus*. The polysaccharides extract (300 mg/kg daily) administered to diabetic mice for 28 days showed best glucose-lowering activity. The results suggested that *C. comatus* serves as good oral hypoglycemic agent for control of diabetes mellitus.[71]

5.11   *Auricularia auricula-judae*

A. auricular-judae is another edible mushroom having immunomodulatory, antitumor, antioxidant, and hypolipidemic activities.[13,75,94] The extract of *A. auricular-judae* administered to diabetic mice resulted in the reduction of plasma glucose and urinary glucose level.[111] Another study revealed that the administration of hot water extract can decrease the blood glucose concentration.[96]

Oral intake of mycelial of *A. auricular* at the rate of 0.5 and 1.0 g/kg of body weight can lower plasma glucose by 35% and 39%, total cholesterol by 18% and 22%, and triglyceride by 12% and 13%, respectively.[62] Polyphenolic compound and polysaccharides found in *A. auricular* prevent hypercholesterolemia and improve the antioxidant level, reducing the total cholesterol level, atherosclerosis, and enhance the status of HDL cholesterol.[13]

5.12 SUMMARY

People are looking for natural treatment with no adverse effects. Many edible mushrooms have antidiabetic properties, for example, *G. lucidum* has ganoderol bioactive compound, which inhibits the β-glucosidase enzyme and prevents the digestion of carbohydrates. *A. bisporus* has the ability to activate the GK activity, which can help to control diabetes. Aqueous and ethanolic extracts of *Pleurotus* spp. show antidiabetic properties. Aqueous fruiting body extracts of *A. ampestris* have antihyperglycemic effects. Like these mushrooms, many others also have antidiabetic properties. *C. comatus* has comatin bioactive compound, which inhibits the nonenzymatic glycosylation reaction and helps to maintain the lower blood glucose level. These bioactive compound expressions can be enhanced in these edible mushrooms. Future research should focus on enhancing the expression of these bioactive compounds. By this approach, we can overexpress the comatin and ganoderol also so that diabetes treatment can be controlled in a natural way.

KEYWORDS

- *Agaricus bisporus*
- antidiabetic
- diabetes mellitus
- *Ganoderma lucidum*
- mushroom

REFERENCES

1. Agardh, C. D.; Stenram, U.; Torffvit, O.; Agardh, E. Effects of Inhibition of Glycation and Oxidative Stress on the Development of Diabetic Nephropathy in Rats. *J. Diabetes Complicat.* **2002**, *16*, 395–400.

2. Agrawal, R. P.; Chopra, A.; Lavekar, G. S.; Padhi, M. M.; Srikanth, N.; Ota, S.; Jain, S. Effect of Oyster Mushroom on Glycaemia: Lipid Profile and Quality of Life in Type 2 Diabetic Patients. *Austr. J. Med. Herbal.* **2010**, *22*, 50–54.

3. Alarcón, J.; Aguila, S.; Arancibia-Avila, P.; Fuentes, O.; Zamorano-Ponce, E.; Hernández, M. Production and Purification of Statins from *Pleurotus ostreatus* (Basidiomycetes) Strains. *Z. Naturforsch. C* **2003**, *58*, 62–66.

4. Albright, A. L. Diabetes. *Exercise Management for Persons with Chronic Diseases and Disabilities*; Human Kinetics (Braun-Brumfield): USA, 1997; pp 94–100.

5. Alexandre, J.; Kahatt, C.; Cvitkovic, F. B.; Faivre, S.; Shibata, S.; et al. A Phase I and Pharmacokinetic Study of Irofulven and Capecitabine Administered Every 2 Weeks in Patients with Advanced Solid Tumors. *Invest. New Drug* **2007**, *25*, 453–462.

6. American Diabetes Association (ADA). Diagnosis and Classification of Diabetes Mellitus. *Diabetes Care* **2008**, *20*, 1183–1197.

7. Anderson, J. W.; Baird, P.; Davis, R. H.; Ferreri, S.; Knudtson, M.; Koraym, A.; Waters, V.; Williams, C. L. Health Benefits of Dietary Fiber. *Nutr. Rev.* **2009**, *67* (4), 188–205.

8. Anonymous. *Total Health Life: High Blood Sugar*; Total Health Institute, 2005. www.totalhealthlife.com/about.html (accessed on July 31, 2018).

9. Badole, S. L.; Shah, S. N.; Patel, N. M.; Thakurdesai, P. A.; Bodhankar, S. L. Hypoglycemic Activity of Aqueous Extract of *Pleurotus pulmonarius* in Alloxan-Induced Diabetic Mice. *Pharm. Biol.* **2006**, *44*, 421–425.

10. Butler, A. E.; Janson, J.; Bonner-Weir, S.; Ritzel, R.; Rizza, R. A.; Butler, P. C. β-Cell Deficit and Increased β-Cell Apoptosis in Humans with Type-2 Diabetes. *Diabetes* **2003**, *52* (1), 102–110.

11. Capes, S. E.; Hunt, D.; Malmberg, K.; Pathak, P.; Gerstein, H. C. Stress Hyperglycemia and Prognosis of Stroke in Nondiabetic and Diabetic Patients: A Systematic Overview. *Stroke* **2001**, *32*, 2426–2432.

12. Chang, S. T. Global Impact of Edible and Medicinal Mushrooms on Human Welfare in the 21st Century, Nongreen Revolution. *Int. J. Med. Mushr.* **1999**, *1*, 1–8.

13. Chen, G.; Luo, Y. C.; Ji, B. P.; Li, B.; Su, W.; Xiao, Z. L.; Zhang, G. Z. Hypocholesterolemic Effects of *Auricularia auricula* Ethanol Extract in ICR Mice Fed a Cholesterol-Enriched Diet. *J. Food Sci. Technol. Mysore* **2011**, *48*, 692–698.

14. Cheng, M.; Li, B. Y.; Li, X. L.; Wang, Q.; Zhang, J. H.; Jing, X. J.; Gao, H. Q. Correlation between Serum Lactadherin and Pulse Wave Velocity and Cardiovascular Risk Factors in Elderly Patients with Type-2 Diabetes Mellitus. *Diabetes Res. Clin. Pract.* **2012**, *95*, 125–131.

15. Chinner, S.; Scherbaum, W. A.; Bornstein, S. R.; Barthel, A. Molecular Mechanisms of Insulin Resistance. *Diab. Med.* **2005**, *22*, 674–682.

16. Chorváthová, V.; Bobek, P.; Ginter, E.; Klvanová, J. Effect of the Oyster Fungus on Glycaemia and Cholesterolaemia in Rats with Insulin Dependent Diabetes. *Physiol. Res.* **1993**, *42*, 175–179.

17. Chung, M. J.; Chung, C. K.; Jeong, Y.; Ham. S. S. Anti-Cancer Activity of Subfractions Containing Pure Compounds of Chaga Mushroom (*Inonotus obliquus*) Extract in Human Cancer Cells and in Balbc/c Mice Bearing Sarcoma-180 Cells. *Nutr. Res. Pract.* **2010**, *4*, 177–182.

18. Dai, Y. C.; Zhou, L. W.; Cui, B. K.; Chen, Y. Q.; Decock, C. Current Advances in *Phellinus sensulato*, Medicinal Species, Functions, Metabolites and Mechanisms. *Appl. Microbiol. Biotechnol.* **2010**, *87* (5), 1587–1593.

19. Daly, M. E.; Vale, C.; Walker, M,; Littlefield, A.; George, K.; Alberti, M. M.; Mathors, J. C. Acute Effects on Insulin Sensitivity and Diurnal Metabolic Profiles of a High-Sucrose Compared with a High-Starch Diet. *Am. J. Clin. Nutr.* **1998**, *67*, 1186–1196.

20. Dhamodharan, G.; Mirunalini, S. A Novel Medicinal Characterization of *Agaricus bisporus* (White Button Mushroom). *Pharmacol. Online* **2010**, *2*, 456–463.

21. Ding, Z. Y.; Lu, Y. J.; Lu, Z. X.; Lv, F. X.; Wang, Y. H.; Bei, X. M.; Wang, F.; Zhang, K. C. Hypoglycaemic Effect of Comatin, an Antidiabetic Substance Separated from *Coprinus comatus* Broth, on Alloxaninduced-Diabetic Rats. *Food Chem.* **2010**, *121*, 39–43.

22. Dong, J. Y.; Zhang, Y. H.; Tong, K.; Quin, L. Q. Depression and Risk of Stroke, a Meta-Analysis of Prospective Studies. *Stroke* **2012**, *43*, 32–37.

23. Fatmawati, S.; Shimizu, K.; Kondo, R. Ganoderol B, a Potent α-Glucosidase Inhibitor Isolated from the Fruiting Body of *Ganoderma lucidum*. *Phytomedicine* **2011**, *18*, 1053–1055.

24. Gao, Y.; Lan, J; Dai, X; Ye, J.; Zhou, S. Phase I/II Study of Ling Zhi Mushroom *Ganoderma lucidum* (W. Curt., Fr.) Lloyd (Aphyllophoromycetideae) Extract in Patients with Type II Diabetes Mellitus. *Int. J. Med. Mushr.* **2004**, *6*, 33–39.

25. Ghaly, I. S.; Ahmed, E. S.; Booles, H. F.; Farag, I. M.; Nada, S. A. Evaluation of Antihyperglycemic Action of Oyster Mushroom (*Pleurotus ostreatus*) and Its Effect on DNA Damage: Chromosome Aberrations and Sperm Abnormalities in Streptozotocin-Induced Diabetic Rats. *Glob. Vet.* **2011**, *7*, 532–544.

26. Ghosh, C. Nutritional Value of Edible Mushroom. In *The Biology and Cultivation of Edible Mushroom*; S. T. Chang, Ed.; Academic Press: New York, 1990; p 842.

27. Goldfine, A. B.; Simonson, D. C.; Folli, F.; Patti, M. E.; Kahn, C. R. In Vivo and In Vitro Studies of Vanadate in Human and Rodent Diabetes Mellitus. *Mol. Cell. Biochem.* **1995**, *153*, 217–231.

28. Guillamón, E.; García-Lafuente, A.; Lozano, M. D.; Arrigo, M.; Rostagno, M. A.; Villares, A.; Martínez, J. A. Edible Mushrooms, Role in the Prevention of Cardiovascular Diseases. *Fitoterapia* **2010**, *81*, 715–723.

29. Guo, Q. C.; Zhang, C. Clinical Observations of Adjunctive Treatment of 20 Diabetic Patients with JinShuiBao Capsule. *J. Admin. Tradit. Chin. Med.* **1995**, *5*, 22–28.

30. Han, C.; Liu, T. A Comparison of Hypoglycemic Activity of Three Species of Basidiomycetes Rich in Vanadium. *Biol. Trace Elem. Res.* **2009**, *127*, 177–182.

31. Han, C.; Yuan, J.; Wang, Y.; Li, L. Hypoglycemic Activity of Fermented Mushroom of *Coprinus comatus* Rich in Vanadium. *J. Trace Elem. Med. Biol.* **2006**, *20*, 191–196.

32. Han, S. B.; Lee, C. W.; Jeon, Y. J.; Hong, N. D.; Yoo, I. D.; Yang, K. H.; et al. The Inhibitory Effect of Polysaccharides Isolated from *Phellinus linteus* on Tumor Growth and Metastasis. *Immunopharmacology* **1999**, *41*, 157–164.

33. Han, S. B.; Lee, C. W.; Kang, J. S.; Yoon, Y. D.; Lee, K. H.; Lee, K.; Park, S. K.; Kim, H. M. Acidic polysaccharide from *Phellinus linteus* Inhibits Melanoma Cell Metastasis by Blocking Cell Adhesion and Invasion. *Int. Immunopharmacol.* **2006,** *6*, 697–702.

34. Hansen, M. B.; Jensen, M. L.; Carstensen, B. Cause of Death among Diabetic Patients in Denmark. *Diabetologia* **2012,** *55*, 294–302.

35. Hao, M.; Head, W. S.; Gunawardana, S. C.; Hasty, A. H.; Piston, D. W. Direct Effect of Cholesterol on Insulin Secretion: A Novel Mechanism for Pancreatic β-Cell Dysfunction. *Diabetes* **2007,** *56* (23), 28–2338.

36. Harris, M. I.; Zimmet, P. Classification of Diabetes Mellitus and Other Categories of Glucose Intolerance. In *International Textbook of Diabetes Mellitus*; K. G. M. M. Alberti, P., Zimmet, R. A., DeFronzo, Keen, H., Eds.; John Wiley and Sons Ltd.: New York; 1997; pp 9–23.

37. Hobbs, C. *Medicinal Mushrooms—An Exploration of Tradition, Healing, and Culture*; Botanica Press: Santa Cruz; 1995; p 251.

38. Hobbs, C. Medicinal Value of *Lentinus edodes* (Berk.) Sing. (Agaricomycetideae). A Literature Review. *Int. J. Med. Mushr.* **2000,** *2*, 287–302.

39. Hobbs, C. R. Medicinal Value of Turkey Tail Fungus *Trametes versicolor* (L., Fr.) Pilát (Aphyllophoromycetideae). *Int. J. Med. Mushr.* **2004,** *6*, 195–218.

40. Hobbs, C. R. The Chemistry, Nutritional Value, Immunopharmacology, and Safety of the Traditional Food of Medicinal Split-Gill Fungus Schizophyllum Commune (Aphyllophoromycetideae): Literature Review. *Int. J. Med. Mushr.* **2005,** *7*, 127–140.

41. Hong, L.; Xun, M.; Wutong, W. Anti-Diabetic Effect of an Alpha Glucan from Fruit Body of Maitake (*Grifola frondosa*) on KK-Ay Mice. *J. Pharm. Pharmacol.* **2007,** *59*, 575–582.

42. Hossain, S.; Hashimoto, M.; Choudhury, E. K.; Alam, N.; Hussain, S.; Hasan, M.; Choudhury, S. K.; Mahmud, I. Dietary Mushroom (*Pleurotus ostreatus*) Ameliorates Atherogenic lipid in Hypocholesterolemic Rats. *Clin. Exp. Pharmacol. Physiol.* **2003,** *30*, 470–475.

43. Hu, S. H.; Wang, J. C.; Lien, J. L.; Liaw, E. T.; Lee, M. Y. Antihyperglycemic Effect of Polysaccharide from Fermented Broth of *Pleurotus citrinopileatus*. *Appl. Microbiol. Biotechnol.* **2006,** *70*, 107–113.

44. Hwang, H. J.; Kim, S. W.; Lim, J. M.; Joo, J. H.; Kim, H. O.; Kim, H. M.; Yun, J. W. Hypoglycemic Effect of Crude Exopolysaccharides Produced by a Medicinal Mushroom *Phellinus baumii* in Streptozotocin-Induced Diabetic Rats. *Life Sci.* **2005,** *76*, 3069–3080.

45. Hyde, K. D.; Bahkali, A. H.; Moslem, M. A. Fungi—An Unusual Source for Cosmetics. *Fung. Div.* **2010,** *43*, 1–9.

46. Jang, J. S.; Lee, J. S.; Lee, J. H.; Kwon, D. S.; Lee, K. E.; Lee, S. Y.; Hong, E. K. Hispidin Produced from *Phellinus linteus* Protects Pancreatic β-Cells from Damage by Hydrogen Peroxide. *Arch. Pharm. Res.* **2010,** *33*, 853–861.

47. Jeong, S. C.; Jeong, Y. T.; Yang, B. K.; Islam, R.; Koyyalamudi, S. R.; Pang, G.; Cho, K. Y.; Song, C. H. White Button Mushroom (*Agaricus bisporus*) Lowers Blood Glucose and Cholesterol Levels in Diabetic and Hypercholesterolemic Rats. *Nutr. Res.* **2010,** *30*, 49–56.

48. Jia, J.; Zhang, X.; Hu, Y. S.; Wu, Y.; Wang, Q. Z.; Li, N. N.; Guo, Q. C.; Dong, X. C. Evaluation of In Vivo Antioxidant Activities of *Ganoderma lucidum* Polysaccharides in STZ-Diabetic Rats. *Food Chem.* **2009,** *115,* 32–36.

49. Jinn, C. W.; Shu, H. H.; Jih, T.; Wang, Ker; Shaw, C.; Yi, C. C. Hypoglycemic Effect of Extract of *Hericium. J. Sci. Food Agric.* **2005,** *58,* 641–646.

50. Jung, M.; Park, M.; Lee, H. C.; Kang, Y. H.; Kang, E. S.; Sang, K. K. Antidiabetic Agents from Medicinal Plants. *Curr. Med. Chem.* **2006,** *13,* 1203–1218.

51. Kan, W. C.; Wang, H. Y.; Chien, C. C.; Li, S. L.; Chen, Y. C.; Chang, L. H.; Cheng, C. H.; Tsai, W. C.; Hwang, J. C. Effects of Extract from Solid-State Fermented *Cordyceps sinensis* on type 2 Diabetes Mellitus. *Evid. Based Complement. Alternat. Med.* **2012,** *743,* 107–110.

52. Kaur, J.; Singh, P.; Sowers, J. R. Diabetes and Cardiovascular Diseases. *Am. J. Ther.* **2002,** *9,* 510–515.

53. Kawagishi, H.; Shimada, A.; Hosokawa, S.; Mori, H.; Sakamoto, H.; Ishiguro, Y.; et al. Erinacines E, F, and G, Stimulators of Nerve Growth Factor (NGF)—Synthesis from the Mycelia of *Hericium erinaceum. Tetrahedron Lett.* **1996,** *37,* 7399–7402.

54. Kidd, P. M. The Use of Mushroom Glucans and Proteoglycans in Cancer Treatment. *Altern. Med. Rev.* **2000,** *5,* 4–27.

55. Kiho, T.; Yamane, A.; Hui, J.; Usui, S.; Ukai, S. Hypoglycemic Activity of a Polysaccharide (CS-F30) from the Cultural Mycelium of *Cordyceps sinensis* and Its Effect on Glucose Metabolism in Mouse Liver. *Phytother. Res.* **2000,** *4,* 647–649.

56. Kim, D. H.; Yang, B. K.; Jeong, S. C.; Hur, N. J.; Das, S.; Yun, J. W.; Choi, J. W.; Lee, Y. S.; Song, C. H. A Preliminary Study on the Hypoglycemic Effect of the Exo-polymers Produced by Five Different Medicinal Mushrooms. *J. Microbiol. Biotechnol.* **2001,** *11,* 167–171.

57. Kim, D. H.; Yang, B. K.; Jeong, S. C.; Park, J. B.; Cho, S. P.; Das, S.; Yun, J. W.; Song, C. H. Production of a Hypoglycemic, Extracellular Polysaccharide from the Submerged Culture of the Mushroom, *Phellinus linteus. Biotechnol. Lett.* **2001,** *23,* 513–517.

58. Kim, H. M.; Kang, J. S.; Kim, J. Y.; Park, S. K.; Kim, H. S.; Lee, Y. J.; Yun, J.; Hong, J. T.; Kim, Y.; Han, S. B. Evaluation of Antidiabetic Activity of Polysaccharide Isolated from *Phellinus linteus* in Non-Obese Diabetic Mouse. *Int. Immunopharmacol.* **2010,** *10,* 72–78.

59. Kim, J. I.; Kang, M. J.; Im, J.; Seo, Y. J.; Lee, Y. M.; Song, J. H.; Lee, J. H.; Kim, M. E. Effect of King Oyster Mushroom (*Pleurotus eryngii*) on Insulin Resistance and Dyslipidemia in db/db Mice. *Food Scibiotechnol.* **2010,** *19,* 239–242.

60. Kim, S. P.; Kang, M. Y.; Kim, J. H.; Nam, S. H.; Friedman, M. Composition and Mechanism of Antitumor Effects of *Hericium erinaceus* Mushroom Extracts in Tumor-Bearing Mice. *J. Agric. Food Chem.* **2011,** *59,* 9861–9869.

61. Kim, O. H.; Yang, B. K.; Hur, N. I.; Das, S.; Yun, J. W.; Choi, Y. S.; Song, C. H. Hypoglycemic Effects of Mycelia Produced from Submerged Culture of *Phellinus linteus* (Berk. et Curt.) Teng (Aphyllophoromycetideae) in Streptozotocin-Induced Diabetic Rats. *Int. J. Med. Mushr.* **2001,** *3,* 21–26.

62. Kim, S. K.; Hong, U. P.; Kim, J. S.; Kim, C. H.; Lee, K. W.; Choi, S. E.; Park, K. H.; Lee, M. W. Antidiabetic Effect of *Auricularia auricula* Mycelia in Streptozotocin-Induced Diabetic Rats. *Nat. Prod. Sci.* **2007,** *13,* 390–393.

63. Kim, Y. W.; Kim, K. H.; Choi, H. J.; Lee, D. S. Anti-Diabetic Activity of Beta-Glucans and Their Enzymatically Hydrolyzed Oligosaccharides from *Agaricus blazei*. *Biotechnol. Lett.* **2005**, *27*, 483–487.
64. Kino, K.; Yamashita, A.; Yamaoka, K.; Watanabe.; J.; Tanaka, S.; Ko, K.; K. Shimizu, K. H. T. Isolation and Characterization of a New Immunomodulatory Protein: Ling Zhi-8 (LZ-8), from *Ganoderma lucidum*. *J. Biol. Chem.* **1989**, *264*, 472–478.
65. Kobayashi, T. Subtype of Insulin-Dependent Diabetes Mellitus (IDDM) in Japan, Slowly Progressive IDDM—The Clinical Characteristics and Pathogenesis of the Syndrome. *Diabetes Res. Clin. Pract.* **1994**, *24*, S95–S99.
66. Kolb, H.; Pozzilli, P. Cow's Milk and Type 1 Diabetes, the Gut Immune System Deserves Attention. *Immunol. Today* **1999**, *20*, 108–110.
67. Kubo, K.; Nanba, H. Antidiabetic Mechanism of Maitake (*Grifola frondosa*). In *Mushroom Biology and Mushroom Products*; Royse, D. J., Ed.; Penn State University: University Park, PA, 1996; pp 215–221.
68. Laaksonen, D. E.; Niskanen, L.; Lakka, H. M.; Lakka, T. A.; Uuistupa, M. Epidemiology and Treatment of the Metabolic Syndrome. *Ann. Med.* **2004**, *36*, 332–346.
69. Lee, Y. S.; Kang, I. J.; Won, M. H.; Lee, J. Y.; Kim, J. K.; Lim, S. S. Inhibition of Protein Tyrosine Phosphatase 1Beta by Hispidin Derivatives Isolated from the Fruiting Body of *Phellinus linteus*. *Nat. Prod. Commun.* **2010**, *5* (12), 1927–1930.
70. Lee, Y. S.; Kang, Y. H.; Jung, J. Y.; Lee, S.; Ohuchi, K.; Shin, K. H.; Kang, I. J.; Park, J. H.; Shin, H. K.; Lim, S. S. Protein Glycation Inhibitors from the Fruiting Body of *Phellinus linteus*. *Biol. Pharm. Bull.* **2008**, *31*, 1968–1972.
71. Li, B.; Lu, F.; Suo, X. M. Glucose Lowering Activity of *Coprinus comatus*. *Agro Food Ind. Hi-Tech.* **2010**, *21*, 15–17.
72. Li, D. H. Diabetes and Pancreatic Cancer. *Mol. Carcinog.* **1997**, *51* (1), 64–67.
73. Li, S. P.; Zhang, G. H.; Zeng, Q.; Huang, Z. G.; Wang, Y. T.; Dong, T. T.; Tsim, K. W. Hypoglycemic Activity of Polysaccharide, with Antioxidation, Isolated from Cultured *Cordyceps* Mycelia. *Phytomedicine* **2006**, *13* (6), 428–433.
74. Lo, H. C.; Tu, S. T.; Lin, K. C.; Lin, S. C. The Anti-Hyperglycemic Activity of the Fruiting Body of *Cordyceps* in Diabetic Rats Induced by Nicotinamide and Streptozotocin. *Life Sci.* **2004**, *74* (23), 2897–2908.
75. Luo, X.; Xu, X.; Yu, M.; Yang, Z.; Zheng, L. Characterisation and Immunostimulatory Activity of an α-(1→6)-D-Glucan from the Cultured *Armillaria tabescens* Mycelia. *Food Chem.* **2008**, *111*, 357–363.
76. Lv. Y.; Han, L.; Yuan, C.; Guo, J. Comparison of Hypoglycemic Activity of Trace Elements Absorbed in Fermented Mushroom of *Coprinus comatus*. *Biol. Trace Elem. Res.* **2009**, *131* (2), 177–185.
77. Ma, H. T.; Hsieh, J. F.; Chen, S. T. Anti-Diabetic Effects of *Ganoderma lucidum*. *Phytochemistry* **2015**, *114*, 109–113.
78. Ma, Z. J.; Fu, Q. Comparison of Hypoglycemic Activity and Toxicity of Vanadium (iv) and Vanadium (v) Absorbed in Fermented Mushroom of *Coprinus comatus*. *Biol. Trace Element Res.* **2009**, *132* (1–3), 278–284.
79. Maritime, A. C.; Sanders, R. A.; Watkins, J. B. Diabetes Mellitus is a Metabolic Disorder Characterized by Hyperglycemia and Insufficiency of Secretion or Action of Endogenous Insulin. *J. Biochem. Mol. Toxicol.* **2003**, *17*, 24–38.

80. Matsuur, H.; Asakawa, C.; Kurimoto, M.; Mizutani, J. Alpha Glycosidase Inhibitor from the Seeds of Balsam Pear (*Momordica charantia*) and the Fruit Bodies of *Grifola frondosa*. *Biosci. Biotechnol. Biochem.* **2002**, *66*, 1576–1578.

81. Mattila, P.; Könkö, K.; Eurola, M.; Pihlava, J. M.; Astola, J.; Vahteristo, L.; Hietaniemi, V.; Kumpulainen, J.; Valtonen, M.; Piironen, V. Contents of Vitamins, Mineral Elements, and Some Phenolic Compounds in Cultivated Mushrooms. *J. Agric. Food Chem.* **2001**, *49*, 2343–2348.

82. Mattila, P.; Salo-Väänänen, P.; Könkö, K.; Aro, H.; Jalava, T. Basic Composition and Amino Acid Contents of Mushrooms Cultivated in Finland. *J. Agric. Food Chem.* **2002**, *50*, 6419–6422.

83. Nolte, M. S.; Karam, J. H. Pancreatic Hormones and Anti-Diabetic Drugs. In *Basic and Clinical Pharmacology*; Katzung, B. G., Ed.; Lange Medical Books (McGraw-Hill): San Francisco, CA, 2001; pp 711–734.

84. Phillips, K. M.; Ruggio, D. M.; Haytowitz, D. B. Folate Composition of 10 Types of Mushrooms Determined by Liquid Chromatography Mass Spectrometry. *Food Chem.* **2011**, *129*, 630–636.

85. Phillips, K. M.; Ruggio, D. M.; Horst, R. L.; Minor, B.; Simon, R. R.; et al. Vitamin D and Sterol Composition of 10 Types of Mushrooms from Retail Suppliers in the United States. *J. Agric. Food Chem.* **2011**, *59* (14), 7841–7853.

86. Porta, M.; Alliione, A. Current Approaches and Perspectives in the Medicinal Treatment of Diabetic Retinopathy. *Pharmacol. Therapeut.* **2004**, *103*, 167–177.

87. Potenza, M. V; Mechanick, J. I. The Metabolic Syndrome, Definition, Global Impact, and Pathophysiology. *Nutr. Clin. Pract.* **2009**, *24*, 560–577.

88. Ravi, B.; Renitta, R. E.; Prabha, M. L.; Issac, R.; Naidu, S. Evaluation of Antidiabetic Potential of Oyster Mushroom (*Pleurotus ostreatus*) in Alloxan-Induced Diabetic Mice. *Immunopharmacol. Immunotoxicol.* **2013**, *35* (1), 101–109.

89. Rubel, R. L.; Santa, H. S. D.; Fernandes, L. C.; Bonatto, S. J. R.; Bello, S.; et al. Hypolipidemic and Antioxidant Properties of *Ganoderma lucidum* (Leyss, Fr) Karst Used as a Dietary Supplement. *World J. Microbiol. Biotechnol.* **2011**, *27*, 1083–1089.

90. Saltiel, A. R.; Kahn, C. R. Insulin Signaling and the Regulation of Glucose and Lipid Metabolism. *Nature* **2001**, *414*, 799–806.

91. Shechter, Y. Insulin-Mimetic Effects of Vanadate: Possible Implications for Future Treatment of Diabetes. *Diabetes* **1990**, *39*, 1–5.

92. Smiderle, F. R.; Olsen, L. M.; Ruthes, A. C.; Czelusniak, P. A.; Santana-Filho, A. P.; Sassaki, G. L.; Gorin, P. A. J.; Iacomini, M. Exopolysaccharides, Proteins and Lipids in *Pleurotus pulmonarius* Submerged Culture Using Different Carbon Sources. *Carbohydr. Polym.* **2012**, *87*, 368–376.

93. Sobngwi, E.; Ndour-Mbave, M.; Boateng, K. A.; Ramaiya, K. L.; Njenga, E. W.; Diop, S. N.; Mbanya, J. C.; Ohwovoriole, A. E. Type-2 Diabetes Control and Complications in Specialized Diabetes Care Centers of Six Sub-Saharan African Countries: The Diabcare Africa Study. *Diabetes Res. Clin. Pract.* **2012**, *95*, 30–36.

94. Song, G.; Du, Q. Isolation of a Polysaccharide with Anticancer Activity from *Auricularia polytricha* Using High-Speed Countercurrent Chromatography with an Aqueous Two-Phase System. *J. Chromatogr. A* **2010**, *1217*, 5930–5934.

95. Suguna, S.; Usha, M. Cultivation of Oyster Mushroom. *J. Food Sci. Technol.* **1995**, *32*, 351–352.

96. Takeujchi, H. He, P.; Mooi, L. Y. Reductive Effect of Hot-Water Extracts from Woody Ear (*Auricularia auricula-judae* Quél.) on Food Intake and Blood Glucose Concentration in Genetically Diabetic KK-Ay Mice. *J. Nutr. Sci. Vitaminol. (Tokyo)* **2004**, *50* (4), 300–304.

97. Teng, B. S.; Wang, C. D.; Yang, H. J.; Wu, J. S.; Zhang, D.; Zheng, M.; Fan, Z. H.; Pan, D.; Zhou, P. A Protein Tyrosine Phosphatase 1B Activity Inhibitor from the Fruiting Bodies of *Ganoderma lucidum* (Fr.) Karst and Its Hypoglycemic Potency on Streptozotocin-Induced Type-2 Diabetic Mice. *J. Agric. Food Chem.* **2011**, *59*, 6492–6500.

98. Thornally, P. J. Glycation in Diabetic Neuropathy: Characteristics, Consequences, Causes and Therapeutic Options. *Int. Rev. Neurobiol.* **2002**, *50*, 37–57.

99. Tourlouki, E.; Matalas, A. L.; Panagiotakos, D. B. Dietary Habits and Cardiovascular Disease Risk in Middle-Aged and Elderly Populations: A Review of Evidence. *Clin. Inter. Aging* **2009**, *4*, 319–330.

100. Ulziijargal, E.; Mau, J. L. Nutrient Compositions of Culinary Medicinal Mushroom Fruiting Bodies and Mycelia. *Int. J. Med. Mushr.* **2011**, *13*, 343–349.

101. Visanthamein, G.; Savita, D. Hypoglycemic and Hypocholesterolemic Effect of Selected Powder. *J. Nutr. Diabet.* **2001**, *38*, 419–427.

102. Wang, J. C. Y.; Hu, S. H.; Wang, J. T.; Chen, K. S.; Chia, Y. C. Hypoglycemic Effect of Extract of *Hericium erinaceus*. *J. Sci. Food Agric.* **2005**, *85*, 641–646.

103. Wani, B. A.; Bodha, R. H.; Wani, A. H.; Nutritional and Medicinal Importance of Mushrooms. *J. Med. Plants Res.* **2010**, *4*, 2598–2604.

104. White, N. H.; Sun, W.; Cleary, P. A.; Danis, R. P.; Davis, M. D.; Hainsworth, D. P.; Hubbard, L. D.; Lachin, J. M.; Nathan, D. M. Prolonged Effect of Intensive Therapy on the Risk of Retinopathy Complications in Patients with Type 1 Diabetes Mellitus, 10 Years after the Diabetes Control and Complications Trial. *Arch. Ophthalmol.* **2008**, *126*, 1707–1715.

105. Winkley, K.; Sallis, H.; Kariyawasam, D.; Leelarathma, L. H.; Chalder, T.; Edmonds, M. E.; Stahl, D.; Ismail, K. Five-Year Follow-Up of a Cohort of People with Their First Diabetic Foot Ulcer, the Persistent Effect of Depression on Mortality. *Diabetologia* **2012**, *55*, 303–310.

106. Xiao, C.; Wu, Q. P.; Tan, J. B.; Cai, W.; Yang, X. B.; Zhang, J. M. Inhibitory Effects on Alpha-Glycosidase and Hypoglycemic Effects of the Crude Polysaccharides Isolated from 11 Edible Fungi. *J. Med. Plants Res.* **2011**, *5*, 6963–6967.

107. Xiao, C.; Wu, Q. P.; Cai, W.; Tan, J. B.; Yang, X. B.; Zhang, J. M. Hypoglycaemic Effects of *Ganoderma lucidum* Polysaccharides in Type 2 Diabetic Mice. *Arch. Pharm. Res.* **2012**, *35*, 1793–1801.

108. Yamac, M.; Zeytinoglu, M.; Kanbak, G.; Bayramoglu, G.; Senturk, H. Hypoglycemic Effect of Crude Exopolysaccharides Produced by Cerrena Unicolor, *Coprinus comatus*, and *Lenzites betulina* Isolates in Streptozotocin-Induced Diabetic Rats. *Pharm. Biol.* **2009**, *47*, 168–174.

109. Ye, L. B.; Zheng, X.; Zhang, J.; Tang, Q.; Yang, Y.; Wang, X.; Li, J.; Liu, Y. F.; Pan, Y. J. Biochemical Characterization of a Proteoglycan Complex from an Edible Mushroom *Ganoderma lucidum* Fruiting Bodies and Its Immunoregulatory Activity. *Food Res. Int.* **2011**, *44*, 367–372.

110. Yousef, M.; Yousri, M.; Aaser, M. Hypoglycemic Effect of Button (*Agaricus bisporus*) and Oyster (*Pleurotus Ostreatus*) Mushrooms on Streptozotocin Induced Diabetic Mice. *Biohealth Sci. Bull.* **2010**, *2* (2), 48–51.

111. Yuan, Z.; He, P.; Cui, J.; Takeuchi, H. Hypoglycemic Effect of Water-Soluble Polysaccharide from *Auricularia auricular-judae* Quél. on Genetically Diabetic KK-Ay Mice. *Biosci. Biotechnol. Biochem.* **1998**, *62*, 1898–1903.

112. Zhang, H. N.; Lin, Z. B. Hypoglycemic Effect of *Ganoderma lucidum* Polysaccharides. *Acta Pharmacol. Sin.* **2004**, *25*, 191–195.

113. Zhang, Z.; Lian, B.; Cui, F.; Huang, D.; Chang, W. Comparison of Regulating Blood Glucose Effects of *Ginkgo biloba* Leaf Extract with and without Biotransformation by *Hericium erinaceus*. *Junwu Xuebao*, **2008**, *27*, 420–430.

114. Zhao, C. S.; Yin, W. T.; Wang, J. Y.; Zhang, Y.; Yu, H.; Cooper, R.; Smidt, C.; Zhu, J. S. *Cordyceps* Cs-4 Improves Glucose Metabolism and Increases Insulin Sensitivity in Normal Rats. *J. Altern. Complement. Med.* **2002**, *8*, 403–405.

115. Zhong, J. J.; Xiao, J. H. Secondary Metabolites from Higher Fungi, Discovery, Bioactivity, and Bioproduction. *Adv. Biochem. Eng. Biotechnol.* **2009**, *113*, 79–150.

116. Zimmet, P. Z.; McCarty, D. J.; de Courten, M. P. The Global Epidemiology of Non-Insulin-Dependent Diabetes Mellitus and the Metabolic Syndrome. *J. Diabetes Complicat.* **1997**, *11* (2), 60–68.

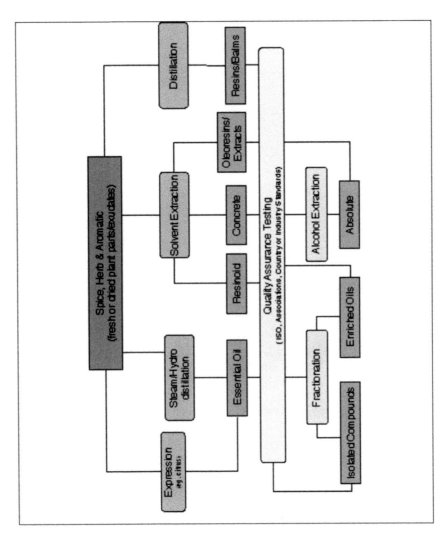

FIGURE 2.1 Extraction of phytomedicines from plants.

Source: http://alf-img2.com/show/plant-extraction-techniques.html.

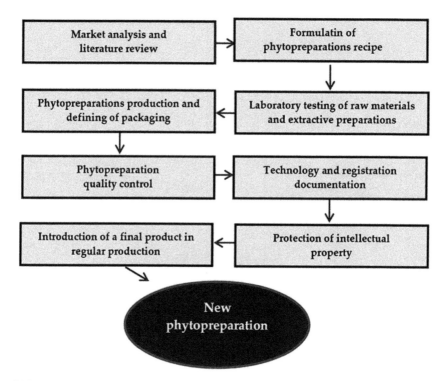

FIGURE 2.2   Schematic representation of the planning and development of new herbal preparation.

Source: Reprinted from Sofija M. Djordjevic (2017). From Medicinal Plant Raw Material to Herbal Remedies. Chapter 16; In: *Aromatic and Medicinal Plants - Back to Nature*; Hany El-Shemy (Ed.); Online; InTech; Open Access; DOI: 10.5772/66618; Available at: https://www.intechopen.com/books/aromatic-and-medicinal-plants-back-to-nature/from-medicinal-plant-raw-material-to-herbal-remedies.

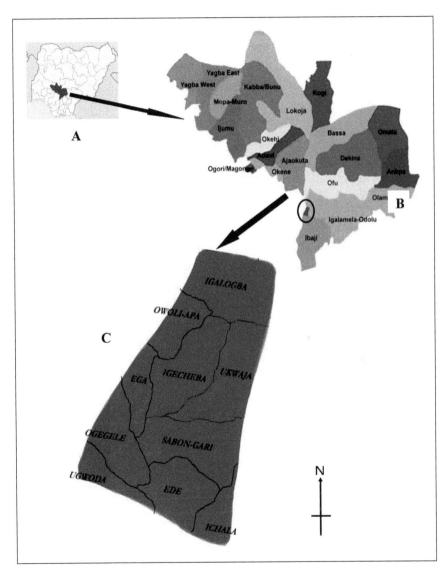

FIGURE 3.1   Location of Idah Local Government and its wards. Map of: A—Nigeria; B—Kogi state; and C—Idah.

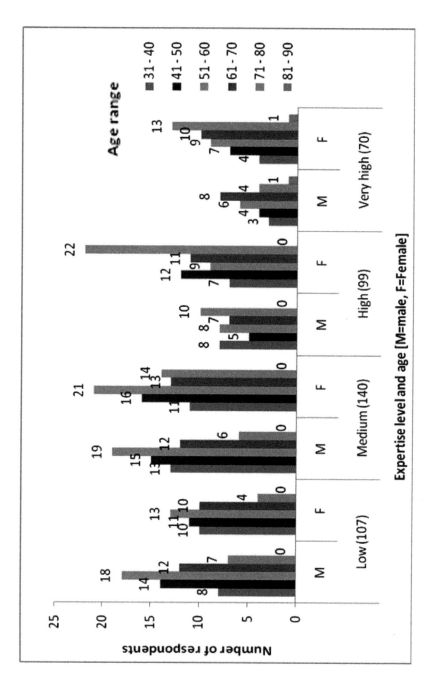

FIGURE 3.3   Respondent's age and level of expertise.

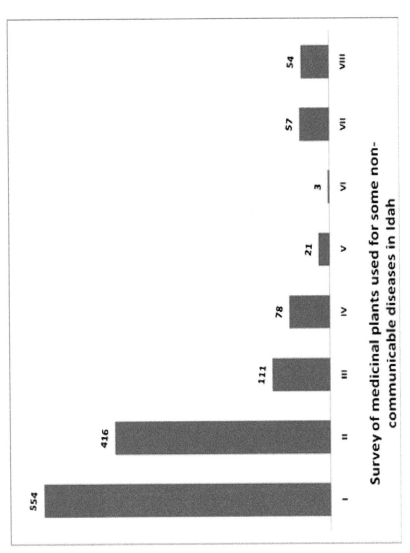

FIGURE 3.4   Population's response with respect to knowledge of medicinal plants.

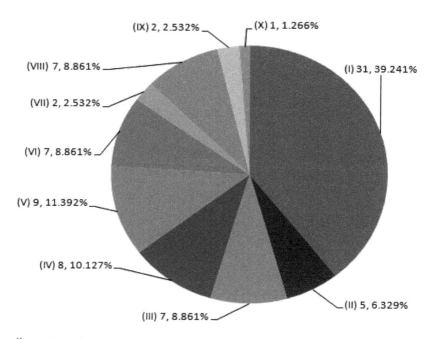

FIGURE 3.5 Modes of extraction and preparation of medicinal plants.

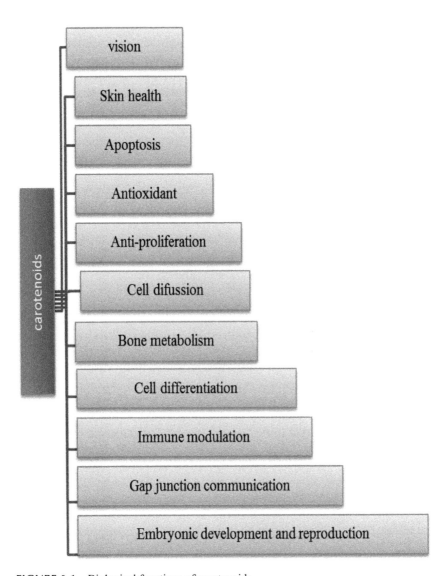

FIGURE 6.1   Biological functions of carotenoids.

CHAPTER 6

NATURAL CAROTENOIDS: WEAPON AGAINST LIFE-STYLE-RELATED DISORDERS

UMAIR SHABBIR, SANA KHALID, MUNAWAR ABBAS, and
HAFIZ ANSAR RASUL SULERIA

ABSTRACT

Carotenoids belong to a group of isoprenoid pigments with abundance in nature. These pigments play a key role in qualitative specification of both fruits and vegetables. Coloring property of these pigments provides an appealing character to foods. Their biological properties allow them to develop a variety of commercialized products. These pigments, in the nutraceutical industry, are known to cure life-style-related disorders, including cardiovascular diseases, cancer, photosensitivity disorders, and immune-sensitivity. This chapter discusses biological properties and the role of carotenoids for prevention and protection from life-style-related disorders. However, there are several research gaps to find out the exact doses for particular disease.

6.1 INTRODUCTION

Pigments are primarily derived from plants and microorganisms such as yeast, cyanobacteria, algae, and fungi.[41] Carotenoids are the second most abundant natural colors after chlorophyll[47] and they are C_{40} lipophilic isoprenoid molecules in nature that contain yellow, red, and orange colors.[81] These pigments serve as antioxidants that act as membrane protector by scavenging peroxyl radicals and O_2. Their antioxidant property is primarily dependent on the structure of pigment. More than 750 structured carotenoids have been identified that are distributed into two fragments: nonoxygenated known as carotenes and oxygenated carotenoids known as xanthophylls.[66] Many of the carotenoids are precursors of vitamin A. Carotenoids are among the bioactive phytochemicals to minimize the threat of chronic diseases, for example, cancer, cardiovascular disorders, cataract, eye health, immunity enhancement, and macular degeneration.[40]

Carotenoids are unable to be synthesized in human body. These are present in different kinds of fruits and vegetables that can be consumed in adequate amounts to fulfill our needs for carotenoids. These carotenoids are involved in the reduction of free radicals to promote health by boosting up the immune system and strengthening endocrine system.[55]

This chapter discusses biological properties and the role of carotenoids for prevention and protection from life-style-related disorders. However, there are several research gaps to find out the exact doses for a particular disease.

6.2 SOURCES OF CAROTENOIDS

Carotenoid availability and their types can be determined by their color, for example, (1) β-carotene and α-carotene are abundantly present in yellow-orange fruits and vegetables; (2) zeinoxanthin and α-cryptoxanthin present in orange fruits, for example, papaya, mandarin, and orange; and (3) lycopene is the main component of tomatoes, grapefruit, and watermelon with bright red color. Lutein (about 45%), β-carotene (about 25–30%), neoxanthin (10–15%), and violaxanthin (10–15%) are present in green leafy vegetables.[53] α-Carotene, lutein, antheraxanthin, zeaxanthin, and β-cryptoxanthin are also found in green vegetables in small amounts. Usually, β-carotene is abundantly found in many vegetables and fruits than to α-carotene.

Da Silva et al.[6] stated that bioactive compounds and carotenoids are rich in traditional and nontraditional tropical fruits that show favorable standpoints for use of fruit species and by-products of these species in products. Recently, many leafy vegetables like *Lactuca indica*, *Moringa oleifera*, and *Oenanthe javanica* have proven as extraordinary sources of carotenoids.

In a study of Thai vegetables, an amount of lutein and β-carotene was documented in *O. javanica* and *L. indica*, correspondingly.[34] Saini et al.[63]

revealed that international nongovernmental organizations (Trees for Life and Educational Concerns for Hunger Organization) have dynamically assisted Moringa leaves as "natural nutrition for the tropics." Citrus fruits and paprika vegetables contain little amounts of apocarotenoids. In pulps of citrus species, 2–7% β-citraurin and red apocarotenoids have been identified.[1] The β-apocarotenoids in total are present at approximately 1.5% of the level of β-carotene.[13] Some of the examples are mentioned in Table 6.1.

TABLE 6.1 Sources of Main Carotenoids with Their Provitamin A and Nonprovitamin A.

Class of carotenoid	Example	Main sources
Provitamin A	β-Carotene	Celery, tomatoes, spinach, lattice, carrot, broccoli, and parsley
	α-Carotene	Green leafy vegetables
	β-Cryptoxanthin	Papayas, peaches, mangoes, and oranges
Nonprovitamin A	Lutein	Green leafy vegetables: kale, spinach, broccoli, parsley, beans, corn, avocado, and Brussels sprouts
	Lycopene	Tomatoes, papaya, grapefruit, and watermelon
	Zeaxanthin	Kale, spinach, broccoli, parsley, beans, corn, avocado, mandarins, peaches, and oranges

Source: Modified from Ref. [6].

Richest sources of carotenoids from plants are ripened fruits or seed pulp of *Momordica cochinchinensis* (common name gac fruit). Vuong et al.[75] documented the concentration of β-carotene and lycopene in gac fruit to be 408 and 83.3 µg/g, respectively. The carotenoids are also present in fruits of Amazonia region that include tucuma (*Astrocaryum aculeatum*), mamey (*Mammea americana*), peach palm, buriti (*Mauritia vinifera*), physalis (*Physalis angulata*), and marimari (*Geoffrola striata*). A total of 60 different carotenoids were identified,[75] with the total carotenoid concentration varying from 38 µg/g in marimari to 514 µg/g in buriti; and β-carotene was maximum among all carotenoids present.[7] The maximum concentration of β-carotene was present in acerola pulp (26.23 µg/g), followed by papaya (20.24 µg/g) and by Surinam cherry (15.64 µg/g) pulp on dry weight basis.

6.3   BIOAVAILABILITY AND METABOLISM OF CAROTENOIDS

The amount of carotenoid, absorbed in body through systematic circulation and available for physiological functions and storage, is called bioavailability.[61] There are several factors affecting the digestion, absorption, movement, and storage of carotenoids.[11,82] Generally, β-carotene extracted from plants has low bioavailability (10–65%) due to resistance of fibers, protein complexes, and carotene to digestion and degradation of plant cell wall to attain sufficient discharge of carotenoids.[58] Research has shown that thermal treatment increases the availability of carotenoids due to bond loosening and disruption of cell walls.[12] Carotenoid bioavailability can also be determined by other factors, including gender, infection, nutritional status, aging, and genetic factors.[82]

Often, carotenoids are derived through biosynthesis of isoprenoid in conjunction with the range of some other naturally occurring substances such as gibberellic acid and steroids. Mevalonic acid is the preliminary product synthetized from all isoprene derivatives, which upon phosphorylation are converted into a phosphorylated isoprene. Therefore, this isoprene results in polymerization. In the chain of polymerization, the position and number of the double bonds are fixed. Biosynthesis and dispossession of xanthophylls, carotenes, and carotenogenesis regulation have been described very well in plants, animals, and humans.[47]

Giuliano et al.[19] reported the metabolism of carotenoids through multigene engineering. Two molecules of geranylgeranyl pyrophosphate are merged into phytoene (40-carbon intermediate in the biosynthesis of all carotenoids) components; this initial reaction is particular to carotenoid type of isoprenoid metabolism. From this phase, small reactions might be set up in various organisms. Respectively, fungi, anoxygenic photosynthetic bacteria, and nonphotosynthetic bacteria can desaturate phytoene three to four times to form lycopene or neurosporene. Comparatively, phytoene converts into lycopene through carotene in two individual groups of reactions by oxygenic photosynthetic organisms. At the level of lycopene or neurosporene, the carotenoid biosynthesis through different pathways to produce the various varieties of carotenoids is initiated in nature.

The bacterio-chlorophyll and lipophilic carotenoid or chlorophyll pigment molecules connect noncovalently through integral membrane proteins in photosynthetic organisms. Carotenoids are protein bounded in nonphotosynthetic organisms and tissues; and this takes place in cell wall or cytoplasmic membranes, fibrils, oil droplets, and crystals. Because animals and humans have no ability to synthesize carotenoids de novo, to combat the requirement carotenoids must be taken through diet. Most of the carotenoids, which are taken through diet, are absorbed through mucosal cells in gastrointestinal tract and remain unchanged in tissues.

Through passive diffusion, carotenoids are absorbed in intestine after being integrated with micelles, which are produced through bile acids and dietary fats. Then, these micellar carotenoids are integrated with chylomicrons and distributed into the lymphatic system.[56] Exclusively, lipoprotein carotenoids are delivered in the plasma. Carotenoids that have oxygen functionality relatively have additional polar in nature than those that are nonfunctional. That is why, lycopene, α-carotene, and β-carotene have the ability to fall in low-density lipoproteins in the circulation, while xanthophylls (e.g., lutein, zeaxanthin, and cryptoxanthins) fall into high-density lipoproteins. By the integration of lipoprotein molecules with receptors, the transportation of carotenoids to extrahepatic tissues and the degradation takes place through lipoprotein lipase.[57]

Surai et al.[68] documented that though more than 40 carotenoids are generally taken through diet, just 6 of these carotenoids and metabolites have been observed in human tissues and organs. In the breast milk of a lactating mother, 34 carotenoids and 8 metabolites have been observed. Mainly, absorption of these carotenoids occurs in intestine; facilitated diffusion has been found additionally to intercede in simple diffusion for absorption of carotenoids in intestine of mammals. The absorption of carotenoids might be because of the uptake to intestinal epithelia through facilitated diffusion and an unfamiliar process of secretion into the intestinal lumen.

After digestion of food containing carotenoids, these are absorbed with fat-soluble molecules in small intestine and then are transferred to the liver. Liver converts them into very low-density lipoproteins (VLDL). Lipase enzymes in the liver change VLDL into different lipoprotein molecules; and as a result of this metabolism, high-density lipoproteins and low-density lipoproteins are formed, which then are transferred to the peripheral tissues. There is still an ambiguity that how much carotenoids are metabolized in liver and how much are converted into other forms.[43] Different components of diet affect the absorption of carotenoids.[70] Among these factors, genetics, sex, hormonal status, and age influence the absorption of carotenoids.[69] These factors are interrelated with each other and their degree of influence on the absorption of carotenoids is still unknown.

Transformation of β-carotene into vitamin A in intestines is very well demonstrated and the metabolic conversion of xanthophylls is less demonstrated in literature. Secondary hydroxyl group through enzymatic oxidation leads to keto-carotenoids that act as a mutual pathway of metabolism of xanthophylls in living organisms.[72]

6.4   RECOMMENDED DIETARY ALLOWANCE OF CAROTENOIDS

Carotenoids play some important roles in cellular function. Figure 6.1 shows activities of carotenoids in animals. Through retinol equivalents (RE), the activity of carotenoids can be evaluated by one RE = 1 μg of retinol by conversion. Eitenmiller et al.[9] stated that transformation of carotenoids and absorption (%) for β-carotene takes into account as 1 RE = 6 μg, while for other carotenoids it is 12 μg (assuming 50% biological activity). For pregnant women, recommended dietary allowance is 800, 1300 RE for breast feeding woman, and 1000 RE for adult man. International units (IU) can also be used to measure the biological activity, in which 1 IU = 0.6 μg of β-carotene and 0.3 μg all-Z-retinol (thus, 1 RE = 3.3 IU). Addition of phenyl group stabilizes biological activity of carotenoids. Synthetic carotenoids prepared by You et al.[83] consisting the aromatic phenyl groups with a parasubstituent at C-13 and C-13' position showed the ability to control the instability of carotenoids. In 2,2-diphenyl-1-picrylhydrazyl (DPPH) and 2,2'-azino-bis(3-ethylbenzothiazoline-6-sulphonic acid) (ABTS) assays, these carotenoids show stronger radical activity than β-carotene.

6.5   BIOLOGICAL FUNCTIONS AND VITAMIN A ACTIVITY

Murthy[46] stated that by scavenging reactive oxygen species (ROS) from extra light, carotenoids have the ability to photo guard the photosynthetic operation in plants such as singlet oxygen and free radicals. Other than coloring properties, carotenoids have main beneficial effect on human health and the use of these pigments in diet is gaining consumer's attention day by day.[84] The antioxidant properties, which are mostly related to biological functioning of carotenoids in humans, are related to their molecular structure. Many of the carotenoids have vitamin A activity.

Among 700 founded carotenoids, nearly 50 have been verified for their provitamin A activity.[27] Von Lintig and Vogt[73] stated that carotenoids show their provitamin A activity by the action of carotene dioxygenase that converts carotenoids into vitamin A in the form of retinol and retinal. Send and Sundholm[65] revealed that carotenoid having minimally one unmodified β-ionone ring possibly has the ability to deliver provitamin A activity. Therefore, provitamin A-active carotenoids are β-cryptoxanthin, β-carotene, α-carotene, and γ-carotene. β-Carotene, that comprises two β-ionone rings, shows 100% provitamin A activity (Fig. 6.2).

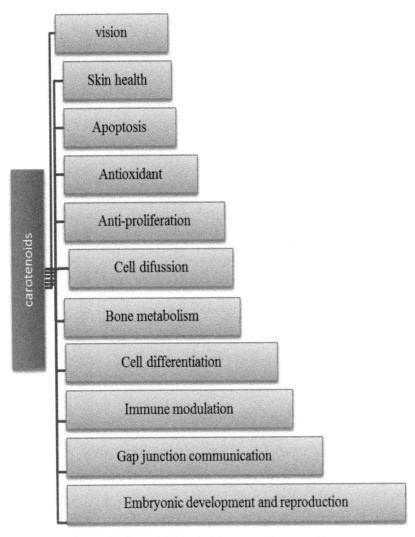

FIGURE 6.1   (See color insert.) Biological functions of carotenoids.

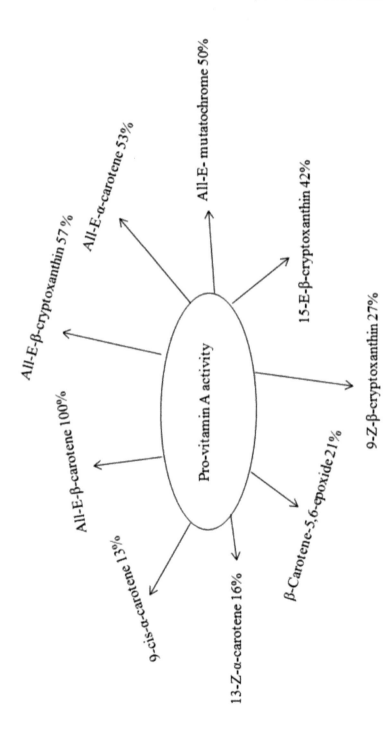

FIGURE 6.2   Provitamin A activity of carotenoids.

As lycopene has less β-ionone ring, therefore, it shows low provitamin A activity. Provitamin A activity of some primary carotenoids is shown in Figure 6.2. Provitamin A activity is the main function of carotenoids and this vitamin A provides antioxidant property through carotenoids by scavenging singlet oxygen and also by deactivation of free radicals, which shows significant functions in the inhibition of various forms of cancer, macular degeneration, cardiovascular diseases, and eye-related disorders.[45]

6.6 BENEFICIAL BIOACTIVITIES OF CAROTENOIDS

6.6.1 CANCER

According to the World Health Organization,[79] nearly one out of six deaths is due to cancer. Therefore, there is a dire need to take some tolerable steps to lessen and amend the factors distressing the danger of cancer. Obviously, one of the simplest steps is ample intake of organic foods that are crowded with biologically active complexes.

Numerous prospective studies have displayed a constructive relationship among the ingestion of carotenoids-rich foods and a reduced hazard of different kinds of cancer.[30] A large assemblage of data on lungs cancer and dietary carotenoids has been brought to limelight. Many studies of carotenoid diet and lung cancer on nonsmokers established a contrary correlation between lung cancer risk and ingestions of foods rich in carotenoids.[5] These illustrations were proven by the β-carotene and Retinol Efficacy Trial study, where mixture of vitamin A supplementation and β-carotene was trialed in both male and female at a high risk of increasing lung cancer (smokers and asbestos workers) and in subjects those who intake larger quantities of alcohol.[21] So far, from thorough scrutinization of the consequences, it has been found that the unpredicted "cancerogenic" properties of carotenoids supplementation may be defined in relations of their strong intervention with the lifestyle disorders of the individuals.[22] The data on studies of prostate cancer support the concept that various carotenoids and carotenoid-rich foods might be involved in the inhibition of the risk of prostate cancer.[48]

Among carotenoids, lycopene is considered as the most potent agent against tumor.[18] The preclinical studies summarize multiple possible ways of lycopene action, claiming, at the same time, its importance in the enrichment of the oxidation stress defense system.[77] Further investigations support the observational studies on the role of lycopene and tomato products in the hindrance of prostate cancer and lungs cancer.[10] Gloria et al.[20] documented that lycopene and β-carotene inhibit cell cycle arrest and cell proliferation in different phases of breast cancer.

Guo et al.[24] stated that antioxidant compounds like α-carotene and β-carotene have been proven beneficial against reducing the threat of cervical cancer. Kim et al.[32] revealed that β-carotenoid acts as a chemotherapeutic agent regulating the metastasis and invasion of neuroblastoma via hypoxia inducible factor-1α. Karppi et al.[29] documented that higher concentration of β-carotene serum was related with low danger of prostate cancer. Moreover, a huge data confirm a correlation among the consumption of fruits and vegetables and reduction of colon cancer development.[42] Freedman et al.[15] documented that taking large amount of foods containing carotenoids resulted in the inhibition of risk of oral and throat cancers by 50%.

6.6.2 CARDIOVASCULAR DISEASES

Inflammation and oxidative stress are one of the main causes for cardiac disorders.[39,60] In oxidative stress, the ROS can lead to oxidation of low-density lipoproteins that show exceptional role in increased risk of atherosclerosis.[33] Carotenoids play a remarkable role in the protection of low-density lipoproteins from oxidation that shows anti-atherogenic properties of carotenoids.[60] Furthermore in in vivo studies, carotenoids inhibited lipid peroxidation mechanisms by which the appearance of carotenoids in cell membranes plays the crucial role to behave as alleviating elements of these structures.[78] Palozza et al.[52] investigated antioxidant properties of these compounds during radical peroxide-induced cholesterol that showed a prominent antioxidant activity in ascending order: β-carotene, lutein, cantaxanthin, and astaxanthin.

Similarly, low levels of α-carotene in serum show reverse relation in development of coronary artery disorder and appearance of arterial plaque (narrowing the veins and arteries), through which α-carotene projected as a potential marker for atherosclerosis in human. Moreover, carotenoids possess higher levels of provitamin A activities that have ability to reduce hazards of angina pectoris disease.[14] The higher levels of lutein and β-cryptoxanthin in plasma have shown low hazards of cardiac infraction for people suffering from cardiac disorders.[33] Gajendragadkar et al.[17] stated that 7 mg dose of lycopene through dietary intake helps to improve the endothelial (vascular) function in cardio vascular patients.

6.6.3   VISION HEALTH

Various studies have demonstrated that zeaxanthin and lutein are major pigments that are responsible for preservation of normal visual functioning and yellowing of the human eye macula.[37] Major carotenoids are in low quantities in the human macula that includes α-carotene, lycopene, β-cryptoxanthin and β-carotene, and zeaxanthin and lutein to absorb blue light and diminish pernicious photo-oxidative effects due to extra blue light thus decreasing eye chromatic aberration.[3] Carotenoids have the ability to protect eye macula from adversative photochemical reactions due to anti-oxidant properties.[36] Visual sensitivity greatly depends on zeaxanthin and lutein concentration in retina of people above the age of 64[26] and is present in lower levels in human and animals with cataracts.[49] Friedman et al.[16] also documented that people with lower levels of lutein and zeaxanthin can face macular degeneration that is the major root of permanent loss of eyesight in people more than age of 65 in developed countries.

Zeaxanthin and lutein spectrum indicate a large absorption band with a peak at 450 nm that is assumed to be participating in absorbing extra blue light before it approaches to photoreceptors; so it prevents the eye macula from being impaired by the blue light.[23] Furthermore, the physical, chemical, and biological properties of both carotenoids possess to save the membrane structure in the eye photoreceptors because of their reactive oxygen scavenging properties.

6.6.4   IMMUNE RESPONSE

Both in vivo and in vitro laboratory studies have confirmed that β-carotene has the ability to protect phagocytic cells from auto-oxidative impairment, encourages effector T-cell functions, boosts T- and B-lymphocyte proliferative responses, enhances the formation of various interleukins, and increases macrophage, along with natural killer cell tumoricidal capacities and cytotoxic T cell. Recent studies have shown that lesser serum concentrations of α-carotene, lycopene, and β-carotene have reported in

HIV-affected women, particularly in those who have lower counts of CD4 helper cells. In HIV and AIDS patients, both selenium and β-carotene are found lesser. They act as antioxidants in both HIV and AIDS conditions and seem to be associated to inhibit HIV replication, cytokine, NF-kB activation, and direct immune modulation.

β-Carotene acts as an immune modulator by enhancing CD4 count and encouraging natural killer cell function. β-Carotene as an antioxidant seems to maintain enzymatic defense system connected in diminishing oxidative impairment.[50] Baralic et al.[2] stated that astaxanthin helps to increase the IgA secretion and reduction of prooxidant–antioxidant balance in humans. In rats, astaxanthin has resulted in improvement of immunological effects by increasing lymph proliferation, phagocytic activity, and has proven itself better in increasing the immunity than other food pigments.[67]

6.6.5 PHOTOSENSITIVE DISORDERS

Carotenoids have the ability to provide skin photo-protection against UV light.[76] They also hold anti-inflammatory effects because of their scavenging action on ROS. Due to these properties, astaxanthin promotes anti-inflammatory properties that maintain vital proteins and lipids of human lymphocytes.[4] Some researchers have revealed that astaxanthin provides protection from CCl_4-induced hepatic impairment by hindering lipid peroxidation, modulating the inflammatory process and cellular antioxidant system.[37] Table 6.2 shows biological functions and medicinal properties of carotenoids, their role in retinitis,[31] hindrance of cataracts,[74] macular degeneration,[71] and gastric infection.[44]

Another property of carotenoids is to act as a preservative that has been used in cosmetics and food products, mixed with some other algal bioactive elements or antioxidants, also in lotions and cream for sun protection.[8] Use of carotenoids shows advantageous influence on patients with skin inflammatory pathology and psoriasis. Lima and Kimball[38] studied lower amounts of carotenoids in skin inflammatory patients with psoriasis disorder. The carotenoids also play significant role in gastric diseases. It has been demonstrated that higher consumption of carotenoids inhibits the expansion of conditions affected by *Helicobacter pylori*,[44] a Gram-negative bacteria class that inhabits the gastric mucosa.[35]

TABLE 6.2 Biological Functions and Medicinal Properties of Carotenoids.

Main carotenoids	Biological function and medicinal properties
Astaxanthin	In liver, prostate tumors, and benign prostatic hyperplasia
	Inflammatory inhibition properties
β-Carotene	Provitamin A function
	Helps in the curing of both acute and chronic coronary syndromes
	In colorectal cancer
	Photo-protection of body skin alongside UV light
Lycopene	In prostate cancer and prostatic hyperplasia
	In the hindrance of acute and chronic coronary syndromes and atherosclerosis
Zeaxanthin	In maintenance of normal vision
	In the hindrance of acute and chronic coronary syndromes
	Helps to prevent hepatocellular carcinoma
	Hindrance of cataracts
	To inhibit macular disintegration linked with age
Lutein	In hindrance of stroke, acute, and chronic coronary syndromes
	In the hindrance of cataracts
	Aids to uphold a typical visual function
	To inhibit macular disintegration linked with age
	To escape gastric infection by *Helicobacter pylori*
	In the hindrance of retinitis

Source: Compiled from Refs. [37,71,74].

6.7 CAROTENOIDS: APPLICATIONS IN FORTIFICATION INDUSTRY

Haila et al.[25] stated that mixing of carotenoids with other foods can bind their activities as antioxidant. There is an extra task to use carotenoids as ingredient in functional foods as they have high melting point, which makes crystals during food storage. In spite of that endogenous carotenoids in foods are generally stable, though carotenoids as food additives are comparatively unpredictable in food systems because they are vulnerable to auto-oxidation, light and oxygen.[80] By mixing of carotenoids with other food material, their functional properties can lose concentration,[59] because the loss of double bond carotenoids might be degraded by reactions.

Additionally, there will be *cis*-configuration when double bonds in carotenoids undertake isomerization.[80] Schieber and Carle[64] revealed that isomerization reactions might be valuable due to *cis*-isomers of carotenoids, for example, lycopene are assumed to be more bioactive and bioavailable.

Rodriguez-Huezo et al.[62] reported that carotenoids are used in the industry as colorants in numerous food supplements and drinks. In various studies, carotenoids reported a chain-breaking antioxidant under particular conditions. However, carotenoids have highly conjugated structure; due to that, it is unstable when exposed to light and oxygen during processing and storage of food. Because of that, there may be a loss of their desirable biological and nutritive properties in addition to the formation of objectionable flavor or aroma components.

Therefore, these components cannot be handled in their crystalline structure but can be in the encapsulated form. Encapsulated form of carotenoids (α-carotene) is used in the industry for food fortification and supplementation. For that purpose, oil in water emulsion proved as a most extensive technique that can be used for encapsulating lipophilic functional compounds, for example, bioactive lipids and flavors. Therefore, there is a solid reason for using carotenoids as functional component in food goods. On the other hand, β-carotenoid is less soluble in oil and insoluble in water at room temperature because of its crystalline form that makes it problematic to integrate in food commodities.[59] Moreover, β-carotenoid is sensitive to temperature elevation, light, and oxygen that limit its applications in food.[51]

Nanotechnology is an encouraging research field, with applications in health industries.[28] Quintanilla-Carvajal et al.[54] reported that with the application of nanotechnology, food industry could have benefits as this technology has prospective to enhance solubility and bioavailability of different functional components such as the "carotenoids."

6.8 SUMMARY

Carotenoids are natural pigments. While providing coloring property, these compounds exhibit health-promoting effects. In fact, carotenoids show the property of antioxidants, directly involved in the mechanism of scavenging the free radicals. Literature has demonstrated high impact of carotenoids on different life style-related disorders, including cardiovascular diseases and cancer. The antioxidant property of carotenoids is the key involved in the prevention of many disorders. Moreover, the exact mechanism and target doze for particular disease are yet to be explored.

KEYWORDS

- antioxidant
- bioavailability
- carotenoids
- metabolism
- nutraceuticals
- phytochemical
- pigments
- provitamin
- vitamin A
- xanthophylls
- zeaxanthin

REFERENCES

1. Agócs, A.; Nagy, V.; Szabó, Z.; Márk, L.; Ohmacht, R.; Deli, J. Comparative Study on the Carotenoid Composition of the Peel and the Pulp of Different Citrus Species. *Innov. Food Sci. Emerg. Technol.* **2007,** *8* (3), 390–394.
2. Baralic, I.; Andjelkovic, M.; Djordjevic, B.; Dikic, N.; Radivojevic, N.; Suzin-Zivkovic, V.; Radojevic-Skodric, S.; Pejic, S. Effect of Astaxanthin Supplementation on Salivary IgA, Oxidative Stress, and Inflammation in Young Soccer Players. *Evid.-Based Complement. Altern. Med.* **2015,** *2*, 1–9.
3. Bates, C. J.; Chen, S. J.; MacDonald, A.; Holden. R. Quantitation of Vitamin E and a Carotenoid Pigment in Cataractous Human Lenses and the Effect of a Dietary Supplement. *Int. J. Vit. Nutr. Res.* **1996,** *66* (4), 316–321.
4. Bolin, A. P.; Macedo, R. C.; Marin, D. P.; Barros, M. P.; Otton, R. Astaxanthin Prevents In Vitro Auto-Oxidative Injury in Human Lymphocytes. *Cell Biol. Toxicol.* **2010,** *26* (5), 457–467.
5. Brennan, P.; Fortes, C.; Butler, J.; Agudo, A.; Benhamou, S.; Darby, S.; Gerken, M.; Jokel, K. H.; Kreuzer, M.; Mallone, S. A Multicenter Case-Control Study of Diet and Lung Cancer among Non-Smokers. *Cancer Causes Control* **2000,** *11* (1), 49–58.

6. Da Silva Messias, R.; Galli, V.; Dos Anjos e Silva, S. D.; Rombaldi, C. V. Carotenoid Biosynthetic and Catabolic Pathways: Gene Expression and Carotenoid Content in Grains of Maize Landraces. *Nutrients* **2014**, *6* (2), 546–563.

7. De Rosso, V. V.; Mercadante, A. Z. Identification and Quantification of Carotenoids, by HPLC–PDA–MS/MS, from Amazonian Fruits. *J. Agric. Food Chem.* **2007**, 55 (13), 5062–5072.

8. DelCampo, J. A.; Moreno, J.; Rodríguez, H.; Vargas, M. A.; Rivas, J.; Guerrero, M. G. Carotenoid Content of Chlorophycean Microalgae, Factors Determining Lutein Accumulation in *Muriellopsis* sp. (*Chlorophyta*). *J. Biotechnol.* **2000**, *76* (1), 51–59.

9. Eitenmiller, R. R.; Landen, W. O.; Ye, L. *Vitamin Analysis for the Health and Food Sciences*; CRC Press (Taylor & Francis Group): Boca Raton, FL, 2007; p 332.

10. Etminan, M.; Takkouche, B.; Caamano-Isorna, F. The Role of Tomato Products and Lycopene in the Prevention of Prostate Cancer, a Meta-Analysis of Observational Studies. *Cancer Epidemiol., Biomark. Prevent.* **2004**, *13* (3), 340–345.

11. Fernandez-Garcia, E.; Carvajal-Lerida, I.; Jaren-Galan, M.; Garrido-Fernandez, J.; Perez-Galvez, A.; Hornero-Mendez, D. Carotenoids Bioavailability from Foods: From Plant Pigments to Efficient Biological Activities. *Food Res. Int.* **2012**, *46* (2), 438–450.

12. Fleshman, M. K.; Lester, G. E.; Ried, K. M.; Kopec, R. E.; Narayanasamy, S.; Curley, R. W. Carotene and Novel Apocarotenoid Concentrations in Orange-Fleshed *Cucumis melo* melons: Determinations of β-Carotene Bioaccessibility and Bioavailability. *J. Agric. Food Chem.* **2011**, *59* (9), 4448–4454.

13. Ford, E. S.; Giles, W. H. Serum Vitamins, Carotenoids and Angina Pectoris, Findings from the National Health and Nutrition Examination Survey, III. *Ann. Epidemiol.* **2000**, *10*, 106–116.

14. Freedman, N. D.; Park, Y.; Subar, A. F.; Hollenbeck, A. R.; Leitzmann, M. F.; Schatzkin, A.; Abnet, C. C. Fruit and Vegetable Intake and Head and Neck Cancer Risk in a Large United States Prospective Cohort Study. *Int. J. Cancer* **2008**, *122* (10), 2330–2336.

15. Freeman, V. L.; Meydani, M. Prostatic Levels of Tocopherols, Carotenoids and Retinol in Relation to Plasma Levels and Self-Reported Usual Dietary Intake. *Am. J. Epidemiol.* **2000**, *151*, 109–118.

16. Friedman, D. S.; Colmain, B. J.; Muñoz, B.; Tomany, S. C.; McCarty, C.; De Jong, P. T.; Nemesure, B.; Mitchell, P.; Kempen, J.; Congdon, N. Prevalence of Age-Related Macular Degeneration in the United States. *Arch. Ophthalmol.* **2004**, *122* (4), 564–572.

17. Gajendragadkar, P. R.; Hubsch, A.; Mäki-Petäjä, K. M.; Serg, M.; Wilkinson, I. B.; Cheriyan, J. Effects of Oral Lycopene Supplementation on Vascular Function in Patients with Cardiovascular Disease and Healthy Volunteers: A Randomized Controlled Trial. *PLoS One* **2014**, *9* (6), 99070.

18. Giovannucci, E. A Review of Epidemiologic Studies of Tomatoes, Lycopene, and Prostate Cancer. *Exp. Biol. Med.* **2002**, *227* (10), 852–859.

19. Giuliano, G. Plant Carotenoids: Genomics Meets Multi-Gene Engineering. *Curr. Opin. Plant Biol.* **2014**, *19*, 111–117.

20. Gloria, N. F.; Soares, N.; Brand, C.; Oliveira, F. L.; Borojevic. R.; Teodoro, A. J. Lycopene and Beta-Carotene Induce Cell-Cycle Arrest and Apoptosis in Human Breast Cancer Cell Lines. *Anticancer Res.* **2014**, *34* (3), 1377–1386.

21. Goodman, G. E.; Thornquist, M. D.; Balmes, J.; Cullen, M. R.; Meysekns, F. L.; Omenn, G. S.; Valanis, B.; Williams, J. H. The Beta-Carotene and Retinol Efficacy Trial, Incidence of Lung Cancer and Cardiovascular Disease Mortality During 6-Year Follow-Up after Stopping β-Carotene and Retinol Supplements. *J. Natl. Cancer Inst.* **2004,** *96* (23), 1743–1750.

22. Góralczyk, R. β-Carotene and Lung Cancer in Smokers: Review of Hypotheses and Status of Research. *Nutr. Cancer* **2009,** *61* (6), 767–774.

23. Greenstein, V. C.; Chiosi, F.; Baker, P.; Seiple, W.; Holopigian, K.; Braunstein, R. E.; Sparrow, J. R. Scotopic Sensitivity and Color Vision with a Blue-Light Absorbing Intraocular Lens. *J. Cataract Refract. Surg.* 2007, *33* (4), 667–672.

24. Guo, L.; Zhu, H.; Lin, C.; Che, J.; Tian, X.; Han, S. Associations between Antioxidant Vitamins and the Risk of Invasive Cervical Cancer in Chinese Women: A Case-Control Study. *Sci. Rep.* **2015,** *5*, 13607–13610.

25. Haila, K. M.; Lievonen, S. M.; Heinonen, M. I. Effects of Lutein, Lycopene, Annatto, and Gamma-Tocopherol on Autoxidation of Triglycerides. *J. Agric. Food Chem.* **1996,** *44* (8), 2096–2100.

26. Hammond, B. R.; Wooten, B.; Snodderly, D. M. Preservation of Visual Sensitivity of Older Individuals: Association with Macular Pigment Density. *Invest. Ophthalmol. Vis. Sci.* **1998,** *39* (2), 397–406.

27. Hurst, W. J. *Methods of Analysis for Functional Foods and Nutraceuticals*; Taylor & Francis Group: Boca Raton, FL, 2008; p 318.

28. Jochen, W.; Paul, T.; McClements, D. J. Functional Materials in Food Nanotechnology. *J. Food Sci.* **2006,** *71* (9), 107–116.

29. Karppi, J.; Kurl, S.; Laukkanen, J. A.; Kauhanen, J. Serum β-Carotene in Relation to Risk of Prostate Cancer: the Kuopio Ischaemic Heart Disease Risk Factor Study. *Nutr. Cancer* **2012,** *64* (3), 361–367.

30. Key, T. J. Fruit and Vegetables and Cancer Risk. *Br. J. Cancer* **2011,** *104* (1), 6–11.

31. Kim, S. H.; Jean, D.; Lim, Y. P.; Lim, C.; An, G. Weight Gain Limitation and Liver Protection by Long-Term Feeding of Astaxanthin in Murines. *J. Korean Soc. Appl. Biol. Chem.* **2009,** *52* (2), 180–185.

32. Kim, Y. S.; Lee, H. A.; Lim, J. Y.; Kim, Y.; Jung, C. H.; Yoo, S. H.; Kim, Y. β-Carotene Inhibits Neuroblastoma Cell Invasion and Metastasis In Vitro and In Vivo by Decreasing Level of Hypoxia-Inducible Factor-1α. *J. Nutr. Biochem.* **2014,** *25* (6), 655–664.

33. Koh, W. P.; Yuan, J.; Wang, R.; Lee, Y. P.; Lee, B. L.; Yu, M. C.; Ong, C. N. Plasma Carotenoids and Risk of Acute Myocardial Infarction in the Singapore Chinese Health Study. *Nutr., Metab. Cardiovasc. Dis.* **2011,** 21 (9), 1–6.

34. Kongkachuichai, R.; Charoensiri, R.; Yakoh, K.; Kringkasemsee, A.; Insung, P. Nutrients Value and Antioxidant Content of Indigenous Vegetables from Southern Thailand. *Food Chem.* **2015,** *173*, 838–846.

35. Kusters, J. G.; Van Vliet, A. H.; Kuipers, E. J. Pathogenesis of *Helicobacter pylori* Infection. *Clin. Microbiol. Rev.* **2006,** *19* (13), 449–490.

36. Landrum, J. T.; Bohne, R. Luteín, Zeaxanthin and the Macular Pigment. *Arch. Biochem. Biophys.* **2001,** *385* (1), 28–40.

37. Le, M.; Xiao-Ming, L. Effects of Lutein and Zeaxanthin on Aspects of Eye Health. *J. Sci. Food Agric.* **2010,** *90* (1), 2–12.

38. Lima, X. T.; Kimball, A. B. Skin Carotenoid Levels in Adult Patients with Psoriasis. *J. Eur. Acad. Dermatol. Venereol.* **2010**, *25* (8), 11–16.
39. Lloyd-Jones, D.; Adams, R.; Carnethon, M. Heart Disease and Stroke Statistics—2009 Update: A Report from the American Heart Association Statistics Committee and Stroke Statistics Subcommittee. *Circulation* **2009**, *119* (3), 480–486.
40. Maldonade, I. R.; Rodriguez-Amaya. D. B.; Scamparini, A. R. P. Carotenoids of Yeasts Isolated from Brazilian Ecosystem. *Food Chem.* **2008**, *107* (1), 145–150.
41. Mata-Gómez, L. C.; Montañez, J. C.; Méndez-Zavala, A.; Aguilar, C. N. Biotechnological Mammalian Carotenoid Metabolism. *Fed. Am. Soc. Exp. Biol.* **2014**, *28*, 4457–4469.
42. McGarr, S. E.; Ridlon, J. M.; Hylemon, P. B. Diet, Anaerobic Bacterial Metabolism, and Colon Cancer: A Review of the Literature. *J. Clin. Gastroenterol.* **2005**, *39* (2), 98–109.
43. Møller, A. P.; Biard, C.; Blount, J. D.; Houston, D. C.; Ninni, P.; Saino, N.; Surai, P. F. Carotenoid-Dependent Signals: Indicators of Foraging Efficiency, Immunocompetence or Detoxification Ability. *Poultry Avian Biol. Rev.* **2000**, *11* (3), 137–159.
44. Molnár, P.; Deli, J.; Tanaka, T.; Kann, Y.; Tani, S.; Gyémánt, N.; Molnár, J.; Kawases, M. Carotenoids with Anti-*Helicobacter pylori* Activity from Golden Delicious Apple. *Phytother. Res.* **2010**, *24* (5), 644–648.
45. Müller, L.; Caris-Veyrat, C.; Lowe, G.; Böhm, V. Lycopene and Its Antioxidant Role in the Prevention of Cardiovascular Diseases—A Critical Review. *Crit. Rev. Food Sci. Nutr.* **2015**, *56* (11), 1868–1879.
46. Murthy, K.; Vanitha, A.; Rajesha, J.; Swamy, M.; Sowmya, P.; Ravishankar, G. A. In vivo Antioxidant Activity of Carotenoids from *Dunaliella salina*, a Green Microalga. *Life Sci.* **2005**, *76* (12), 1381–1390.
47. Nisar, N.; Li, L.; Lu, S.; Khin, N. C.; Pogson, B. J. Carotenoid Metabolism in Plants. *Mol. Plant* **2015**, *8* (1), 68–82.
48. Nordström, T.; Van Blarigan, E. L.; Ngo, V.; Roy, R.; Weinberg, V.; Song, X.; Simko, J.; Carroll, P. R.; Chan, J. M.; Paris, P. L. Associations between Circulating Carotenoids, Genomic Instability and the Risk of High-Grade Prostate Cancer. *Prostate* 2016, *76* (4), 339–348.
49. Olmedilla, B.; Granado, F.; Blanco, I.; Vaquero, M.; Cajigal, C. Lutein in Patients with Cataracts and Age-Related Macular Degeneration: A Long Term Supplementation Study. *J. Sci. Food Agric.* **2001**, *81* (9), 904–909.
50. Omayma, A.; Eldahshan Abdel Nasser, B. Carotenoids. *J. Pharmacogn. Phytochem.* **2013**, *2*, 2278–4136.
51. Orset, S.; Leach, G. C.; Morais, R.; Young, A. J. Spray-Drying of the Microalga *Dunaliella salina*: Effects on Beta-Carotene Content and Isomer Composition. *J. Agric. Food Chem.* **1999**, *47* (11), 4782–4790.
52. Palozza, P.; Barone, E.; Mancuso, C.; Picci, N. The protective Role of Carotenoids against 7-Keto-Cholesterol Formation in Solution. *Mol. Cell. Biochem.* **2008**, *309* (1–2), 61–68.
53. Priyadarshani, A. M. B.; Jansz, E. R. A Critical Review on Carotenoid Research in Sri Lankan Context and Its Outcomes. *Crit. Rev. Food Sci. Nutr.* **2014**, *54* (5), 561–571.

54. Quintanilla, M. X.; Camacho, B. H.; Meraz-Torres, L. S.; Chanona, J. J.; Alamilla, L.; Jimenéz-Aparicio, A.; Gutiérrez-López, G. F. Nanoencapsulation: A New Trend in Food Engineering Processing. *Food Eng. Rev.* **2010**, *2* (1), 39–50.

55. Rahiman, R.; Ali, M. A. M.; Rahman, M. S. Carotenoids Concentration Detection Investigation: A Review of Current Status and Future Trend. *Int. J. Biosci., Biochem. Bioinform.* **2013**, *3*, 466–470.

56. Rao, A. V.; Rao, L. G. Carotenoids and Human Health. *Pharmacol. Res.* **2007**, *55* (3), 207–216.

57. Reboul, E.; Borel, P. Proteins Involved in Uptake, Intracellular Transport and Basolateral Secretion of Fat-Soluble Vitamins and Carotenoids by Mammalian Enterocytes. *Progress Lipid Res.* **2011**, *50* (4), 388–402.

58. Rein, M. J.; Renouf, M.; Cruz-Hernandez, C.; Actis-Goretta, L.; Thakkar, S. K.; da Silva Pinto, M. Bioavailability of Bioactive Food Compounds: A Challenging Journey to Bioefficacy. *Br. J. Clin. Pharmacol.* **2013**, *75* (3), 588–602.

59. Ribeiro, H. S.; Cruz, R. C. D. Biliquid Foams Containing Carotenoids. *Eng. Life Sci.* **2005**, *5* (1), 84–88.

60. Riccioni, G. Carotenoids and Cardiovascular Disease. *Curr. Atheroscl. Rep.* **2009**, *11*, 434–439.

61. Rodriguez-Amaya, D. B. Status of Carotenoid Analytical Methods and In Vitro Assays for the Assessment of Food Quality and Health Effects. *Curr. Opin. Food Sci.* **2005**, *1*, 56–63.

62. Rodriguez-Huezo, M. E.; Pedroza-Islas, R.; Prado-Barragan, L. A.; Beristain, C. I.; Vernon-Carter, E. J. Microencapsulation by Spray Drying of Multiple Emulsions Containing Carotenoids. *J. Food Sci.* **2004**, *69* (7), 51–359.

63. Saini, R. K.; Nile, S. H.; Park, S. W. Carotenoids from Fruits and Vegetables: Chemistry, Analysis, Occurrence, Bioavailability and Biological Activities. *Food Res. Int.* **2015**, *76*, 735–750.

64. Schieber, A.; Carle, R. Occurrence of Carotenoid *cis*-Isomers in Food: Technological, Analytical, and Nutritional Implications. *Trends Food Sci. Technol.* **2005**, *16* (9), 416–422.

65. Send, R.; Sundholm, D. The Role of the β-Ionone Ring in the Photochemical Reaction of Rhodopsin. *J. Phys. Chem.* **2007**, *111* (1), 27–33.

66. Stange, C.; Flores, C. *Advances in Photosynthesis—Fundamental Aspects.* InTech, 2012; p 412. https://www.intechopen.com/books/advances-in-photosynthesis-fundamental-aspects

67. Sun, W.; Xing, L.; Lin, H.; Leng, K.; Zhai, Y.; Liu, X. Assessment and Comparison of In Vitro Immunoregulatory Activity of Three Astaxanthin Stereoisomers. *J. Ocean Univ. China* **2016**, *15* (2), 283–287.

68. Surai, P. F.; Bortolotti, G. R.; Fidgett, A. L.; Blount, J. D.; Speake, B. K. Effects of Piscivory on the Fatty Acid Profiles and Antioxidants of Avian Yolk: Studies on Eggs of the Gannet, Skua, Pelican and Cormorant. *J. Zool.* **2001**, *255* (3), 305–312.

69. Surai, P. F.; Speake, B. K.; Sparks, N. H. C. Carotenoids in Avian Nutrition and Embryonic Development. Absorption, Availability and Levels in Plasma and Egg Yolk. *J. Poultry Sci.* **2001**, *38* (1), 1–27.

70. Surai, P. F. *Natural Antioxidants in Avian Nutrition and Reproduction*; Nottingham University Press: Nottingham, UK, 2002; p 313.

71. Tan, J. S.; Wang, J. J.; Flood, V.; Rochtchina, E.; Smith, W.; Mitchell, P. Dietary Antioxidants and the Long-Term Incidence of Age-Related Macular Degeneration: The Blue Mountains Eye Study. *Ophthalmology* **2008,** *115* (2), 334–341.

72. Tanaka, T.; Shnimizu, M.; Moriwaki, H. Cancer Chemoprevention by Carotenoids. *Molecules* **2012,** *17* (3), 3202–3242.

73. Von Lintig, J.; Vogt, K. Filling the Gap in Vitamin a Research Molecular Identification of an Enzyme Cleaving β-Carotene to Retinal. *J. Biol. Chem.* **2000,** *275* (16), 11915–11920.

74. Vu, H. T.; Robman, L.; Hodge, A.; McCarty, C. A.; Taylor. H. R. Lutein and Zeaxanthin and the Risk of Cataract: The Melbourne Visual Impairment Project. *Invest. Ophthalmol. Vis. Sci.* **2006,** *47* (9), 3783–3786.

75. Vuong, L. T.; Franke, A. A.; Custer, L. J.; Murphy, S. P. *Momordica cochinchinensis* Spreng. (gac) Fruit Carotenoids Reevaluated. *J. Food Compos. Anal.* **2006,** *19* (6–7), 664–668.

76. Wertz, K.; Hunziker-Buchwald, P.; Seifert, N.; Riss, G.; Neeb, M.; Steiner, G.; Goralczyk, R. β-Carotene Interferes with Ultraviolet Light A-Induced Gene Expression by Multiple Pathways. *J. Invest. Dermatol.* **2005,** *124* (2), 428–434.

77. Wertz, K.; Siler, U.; Góralczyk, R. Lycopene: Modes of Action to Promote Prostate Health. *Arch. Biochem. Biophys.* **2004,** *430* (1), 127–134.

78. Wieslaw, I. *Carotenoids in Health and Disease*; CRC Press: Boca Raton, FL, 2004; p 418.

79. World Health Organization (WHO). *Cancer.* http://www.who.int/mediacentre/factsheets/fs297/ (accessed on March 03, 2017).

80. Xianquan, S.; Shi, J.; Kakuda, Y.; Yueming, J. Stability of Lycopene During Food Processing and Storage. *J. Med. Food* **2005,** *8* (4), 413–422.

81. Yatsunami, R.; Ando, A.; Yang, Y.; Takaichi, S.; Kohno, M.; Matsumura, Y.; Ikeda, H.; Fukui, T.; Nakasone, K.; Fujita, N.; Sekine, M. Identification of Carotenoids from the Extremely Halophilic Archaeon *Haloarcula Japonica*. *Front. Microbiol.* **2014,** *5,* 100–105.

82. Yeum, K, J.; Russell, R. M. Carotenoid Bioavailability and Bioconversion. *Annu. Rev. Nutr.* **2002,** *22* (1), 483–504.

83. You, J. S.; Jeon, S.; Byun, Y. J.; Koo, S. Choi S. S. Enhanced Biological Activity of Carotenoids Stabilized by Phenyl Groups. *Food Chem.* **2015,** *177,* 339–345.

84. Zhang, J.; Sun, Z.; Sun, P.; Chen, T.; Chen, F. Microalgal Carotenoids: Beneficial Effects and Potential in Human Health. *Food Funct.* **2014,** *5* (3), 413–425.

CHAPTER 7

EDIBLE VACCINE: JOURNEY FROM MUSHROOM TO SYRINGE

VIVEK K. CHATURVEDI, SUSHIL K. DUBEY, N. TABASSUM, and
M. P. SINGH

ABSTRACT

Vaccines exhibit great potential to fight against infectious disease.
Edible vaccines make contact with the lining of digestive tract by
which it activates the mucosal as well as systemic immunity. These
properties of edible vaccines provide more effective protection from
the dangerous microorganisms which cause deadly diseases. However,
among properties of edible vaccines, there are many other issues, which
are still to be addressed. For example, the vaccine production from the
plants is in low amount. While solving such problems, the researchers
also have to make sure that the vaccine produced by the fruits should
have a predictable dose.

7.1 INTRODUCTION

Edible vaccines (EVs) are milestones in medical biotechnology to provide
protection against variety of diseases. They provide exciting possibilities
for reducing the load of diseases especially in developing country, where
storage and administration of vaccines are the major issues. These vaccines
suppress many diseases like diarrhea, type-1 diabetes, multiple sclerosis,
rheumatoid arthritis, autoimmune disorders, etc. The genes encoding
bacterial and viral pathogenic antigens can be expressed in plants in a form
in which they possess native immunogenic properties. Since it is a nonin-
vasive method of immunization, therefore, the chances of infection are
also decreased. There is always a risk associated with vaccine production

from mammalian cell culture, which can get contaminated with animal viruses. While in case of EVs, it is not possible because plant viruses do not infect humans. Thus, they are harmless and safety is guaranteed. The oral administration of EVs provides "mucosal immunity" at various sites by secreting antibodies. The production is highly efficient, easily scaled up, and reduces cost of transportation and refrigeration.

In quest of better quality of life of human beings, various EVs have been developed to prevent many life-threatening diseases. Vaccination allows the individual's body to generate an adaptive immune response against specific antigen being administered. Diseases like polio, malaria, and many others that were once considered lethal can now be treated by available vaccines.[5] One of the most innovative approaches for immunization by EVs is that it can be eaten and are derived from natural sources. There are many EVs that have been developed, for example, hepatitis B and C, cholera, measles, etc. The main advantage of EVs is that the plants can express more than one transgene, which is very helpful in diseases requiring booster doses or multiple antigens to induce immunity.[57]

For production of edible vaccines, the selected gene of desired protein is introduced to the plants. The plant developed by transformation of foreign gene is called transgenic plants. The process of EV development is similar to the process of subunit vaccine development because it also contains desire antigen, not whole pathogen.[70] Plants used for oral vaccine production should be produced in edible parts and have some important properties: (1) firstly, the vaccine antigens are heat sensitive, therefore the plant part should be consumed uncooked; (2) secondly, the vaccine content in plant is in very small percentage so plant part should be rich in protein content; and (3) lastly, it should not produce toxic compounds and can grow widely. Different antigens have been expressed in different plants like maize, rice, potato, tobacco, etc.[71]

Mushrooms are rich in proteins and many micronutrients. It contains large number of biologically active compounds. Some of these compound complexes are capable to stimulate the nonspecific immune response. The immunomodulatory properties of the mushrooms are due to many bioactive compounds isolated from mushrooms, for example, β-d-glucan, lectins, and many kinds of polysaccharides. Lectins are glycoproteins to promote the cell agglutination in the body. Lectins have been isolated from many edible mushrooms. Lectin also has many biological activities like antiviral, immunomodulatory, antitumor, etc.[12]

This chapter highlights the existing conventional methods for the production of edible vaccines, pros and cons, and research studies to improve EV production from mushrooms. It also presents challenges during the production process.

7.2   THE DEVELOPMENT OF EV

For the development of EV, the selected foreign genes are incorporated into the plants and after that the transformed plants are able to generate the desired proteins.[6] The development of plant by this method is called as transgenic plant and the process is called as transformation.[2] A general method for the development of EV is presented in Figure 7.1. Several such approaches are discussed in this section.

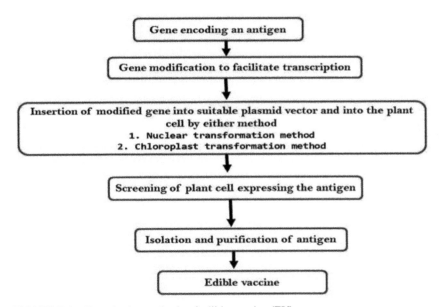

FIGURE 7.1   Steps in the synthesis of edible vaccine (EV).

A gene of interest is inserted into the Ti plasmid of *Agrobacterium*. This whole construct is allowed to be transformed into the plant cells.[20,40] This approach is time-consuming with lower yield. However, in case of dicotelydenous plants like potato, tomato, and tobacco,[11] the result is satisfactory. Researchers have proved experimentally that this method is somehow good for some plants.[55]

7.2.1   BIOLISTIC METHOD

In this method, gene gun is used for delivery of desired gene. DNA is coated with gold, tungsten, etc., and is bombarded into the cells or tissue.[77] Then these transformed cells will develop into a whole plant. Biolistic method is very attractive because it has high regenerative ability.[19] This method is expensive because it requires sophisticated devices like particle gun, gold particle, etc.

7.2.2   ELECTROPORATION

In this method, the DNA is incorporated into the cell through pores developed by high voltage of electric pulse. In this approach, mild enzymatic treatment is needed for loosening the cell wall because the cell wall provides effective barrier to prevent the entry of foreign DNA into the cell cytoplasm.[32]

7.3   TRANSGENIC PLANTS FOR PRODUCTION OF EV

There are many plants, which have been used to express antigens for rotavirus, cholera, gastroenteritis, autoimmune diseases, or rabies.[49] There are several research studies on potato cultivation, but there are several drawbacks of the potatoes as EVs because it cannot be eaten raw and EVs are heat sensitive. Many plants produce fruits and can be consumed in raw form, like tomatoes, corn, tobacco, bananas, carrots, and peanuts. Such plants can be best option for the EV production because these plants have been tested for the genetic transformation.[17]

7.3.1   POTATOES

The first attempt for the production of EV in potatoes *(Solanum tuberosum)* was carried by Mason et al.[52] to fight enteritis, caused by *Escherichia coli* strain LT-B in mice. Meanwhile during subsequent year, antigen against Norwalk virus capsid antigen and for *Vibro cholera* enterotoxin was developed and its efficacy was tested in rats and in humans.[76] In 2005,

Thanavala et al.[78] suggested that these transgenic potatoes play role as an oral reinforcement to the hepatitis B vaccine in humans. When wild rabbits were immunized with these transgenic potatoes, then it provided protection against the infection of rabbit hemorrhagic virus (RHDV).[8]

7.3.2   LETTUCE

The β-subunits of *E. coli*, which is a thermo labile protein, causes enteric in human and in animals is expressed into the lettuce (*Lactuca sativa*).[39] In 2005, swine fear hog pest virus's glycoprotein was expressed in the lettuce.[46] This food is mostly used as raw, therefore it has great potential as an EV.

7.3.3   TOMATOES

In tomato (*Solanum lycopersicum*), the antigen of corona virus, which causes acute respiratory syndrome (SARS), had been developed.[64] The expression of CT-B protein of *Vibrio cholera* in leaves, stem, fruits, etc. had been expressed in tomatoes, which are confirmed by enzyme-linked immunosorbent assay (ELISA) and Western blot analysis.[33] By real-time quantitative reverse transcription-polymerase chain reaction (qRT-PCR) and ELISA, it is confirmed that the expression of HBsAg is stable in tomatoes (Mega variety).[48,73] For Alzheimer's disease, human β-amyloid antigen was expressed in the tomato.[88]

7.3.4   ALFALFA

The eBRV4 antigen from VP4 of hog rotavirus had been expressed in Alfalfa (*Medicago sativa*).[83] In 2005, glycoprotein E2 of hog pest virus was expressed[46] and the protein σC of capsid virus, which causes infection in poultry, was expressed in 2009. Meanwhile, some other plants like *A. thaliana* were used to express the same the same protein.[27,84] Researchers also showed that in alfalfa, Eeg95-EgA31 of *Echinococcus granulosus* can be expressed efficiently. Researchers purified this protein from leaves and delivered it to the target organisms.[86]

7.3.5   RICE

In 2007, researchers successfully attempted the expression of B subunits of *E. coli* in rice (*Oryza sativa*).[61] VP2 antigen was also expressed in rice, which provides safety from infectious bursitis by producing antibodies when challenged to chicken.[84] The expression of HBsAg in rice seeds was confirmed in 2008 by using PCR and Southern blot techniques.[65] Moreover in 2008, B-subunit of *E. coli* was transformed into the rice by using the bioballistic approach and its expression was confirmed by the PCR.[60] The expected rice production in 2016–2018 is 490 million metric tons in the world wide.[81] Therefore, it can be concluded that the development of EV using rice will help more people in the worldwide to cure from different diseases. Rice is available easily to everyone and is an important food.

7.3.6   BANANA

In banana plant, the expression of HBsAg has been done successfully. Researchers used four expression cassettes, that is, PHB, pEFEHER, pEFEHBS, and PHER. The expression level of HBsAg was analyzed by PCR, Southern hybridization, and reverse transcription PCR.[43] It has been shown that the expression level in plants can rise up to 19.92 ng/g.[23] The major disadvantage with banana is, it takes too much time to develop into a mature plant.

7.3.7   CARROTS

Transgenic carrots (*Daucus carota*) contain antigen of B-subunits of *E. coli*, which produce thermolabile toxin, and can induce the production of IgA and IgG immunoglobuline.[67] In 2010, researchers developed a transgenic carrot plant, which expresses the UreB-subunit of *Helicobacter pylori*.[90] HIV antigen expression was also attempted in carrot along with *A. thaliana*. The use of transgenic carrots for the treatment of HIV showed positive effective results because carrots itself are good source of carotenoids, which increase the lymphocytes, monocytes counts.[16] Thus, people having weak immunity can be benefitted from this EV.

7.3.8 TOBACCO

Tobacco (*Nicotiana benthamiana*) cannot be eaten like other edible plants. This plant is used as a model to check the expression of foreign gene, so this plant is also used for the development of EVs.[49] In 1996, tobacco was studied along with potato to develop the EV. Norwalk virus capsid protein was expressed in tobacco, which cures from gastroenteritis.[25] In 2012, researchers demonstrated the expression of HPAIV H5N1 of avian flu virus. When the rats were challenged with this antigen, then it stimulated the expression of IgG immunoglobuline.[36,72] Recently, researchers reported the expression of anthrax and *Eimeria tenella* antigen in tobacco plants.[21,69]

7.4 IMMUNE RESPONSE GENERATED BY EV

Long-term immunization by vaccines is developed by the memory cells, generated when exposed with antigens. When the antigen is exposed to the immune cells, then two types of cells are generated: one is effectors cell and second is memory cell. The effectors cells provide immediate immunity and memory cells are stored for the future; and if same antigen is encountered then memory cells can produce antibodies more rapidly. Vaccines generate antigen-specific response by which specific antibodies are generated, which neutralizes the pathogenic antigens. Antibodies are generated by the B-lymphocyte cells, which convert into the plasma cells on pathogenic antigen interaction.[14]

In case of EVs, it generates IgA immunoglobuline by oral and gut mucosa and generates systemic immune response also.[82] The mucosal immune system is different form of humoral immune system and not activated by the parenteral administration of antigens. In mucosal immune system, it needs antigen interaction with mucosal surface. Mucosal surfaces are the first line of defense of immune system. It provides safety from transmissible diseases, which can infect through oral, respiratory, urinary, and genital routes. The major advantage of EV is that it can be used alone or with the combination of other vaccines also.[74]

Research has reported that if certain protein is administrated then sometimes body starts to downregulate its response. Researchers need to standardize the safe dose and its handling so that EV could provide better immunity.[44]

7.5   EDIBLE MUSHROOM AS A VACCINE

Edible mushrooms are nutritious food substance containing medicinal properties. People in China and Romans resemble mushrooms to be God's diet: "It protects from childhood to youth from malnutrition due to the availability of proteins, vitamins, minerals, fats, and carbohydrate." In mushroom, protein content is found much higher and varies from 22% to 35% than any other plants. The mushroom contains 0.3–0.4% of low fat, 4–5% of carbohydrate, and high amount of essential fatty acids.[1] It is considered as a decent diet for diabetic patients because of low levels of sugar and 0.5% of the starch. Mushroom contains numerous bioactive mixes, for example, polysaccharide–protein complexes and polysaccharides. These bioactive mixes (phenolics, steroids, terpenes, basic cell wall components, and polysaccharides β glucan family) play major role in various diseases like cancer, diabetes, bone health, cardiovascular diseases, liver dysfunction, and obesity.[68]

The idea of mushroom-based EV technology has shown low cost production results, and is rich in nutrition and effective therapeutic therapy against many diseases. These include high biosynthetic capacity with production of immunomodulatory compounds, and the availability of genetic transformation methods. Not many reports have been distributed on the utilization of edible mushroom as a creation stage for biopharmaceuticals. Among the *Pleurotus* family, *P. eryngii* species has been designed to assess its potential for generation of biopharmaceuticals, Interleukin-32.[13] Other researcher changed mushroom by an articulation of human development hormone quality (hGH) in *P. eryngii*.[38]

Human growth hormone (hGH2) gene has successfully been transformed in mushroom, showing that mushroom is very useful plant for production of transgenic edible vaccine. *Pleurotus eryngii* (King oyster mushroom) transformed by the recombinant vector (pPEVbGH) containing Bovine growth hormones (bGH) *via Agrobacterium tumefaciens*-mediated transformation showed control of cauliflower mosaic virus (CaMV).[63] Reactive oxygen species (ROS), free radicals, or peroxidases are formed during the metabolism of oxygen. ROS play an important role by regulating cellular processes like cell growth, ranging from cell proliferation to apoptosis, which causes DNA and protein damage. To reduce oxidative stress, methionine sulfoxidereductase A (PoMsrA) gene overexpression in *Pleurotus ostreatus via Agrobacterium*-mediated transformation, *Pleurotus ostreatus* behave like EV, which reduces ROS under stress conditions.[87]

7.6 ADVANTAGES OF EV COMPARED WITH INJECTABLE VACCINES

EVs are safer than injectable or "oral vaccine like polio vaccine." At developed industrial scale, preparation of injectable vaccine is highly expensive, requires their storage for temperature maintenance, sterile injections, and specially-built manufacturing facilities. For production and industrial development of EVs, costly equipment and machines are not necessary as they could grow on rich land and stored at room temperature. By expression of many antigen (transgene), EVs can be delivered with better efficacy. These EVs enable few antigens to approach insusceptible framework particularly M-cells, similar to a trivalent antibody against cholera, ETEC (enterotoxigenic *E. coli*), and rotavirus; inspire a critical resistant reaction to every one of the three antigens.[89]

Those vaccines, which are integrated into mammalian cells, are by and large debased with pathogenic organisms, yet the plant-based immunizations are definitely not contaminated by microbes.[71] Complication in injection in comparison to oral administration of EV is much easy to be taken by children. Needles utilized as a part of customary vaccination process likewise cause natural pollution and convey the danger of spreading second-hand sicknesses. In comparison of conventional injectable vaccine, plant-based EV induces mucosal immune responses.[40]

7.7 CLINICAL USE OF EV

7.7.1 HEPATITIS B

The first-generation vaccine purified from the inactivated serum of carriers having "subviral particles (SVPs) of HBV" was prepared. The efficacy of this vaccine was very high,[26,41] but its amount was insufficient for immunization as well as their production rate is costly. The small HBsAg (surface antigen), expressed for the first time in plants,[53] was purified and given intraperitoneally.[79] The surface antigen of HBV transgene was communicated in transgenic potato,[35] tomato,[30] and momentarily in *N. benthamiana*.[29,30] The substantial surface antigen was expressed in tomato[48] and the two antigens in tobacco and lettuce. For a timeframe, recombined anti-HBV vaccines give significant prophylaxis and control of HepB,[56] because of decreased cost per dose alongside expanded accessibility and mass vaccination programs.[18] Selected examples of transgenic plants are given in Table 7.1.

TABLE 7.1 Selected Examples of Vaccine Produced in Transgenic Plant and Their Applications.

Transgenic plant	Vaccine (recombinant protein)	Protection against	Reference
Banana	HBsAg	Hepatitis-B	43
Pleurotus eryngii (Edible mushroom)	Interleukin-32	Inflammatory and autoimmune diseases	13
Potato	Cholera toxin B-subunit	*Vibrio cholerae*	3
Potato	*Vibrio cholera* toxin B-subunit	Autoimmune diabetes	3
Potato/rice	α1 interferon	Viral protection, anticancer	59
Rice	Human cytomegalovirus glycoprotein B	Human cytomegalovirus	75
Rice	amyloid β	Alzheimer's disease	58
Rice	urease subunit B	*Helicobacter pylori*	22
Tobacco	Foot and mouth virus (VPI)	Foot and mouth virus	34
Tobacco	Heat labile Enterotoxin B	*E. coli*	37
Tobacco	Herpes virus B surface antigen	Herpes simplex virus	48
Tobacco	Hemagglutinin	Influenza	45
Tobacco	Malarial B cell epitope	Malaria	80
Tobacco/potato	Capsid protein	Norwalk virus	31
Tobacco/potato	Envelope surface protein	Hepatitis B	23
Tomato	Rabies virus glycoprotein	Rabies virus	54
Tomato	Tat gene	HIV-1	66

7.7.2 MEASLES

Around the world, measles cause hard of hearing or create encephalitis. The vaccine presently being used produces 95% sero-transformation in people.[15] Measles live-attenuated vaccine does not create oral inoculation impact and is additionally obliterated by warm ambient. Consequently, refrigeration is essential for its stockpiling. For the advancement of eatable vaccine, MV-H antigen was chosen, which can be transformed in tobacco plant by plasmid/vector *A. tumefactions*.[28] Analysts additionally contemplated that transgenic carrot plant can be utilized to convey viral antigens for the improvement of measles vaccine.[51]

7.7.3 DIABETES

It is an immune system infection in which insulin-delivering beta cells are destroyed by the body's own particular invulnerable framework. In mice, diabetes could be forestalled by encouraging them with designed transgenic plants to create a diabetes-related protein.[7] The entire thought depended on "oral resilience" where the immune system framework is killed right on time by instructing the body to endure the "antigenic proteins."[50]

7.7.4 MALARIA

Malaria is caused by protozoan parasite of genus *Plasmodium*. The antigens, which are being researched for the improvement of EVs for jungle fever (malaria) are merozoite surface protein MsP4 and MsP5 from *Plasmodium falciparum* and MsP4/MsP5 from *Plasmodium yoelii*.[6,57] Oral administration of recombinant MsP4, MsP5, and MsP4/5 coregulated with cholera poison B (CTB) as mucosal adjuvant demonstrated compelling counteracting agent reaction against blood arrange parasite mice.[62]

7.8 PROS AND CONS OF EV

7.8.1 PROS OF EV

- Antigen protection through bioencapsulation.[32,42,57,62,70]
- Cold storage conditions are not required for these vaccines as they are heat stable.
- Costly equipments and machines are not required for their production, as they can easily grow on rich soils and does not require fermenters unlike traditional vaccines.
- Economical in terms of mass production.
- Enhanced compliance (especially in children).
- For proteins administered orally, low downstream processing is required.

- Local crop can be engineered to produce vaccine, thereby eliminating the process of transportation and distribution.
- Needle-free vaccine, thereby making it cost-effective in terms of medical equipment and in turn the use of needle and syringes increases the risk of second hand diseases.
- New or various transgenes can be presented by the sexual intersection of plants.
- Plant-derived vaccine (immunizations) could be the hotspot for new antibodies consolidating various antigens, called as second-generation vaccine, which enable a few antigens to approach M cells at the same time.
- Plant-inferred antigens gather unexpectedly into oligomers and into infection like particles.
- Processing and purification can be eliminated because these vaccines can be administered by eating the part of the plant.
- Seroconversion in the presence of maternal antibodies.
- Simple for large-scale manufacturing framework by rearing when contrasted with an animal framework.
- It can be delivered orally; therefore, medical professionals are not necessary.
- The body's first line of defenses is triggered by them.
- The seeds of transgenic plants contain lesser dampness content and can be effectively dried; in this manner, they offer more prominent stockpiling openings. Furthermore, plants with oil or their fluid concentrates have more storage opportunities.
- There is reduced risk of product contamination by mammalian viruses, blood-borne pathogens, bacterial toxins, oncogenes.
- They act as delivery vehicle to transport the desired protein in the food to produce immunization. There is no need of adjuvant for immune response.
- They contain heat-killed pathogens, therefore are safe because there is no risk of proteins to reform into infectious organism.

7.8.2 CONS OF EV

- Care to be taken for appropriate recognizing the "vaccine (transgenic) fruits" and "ordinary fruits" to maintain a strategic to avoid misadministration.[32,42,57,62,70]

- Certain foods cannot be eaten crude (e.g., potato) in this manner needs cooking, thus is in charge of denaturation or debilitating the protein exhibit in it.
- Check of allergens in the plant and plant items.
- Cross pollination and their problems.
- Development of oral tolerance to the particular vaccine.
- Doses are simultaneously differing from age to age, plant to plant, protein content, and weight.
- Effects on insects and soil microbes.
- Identification of proper dosage and decide which plant part can produce proper dose, whether plant products, pill, intramuscular or intravenous injection of purified antigen.
- In plants and human their glycosylation pattern is different, resulting which could affect the functions of vaccines.
- May also have side effects due to the interaction between vaccine and vector.
- Plant-derived vaccines and their delivery in the form of capsule or pill.
- Selecting suitable plant which will give ideal antigen expression could be difficult task, time-consuming, and expensive.
- The effect of plant-derived vaccines on nature and human well-being, as it could get blended with human food supply or the earth.
- The major disadvantages of EVs are getting infectious microbial infestation, for example, potatoes containing vaccine can last longer if stored at 4°C while a tomato cannot last long time.
- Their organization requires techniques for the institutionalization of plant material/item as low dosages may deliver the lesser number of antibodies and high measurements can be in charge of safe resilience.

7.9 SUMMARY

The EVs are not only cost effective and easily available but also they can be safe and functional also. With the development of suitable vaccine, there can be safe and effective delivery system for complex diseases like HIV, malaria, diabetes, etc. The production of EVs in plants or vegetables will be feasible on social acceptance as well as technical merits. Further research is required to increase the expression level of desired genes and to make these EVs meet the benchmarks of value regarding virtue, potency, wellbeing, and adequacy. Also, further research is necessary how to bring these EVs from labs to hospitals so that it can be used to treat diseases.

KEYWORDS

- **Alzheimer**
- **edible mushroom**
- **edible vaccine**
- **interleukin 32**
- ***Medicago sativa***
- ***Plasmodium falciparum***
- ***Pleurotuseryngii***

REFERENCES

1. Agarwal, S.; Vaseem, H.; Kushwaha, A.; Gupta, K. K.; Maurya, S.; Chaturvedi, V. K.; Pathak, R. K.; Singh, M. P. Yield, Biological Efficiency and Nutritional Value of *Pleurotussajor-caju* Cultivated on Floral and Agro-waste. *Cell Mol. Biol.* **2016,** *62* (3), 1–5.
2. Altpeter, F.; Baisakh, N.; Beachy, R. Particle Bombardment and the Genetic Enhancement of Crops: Myths and Realities. *Mol. Breed.* **2005,** *15* (3), 305–327.
3. Arakawa, T.; Chong, D. K. X.; Merritt, J. L.; Langridge, W. H. R. Expression of Cholera Toxin B -Subunit Oligomers in Transgenic Potato Plants. *Transgenic Res.* **1997,** *6* (6), 403–413.
4. Arakawa, T.; Chong, D. K. X.; Langridge, W. H. R. Efficacy of a Food Plant-based Oral Cholera Toxin B Subunit Vaccine. *Nat. Biotechnol.* **1998,** *16* (3), 292–297.
5. Aryamvally, A.; Gunasekaran, V.; Narenthiran, K. R. Strategies Toward Edible Vaccines: An Overview. *J. Dietary Suppl.* **2017,** *14* (1), 1–16.
6. Bhairy, S.; Hirlekar, R. Edible Vaccines: An Advancement in Oral Immunization. *Asian J. Pharm. Clin. Res.* **2017,** *10* (2), 71–77.
7. Blanas, E.; Carbone, F. R.; Allison, J.; Miller, J. F.; Heath, W. R. Induction of Autoimmune Diabetes by Oral Administration of Auto Antigen. *Science* **1996,** *274* (5293), 1707–1709.
8. Castañon, S.; Marín, M. S.; Martín-Alonso, J. M.; Boga, J. A.; Casais, R.; Humara, J. M.; Ordás, R. J.; Parra, F. Immunization with Potato Plants Expressing Vp60 Protein Protects Against Rabbit Hemorrhagic Disease Virus. *J. Virol.* **1999,** *73* (5), 4452–4455.

9. Cebadera, M. Plantas Modificadas Genéticamente Como Vacunas Comestibles: Aspectos Científicos Socioeconómicos (Genetically Modified Plants as Edible Vaccines: Social–Economic Scientific Aspects). Ph.D. Thesis, Universidad Complutense Madrid, Madrid, España, 2012, p. 216.

10. Charmi, P. S.; Trivedi, N.; Vachhani, U. D.; Joshi, V. J. Edible Vaccine: A Better Way for Immunization. *Int. J. Curr. Pharm. Res.* **2011**, *3* (1), 53–56.

11. Chikwamba, R.; Cunnic, J.; Hathway, D.; McMurary, J.; Mason, H. Functional Antigen in a Practical Crop: LT-B Producing Maize Protects Mice Against *E. coli* Heat Liable Enterotoxin (LT) and Chorea Toxin (CT). *Transgenic Res.* **2002**, *11*, 479–493.

12. Chunchao, H.; Guo, J. Y. Hypothesis: Supplementation with Mushroom Derived Active Compound Modulates Immunity and Increases Survival in Response to Influenza Virus (H1N1) infection. *Evid. Based Complement. Alternat. Med.* **2011**, *11*, 1–3.

13. Chung, S. J.; Kim, S.; Sapkota, K.; Choi, B. K.; Shin, C.; Kim. S. J. Expression of Recombinant Human Interleukin-32 in Pleurotuseryngii. *Ann. Microbiol.* **2011**, *61* (2), 331–338.

14. Claire, A. S. Vaccine Immunology. Chapter 2; In: *Vaccines, Section 1:General Aspects of Vaccination*; 6th ed., Elsevier Inc.: New York, 2013, pp 17–36.

15. Cutts, F.; Henao-Restrepo, A. M.; Olive, J. Measles Elimination: Progress and Challenges, *Vaccine* **1999**, *17*, 47–52.

16. Ekam, V. S.; Udosen, E. O.; Chighu, A. E. Comparative Effect of Carotenoid Complex from Goldenneo-life Dynamite and Carrot Extracted Carotenoids on Immune Parameters in Albino Wistar Rats. *Niger. J. Physiol. Sci.* **2006**, *21* (1–2), 1–4.

17. Glick, B. R.; Pasternak, J. J.; Patten, Ch. L. *Molecular Biotechnology. Principles and Applications of Recombinant DNA*. ASM Press: Herndon, VA, USA, 2010, p. 999.

18. Goldstein, S. T.; Fiore, A. E. Toward the Global Elimination of Hepatitis B Virus Transmission. *J. Pediatr.* **2001**, *139*, 343–345.

19. Gómez, E.; Zoth, S. C.; Berinstein, A. Plant-based Vaccines for Potential Human Application. *Human Vaccines* **2009**, *5* (11), 738–744.

20. Gómez, E.; Zoth, S. C.; Carrillo, E.; Berinstein, A. Developments in Plant-based Vaccines Against Diseases of Concern in Developing Countries. *Open Infectious Dis. J.* **2010**, *4*, 55–62.

21. Gorantala, J.; Grover, G.; Rahi, A.; Chaudhary, P.; Rajwanshi, R.; Sarin, L. B.; Bhatnagar, R. Generation of Protective Immune Response Against Anthrax by Oral Immunization with Protective Antigen Plant-based Vaccine. *J. Biotechnol.* **2014**, *176* (1), 1–10.

22. Gu, Q.; Han, N.; Liu, J.; Zhu, M. Expression of Helicobacter pylori Urease Subunit B Gene in Transgenic Rice. *Biotechnol Lett.* **2006**, *28* (20), 1661–1666.

23. Guan, Z. J.; Guo, B.; Huo, Y. L.; Guan, Z. P.; Wei, Y. H. Overview of Expression of Hepatitis B Surface Antigen in Transgenic Plants. *Vaccine* **2010**, *28* (46), 7351–7362.

24. Guana, B.; Bin, G.; Yan-lin, H.; Zheng-ping, G.; Ya-hui, W. A Overview of Expression of Hepatitis B Surface Antigen in Transgenic Plants *Zheng-jun. Vaccine* **2010**, *28*, 7351–7362.

25. Hahn, B. S.; Jeon, I. S.; Jung, Y. J.; Kim, J. B.; Park, J. S.; Ha, S. H.; Kim, K. H.; Kim, H. M.; Yang, J. S.; Kim, Y. H. Expression of Hemagglutinin-neuraminidase Protein of Newcastle Disease Virus in Transgenic Tobacco. *Plant Biotechnol. Rep.* **2007**, *1* (2), 85–92.

26. Hilleman, M. R. Critical Overview and Outlook: Pathogenesis, Prevention, and Treatment of Hepatitis and Hepatocarcinoma Caused by Hepatitis B Virus. *Vaccine* **2003**, *21*, 4626–4649.

27. Huang, L. K.; Liao, S. C.; Chang, C. C.; Liu, H. J. Expression of Avian Reovirus -C Protein in Transgenic Plants. *J. Virol. Methods* **2006**, *134*, 217–222.

28. Huang, Z.; Dry, I.; Webster, D. Plant Derived Measles Virus Hemagglutinin Protein Induces Neutralizing Antibodies in Mice. *Vaccine* **2001**, *19*, 2163–2171.

29. Huang, Z.; Elkin, G.; Maloney, B. J.; Buehner, N.; Arntzen, C. J.; Thanavala, Y.; Mason, H. S. Virus-like Particles Expression and Assembly in Plants: Hepatitis B and Norwalk Viruses. *Vaccine* **2005**, *23*, 1851–1858.

30. Huang, Z.; LePore, K.; Elkin, G.; Thanavala, Y.; Mason, H. S. High-yield Rapid Production of Hepatitis B Surface Antigen in Plant Leaf by a Viral Expression System. *Plant Biotechnol. J.* **2008**, *6*, 202–209.

31. Mason, H. S.; Ball, J. M.; Shi, J. J.; Jiang, X.; Estes, M. K.; Arntzen, C. J. Expression of Norwalk Virus Capsid Protein in Transgenic Tobacco and Potato and Its Oral Immunogenicity in Mice. *PNAS* 1996, *93* (11), 5335–5340

32. Jan, N.; Shafi, F.; Hameed, O. B. An Overview on Edible Vaccines and Immunization. *Austin J. Nutr. Food Sci.* **2016**, *4* (2), 1–5.

33. Jiang, X. L.; He, Z. M.; Peng, Z. Q.; Qi, Y.; Chen, Q.; Yu, S. Y. Cholera Toxin B protein in Transgenic Tomato Fruit Induces Systemic Immune Response in Mice. *Transgenic Res.* **2007**, *16* (2), 169–175.

34. Joensuu, J. J.; Brown, K. D.; Conley, A. J.; Clavijo, A.; Menassa, R. Expression and Purification of an Anti-foot Mouth Disease Virus Single Chain Variable Antibody Fragment in Tobacco Plants. *Transgenic Res.* **2009**, *18* (5), 685–696.

35. Joung, Y. H.; Youm, J. W.; Jeon, J. H.; Lee, B. C.; Ryu, C. J.; Hong, H. J.; Kim, H. C.; Joung, H.; Kim, H. S. Expression of the Hepatitis B Surface S and preS2 Antigens in Tubers of *Solanum tuberosum. Plant Cell Rep.* **2004**, *22*, 925–930.

36. Kanagarajan, S.; Tolf, C.; Lundgren, A.; Waldenstrom, J.; Brodelius, P. E. Transient Expression of Hemagglutinin Antigen from Low Pathogenic Avian Influenza A (H7N7) in Nicotianabenthamiana. *PLoS One* **2012**, *7*, 682–693.

37. Kang, T. J.; Han, S. C.; Jang, M. O;, Kang, K. H.; Jang, Y. S.; Yang, M. S. Enhanced Expression of B-subunit of *Escherichia coli* Heat Labile Enterotoxin in Tobacco by Optimization of Coding Sequences. *Appl. Biochem. Biotechnol.* **2004**, *117* (3), 175–187.

38. Kim, S.; Sapkota, K.; Choi, S. B.; Kim, S. J. Expression of Human Growth Hormone Gene in *Pleurotuseryngii. Central Eur. J. Biol.* **2010**, *5* (6), 791–799.

39. Kim, T. G.; Kim, M. Y.; Kim, B. G.; Kang, T. J. Synthesis and Assembly of *Escherichia coli* Heat-labile Enterotoxin B Subunit in Transgenic Lettuce (*Lactucasativa*). *Protein Expr. Purif.* **2007**, *51* (1), 22–27.

40. Kim, T. G.; Yang, M. S. Current Trends in Edible Vaccine Development Using Transgenic Plants. *Biotechnol. Bioproc. Eng.* **2010**, *15*, 61–65.

41. Krugman, S. The Newly Licensed Hepatitis B Vaccine. Characteristics and Indications for Use. *JAMA* **1982**, *247*, 2012–2015.

42. Kumar, B. V.; Raja, T. K.; Wani, M. R. Transgenic Plants as Green Factories for Vaccine Production. *Afr. J. Biotechnol.* **2013**, *12* (43), 6147–6158.

43. Kumar, G. B.; Ganapathi, T. R.; Revathi, C. J.; Srinivas, L.; Bapat, V. A. Expression of Hepatitis B Surface Antigen in Transgenic Banana Plants. *Planta* **2005**, *222* (3), 484–493.

44. Langridge, W. H. Edible Vaccines. *Sci. Am.* **2000,** *283* (3), 66–71.

45. Le, M. F.; Mercier, G.; Chan, P.; Burel, C. Biochemical Composition of Hemagglutinin Based Influenza Virus-like Particle Vaccine Produced by Transient Expression in Tobacco Plants. *Plant Biotechnol. J.* **2015,** *13* (5), 717–725.

46. Legocki, A. B.; Miedzinska, K.; Czaplin, S. M.; Płucieniczak, A.; Wedrychowicz, H. Immunoprotective Properties of Transgenic Plants Expressing E2 Glycoprotein from CSFV and Cysteine Protease from Fasciola Hepatica. *Vaccine* **2005,** *23,* 1844–1846.

47. Lindh, I.; Brave, A.; Hallengard, D.; Hadad, R.; Kalbina, I.; Strid, A.; Andersson, S. Oral Delivery of Plant-derived HIV-1 p24 Antigen in Low Doses Shows a Superior Priming Effect in Mice Compared to High Doses Ingrid. *Vaccine,* **2014,** *32* (20), 2288–2293.

48. Lou, X. M.; Yao, Q. H.; Zhang, Z.; Peng, R. H.; Xiong, A. S.; Wang, H. K. Expression of the Human Hepatitis B Virus Large Surface Antigen Gene in Transgenic Tomato Plants. *Clin. Vaccine Immunol.* **2007,** *14,* 464–469.

49. Ma, J. K.; Drake, P. M.; Christou, P. The Production of Recombinant Pharmaceutical Proteins in Plants. *Nature* **2003,** *4,* 794–805.

50. Ma, S. W.; Zhao, D. L.; Yin, Z. Q. Mukherjee, R.; Singh, B. et al. Transgenic Plants Expressing Auto Antigens Fed to Mice to Induce Oral Immune Tolerance. *Nat. Med.* **1997,** *3,* 793–796.

51. Marquet-Blouin, E.; Bouche, F. B.; Steinmetz, A.; Muller, C. P. Neutralizing Immunogenicity of Transgenic Carrot (*Daucus Carota*) L -derived Measles Virus Hemagglutinin. *Plant Mol. Biol.* **2003,** *51,* 458–469.

52. Mason, H. S.; Haq, T. A.; Clements, J. D. Edible Vaccine Protects Mice Against *Escherichia coli* Heat-labile Enterotoxin (LT): Potatoes Expressing a Synthetic LT-B gene. *Vaccine* **1998,** *16* (13), 1336–1343.

53. Mason, H. S.; Lam, D. M. K.; Arntzen, C. J. Expression of Hepatitis B Surface Antigen in Transgenic Plants. *Proc. Natl. Acad. Sci. USA.* **1992,** *89,* 11745–11749.

54. McGarvey, P. B.; Hammond, J.; Dienelt, M. M.; Hooper, D. C. Expression of the Rabies Virus Glycoprotein in Transgenic Tomatoes. *Biotechnology* **1995,** *13* (13), 1484–1487.

55. Mercenier, A.; Wiedermann, U.; Breiteneder, H. Edible Genetically Modified Microorganisms and Plants for Improved Health. *Curr. Opin. Biotechnol.* **2001,** *12* (5), 510–515.

56. Michel, M. -L.; Tiollais, P. Hepatitis B Vaccines: Protective Efficacy and Therapeutic Potential. *Pathol. Biol.* **2010,** *58,* 288–295.

57. Mishra, M.; Gupta, P. N.; Khatri, K.; Goyal, A. K.; Vyas, S. P. Edible Vaccines: A New Approach to Oral Immunization. *Ind. J. Biotech.* **2007,** *7,* 283–294.

58. Nojima, J,; Ishii-Katsuno, R.; Futai, E.; Sasagawa, N.; Watanabe, Y.; Yoshida, T.; Ishiura, S. Production of Anti-amyloid β Antibodies in Mice Fed Rice Expressing Amyloid β. *Biosci. Biotechnol. Biochem.* **2011,** *75* (2), 396–400.

59. Noriho, F.; Noriko, T.; Yasushi, O.; Ryo, F. Production of Biologically Active Atlantic Salmon Interferon in Transgenic Potato and Rice Plants. *J. Biosci. Bioeng.* **2010,** *110* (2), 201–207.

60. Oszvald, M.; Kang, T. J.; Tomoskozi, S.; Jenes, B.; Kim, T. G.; Cha, Y. S.; Tamas, L.; Yang, M. S. Expression of Cholera Toxin B Subunit in Transgenic Rice Endosperm. *Mol. Biotechnol.* **2008,** *40* (3), 261–268.

61. Oszvald, M.; Kang, T. J.; Tomoskozi, S.; Tamas, C.; Tamas, L.; Kim, T. G.; Yang, M. S. Expression of a Synthetic Neutralizing Epitope of Porcine Epidemic Diarrhoea

Virus Fused with Synthetic B Subunit of *Escherichia coli* Heat Labile Enterotoxin in Rice Endosperm. *Mol. Biotechnol.* **2007**, *35* (3), 215–223.

62. Pant, G.; Sanjana, W. K. Edible Vaccines: A Boon to Medical Science. *Int. J. Curr. Agric. Res.* **2014**, *3* (5), 76–80.

63. Park, H. S.; Choi, J. W. Functional Expression of Bovine Growth Hormone Gene in *Pleurotuseryngii*. *Biotechnol. Bioproc. Eng.* **2014**, *19* (1), 33–42.

64. Pogrebnyak, N.; Golovkin, M.; Andrianov, V.; Spitsin, S.; Smirnov, Y.; Egolf, R.; Koprowski, H. Severe Acute Respiratory Syndrome (SARS) S Protein Production in Plants: Development of Recombinant Vaccine. *Proc. Natl. Acad. Sci.* **2005**, *102* (25), 9062–9067.

65. Qian, B. J.; Shen, H. F.; Liang, W. Q.; Guo, X. M.; Zhang, C.; Wang, Y.; Li, G.; Wu, A.; Cao, K.; Zhang, D. Immunogenicity of Recombinant Hepatitis B Virus Surface Antigen Fused with Pres1 Epitopes Expressed in Rice Seeds. *Transgenic Res.* **2008**, *17* (4), 621–631.

66. Ramírez, Y. J.; Tasciotti, E.; Gutierrez-Ortega, A.; Donayre, T. A. J.; Olivera, F. M. T.; Giacca, M.; Gómez, L. M. A. Fruit-specific Expression of the Human Immunodeficiency Virus Type 1 Tat Gene in Tomato Plants and its Immunogenic Potential in Mice. *Clin. Vaccine Immuno.* **2007**, *14* (6), 685–692.

67. Rosales-Mendoza, S.; Alpuche-Solís, A.; Soria-Guerra, R. Expression of an *Escherichia coli* Antigenic Fusion Protein Comprising the Heat Labile Toxin B Subunit and the Heat Stable Toxin and its Assembly as a Functional Oligomer in Transplastomic Tobacco Plants. *Plant J.* **2008**, *57* (1), 45–54.

68. Roupas, P.; Keogh, J.; Noakes, M.; Margetts, C.; Taylor, P. The Role of Edible Mushrooms in Health: Evaluation of the Evidence. *J. Funct. Foods* **2012**, *4*, 687–709.

69. Sathish, K.; Sriraman, R.; Subramanian, B. M.; Rao, N. H.; Kasa, B.; Donikeni, J. Plant Expressed Coccidial Antigens as Potential Vaccine Candidates in Protecting Chicken Against Coccidiosis. *Vaccine* **2012**, *30* (3), 4460–4464.

70. Shah, C. P.; Trivedi, M. N.; Vachhani, U. D.; Joshi, V. J. Edible Vaccine: A Better Way for Immunization. *Int. J. Curr. Pharm. Res.* **2011**, *3* (1), 5356–5359.

71. Sharma, M.; Sood, B. A Banana or a Syringe: Journey to Edible Vaccines. *World J. Microbiol. Biotechnol.* **2011**, *27* (3), 471–477.

72. Shoji, Y.; Farrance, C. E.; Bautista, J. A Plant-based System for Rapid Production of Influenza Vaccine Antigens. *Influ. Other Respir. Viruses* **2012**, *6* (3), 204–210.

73. Srinivas, L.; Kumar, G.; Ganapathi, T. R.; Revathi, C. J.; Bapat, V. A. Transient and Stable Expression of Hepatitis B Surface Antigen in Tomato (*Lycopersiconesculentum*). *Plant Biotechnol. Rep.* **2008**, *2* (1), 1–6.

74. Streatfield, S. J.; Jilka, J. M.; Hood, E. E.; Turner, D. D.; Bailey, M. R.; Mayor, J. M.; Woodard, S. L.; Beifuss, K. K.; Horn, M. E. Plant-based Vaccines: Unique Advantages. *Vaccine* **2001**, *19*, 2742–2748.

75. Tackaberry, E. S.; Prior, F. A.; Rowlandson, K. Sustained Expression of Human Cytomegalovirus Glycoprotein B (UL55) in the Seeds of Homozygous Rice Plants. *Mol. Biotechnol.* **2008**, *40* (1), 1–12.

76. Tacket, C. O.; Mason, H. S.; Losonsky, G. Immunogenicity in Humans of a Recombinant Bacterial Antigen Delivered in a Transgenic Potato. *Nat. Med.* **1998**, *4* (5), 607–609.

77. Taylor, N. J; Fauquet, C. M. Microparticle Bombardment as a Tool in Plant Science and Agricultural Biotechnology. *DNA Cell Biol.* **2002,** *21* (12), 963–977.
78. Thanavala, Y.; Lugade, A. Oral Transgenic Plant-based Vaccine for Hepatitis B. *Immunol. Res.* **2010,** *46* (1), 4–11.
79. Thanavala, Y.; Yang, Y. F.; Lyons, P.; Mason, H. S. Arntzen, C. J. Immunogenicity of Transgenic Plant-derived Hepatitis B Surface Antigen. *Proc. Natl. Acad. Sci. USA.* **1995,** *92*, 3358–3361.
80. Turpen, T. H.; Reinl, S. J.; Charoenvit, Y.; Hoffman, S. L.; Fallarme, V.; Grill, L. K. Malarial Epitopes Expressed on the Surface of Recombinant Tobacco Mosaic Virus. *Biotechnology* **1995,** *13* (1), 53–57.
81. USDA. Rice World Markets and Trade. Foreign Agricultural Service/USDA. Office of Global Analysis, Washington DC, January 2017, p 28.
82. Walmsley, A. M.; Arntzen, C. J. Plants for Delivery of Edible Vaccines. *Curr. Opin. Biotechnol.* **2000,** *11* (2), 126–129.
83. Wigdorovitz, A.; Mozovoj, M.; Santos, M.; Parreno, V. Protective Lactogenic Immunity Conferred by an Edible Peptide Vaccine to Bovine Rotavirus Produced in Transgenic Plants. *J. Gen. Virol.* **2004,** *85* (7), 1825–1832.
84. Wu, J.; Yu, L.; Li, L.; Hu, J.; Zhou, J.; Zhou, X. Oral Immunization with Transgenic Rice Seeds Expressing VP2 Protein of Infectious Bursal Disease Virus Induces Protective Immune Responses in Chickens. *Plant Biotechnol. J.* **2007,** *5* (5), 570–578.
85. Xiao-Ming, L. X. M.; Quan-Hong, Y.; Zhen, Z. Expression of the Human Hepatitis B Virus Large Surface Antigen Gene in Transgenic Tomato Plants. *Clin. Vaccine Immunol.* **2007,** *14* (4), 464–469.
86. Yan-Ju, Y. E.; Wen-Gui, L. I. Immunoprotection of Transgenic Alfalfa (*Medicago-sativa*) Containing Eg95-EgA31 Fusion Gene of Echinococcusgranulosus Against Egprotoscoleces. *J. Trop. Med.* **2010,** *3*, 10–13.
87. Yin, C.; Zheng, L.; Zhu, J.; Chen, L.; Ma, A. Enhancing Stress Tolerance by Overexpression of a Methionine Sulfoxide Reductase A (MsrA) Gene in *Pleurotus ostreatus. Appl. Microbiol. Biotechnol.* **2015,** *99* (7), 3115–3126.
88. Youm, J. W.; Jeon, J. H.; Kim, H.; Kim, Y. H.; Ko, K.; Joung, H.; Kim, H. Transgenic Tomatoes Expressing Human Beta-amyloid for Use as a Vaccine Against Alzheimer's Disease. *Biotechnol. Lett.* **2008,** *30* (10), 1839–1845.
89. Yu, J; Langridge, W. H. R. A Plant-based Multicomponent Vaccine Protects Mice from Enteric Diseases. *Nat. Biotechnol.* 2001, *19* (6), 548–552.
90. Zhang, H.; Liu, M.; Li, Y.; Zhao, Y.; He, H.; Yang, G.; Zheng, C. Oral Immunogenicity and Protective Efficacy in Mice of a Carrot-derived Vaccine Candidate Expressing UREB Subunit Against Helicobacter Pylori. *Protein Expr. Purif.* **2010,** *69* (2), 127–131.
91. Zhang, X.; Buehner, N.; Hutson, A.; Estes, M.; Manson, H. Tomato is a Highly Effective Vehicle for Expression and Oral Immunization with Norwalk Virus Capsid Protein. *Plant Biotechnol. J.* **2006,** *4* (4), 419–432.

PART III
Scope and Applications of Plant-Based Products

CHAPTER 8

NATURAL PRODUCTS AS ECONOMICAL AGENTS FOR ANTIOXIDANT ACTIVITY

NIDA NAZAR, ABDULLAH IJAZ HUSSAIN,
SYED MAKHDOOM HUSSAIN, and POONAM SINGH NIGAM

ABSTRACT

Plants are major source of natural products, either as standardized extracts or as pure compounds from different parts, due to unmatched bioavailability of chemical diversity. Many flowering and nonflowering plants and their parts (leaves, flowers, seeds, and stems) have been studied for their medicinal properties due to plant-based biocompounds, which may be derived from natural resources (biological/plants), synthetic chemicals, or recombinant DNA technology. Bioactive molecules derived from biological resources have significant importance in different industries, especially the pharmaceuticals and foods. The specific fractions rich in required bioactivities isolated from the plant extracts could be useful in the nutraceutical and pharmaceutical industry. The bioactivity in extracts prepared from plants using polar and nonpolar solvents, and also in the essential oils of certain seasonal plants under different environmental conditions has been studied by many researchers for the assessment of antioxidant properties. Seasonal variations have also been found in bioactivities of the extracts from different plant parts. This chapter reviews bioactivities of plant-based antioxidants. The information in this chapter will be helpful for researchers and pharmacists for future studies.

8.1 INTRODUCTION

It has been reported that approximately 400,000 to 500,000 species of plants are available on this planet. Various species of flowering and nonflowering plants have been recognized as valuable resources due to the fact that these plants possess a huge renewable wealth of novel bioactive molecules.[70,71,74] Historically, many plants possess medicinal properties and major contributions in the area of human health and well-beings. As medicinal source, the use of plants is in the form of extract of specific plant part or combinations. Plant-based biomolecules have been studied for their wide range of bioactivities especially antioxidant and antimicrobial (antibacterial and antifungal) activities.[24,32,58] In addition, food bioactive secondary metabolic compounds (such as phenolics, carotenoids, alkaloids, and terpenoids) also act as potential chemopreventive agents against different diseases.[23,112] Moreover, these compounds have gained more attention due to their vast bioactive properties, in life sciences as necessary active part of our daily diet.

Hence, these bioactive molecules represent vast untapped source for medicines and other healthcare products. In recent years, various natural sources (herbs, fruits, and vegetables) have gained interest for the production of antioxidants and the manufacturing of antioxidants supplements. This concern is mainly rising due to consumption of synthetic antioxidant compounds such as butylates hydroxyanisole (BHA) and butylated hydroxytoluene (BHT), which have proven toxic for pharmaceutical industry, food industry, and human health. Moreover, due to their enormous therapeutic potential, various plants have been investigated and studied by many researchers.[2,54,58,70,71,74]

The aim of this chapter is to explore the use of selected bioactive molecules of plant origin that are efficient against pathogens. Moreover, these natural compounds are recognized for the treatment of several health disorders by mitigating most of the side effects. In parallel, plant-based extracts have gained vast use as rich source of organic compounds as described in detail in a special publication edited by Colla and Rouphael,[31] who used term of botanicals under the heading biostimulants for plant-based extracts. In plants, their exogenous use is mainly for the growth enhancement under stressful conditions, especially by enhancing the plant at oxidative potential.

8.2   BIOACTIVE COMPOUNDS

Natural products or biological resources are main sources of bioactive molecules or phytochemicals.[23] All medicinal plants contain phytochemicals components that are beneficial for healing and curing of human diseases.[35,108] Potential phytochemicals are found naturally in leaves, stem, roots, and fruits/seeds of a medicinal plant and show strong defense mechanisms and safeguards against numerous diseases. Phytochemicals are classified as primary and secondary compounds. Primary ingredients are proteins, chlorophyll, and common sugars while terpenoid, alkaloids, and phenolic compounds are secondary compounds (or metabolites).[88]

8.2.1   POLYPHENOLS

Phenolics are most overflowing secondary metabolites in plants that are largely circulated in the kingdom of planta. Plant polyphenols have gained considerable importance due to their potential as antioxidant and their obvious effects in the deterrence of several oxidative stress-related syndromes such as cancer. Phenolics are extensive compounds of plant foods (vegetables, fruits, cereals, legumes, olive, etc.) and beverages (beer, coffee, tea, wine, etc.), as well as somewhat liable for the general organoleptic properties of plant-based foods.[35]

Over 8000 phenolic structures have presently been identified. Phenolics are aromatic compounds having one or more hydroxyl groups along with one or more aromatic rings. Single substituted phenolic rings are classified as simplest bioactive phytochemicals. The main classes of phenylpropane-derived group of compounds are caffeic and cinnamic acids. Figure 8.1a represents caffeic acid, which work as antifungal,[38] antibacterial, and anti-viral[148] agent, present in most common herbs such as thyme and tarragon. It is also present in high concentration in tea that have in daily use in most of the countries. Pyrogallol and catechol are hydroxylated phenols and are toxic against microorganisms. Two and three hydroxyl groups are present in catechol and pyrogallol structures, respectively.

Phenolic compounds at lower level of oxidation—retaining a C_3 side chain do not have any oxygen—are categorized as essential oils and have strong antimicrobial activities. For example, clove oil has encompassed as well-characterized representatives as it is good source of eugenol. Contrary to bacteria and fungi, eugenol is also bacteriostatic.[32,35,61]

8.2.2 QUINONES

Quinones are class of compounds having aromatic rings with ketone substitutions as shown in Figure 8.1b. These compounds are reactive and abundant in nature. Being colored in nature during cutting or damaging of food fruit and vegetables, these compounds are accountable for the browning reactions. For example, in human skin during melanin biosynthetic pathway it acts as an intermediate.[130] It retains dyeing property due to presence in henna (tree).[43]

Through oxidation and reduction reactions, its alternation between hydroquinone and quinine takes place easily. A complex of naphthoquinone is called vitamin K. It has been reported that in body tissues, the ease of oxidation may be associated to its antihemorrhagic property.[61]

Besides a source of free radicals, nucleophilic amino acids form irreversible complex with quinines in proteins molecules, which are frequently responsible for the protein inactivation and loss of metabolic functions.[141] Consequently, quinine has pronounced potential range of antimicrobial effect. Membrane-bound enzymes, surface-exposed adhesions, and cell wall polypeptides are apparently targeted in the microbial cell. Substrate unattainable to the microorganism may also diminish by quinones, which have comprehensively been examined the conceivable toxic effects.[32]

8.2.3 FLAVONES, FLAVONOLS, AND FLAVONOIDS

Phenolic compounds having one carbonyl group in their structure are termed as flavones shown in Figure 8.1c (while quinones have two carbonyl groups) and the flavonols are resulted due to addition of three hydroxyl groups shown in Figure 8.1d.[43] Aromatic rings interconnected through C6-C3 linkage are identified as flavonoids as shown in Figure 8.1e, which are also hydroxylated phenolic compounds. Their synthesis also takes place against microbial infection.[36] It is also important to know that in vitro response to extensive assortment of microorganisms, they have been effective antimicrobial substances. Their complex formation ability with extracellular entities, such as proteins as well as with soluble proteins is perhaps responsible for their bioactivities. Microbial membrane may possibly be interrupted by more lipophilic flavonoids.[146]

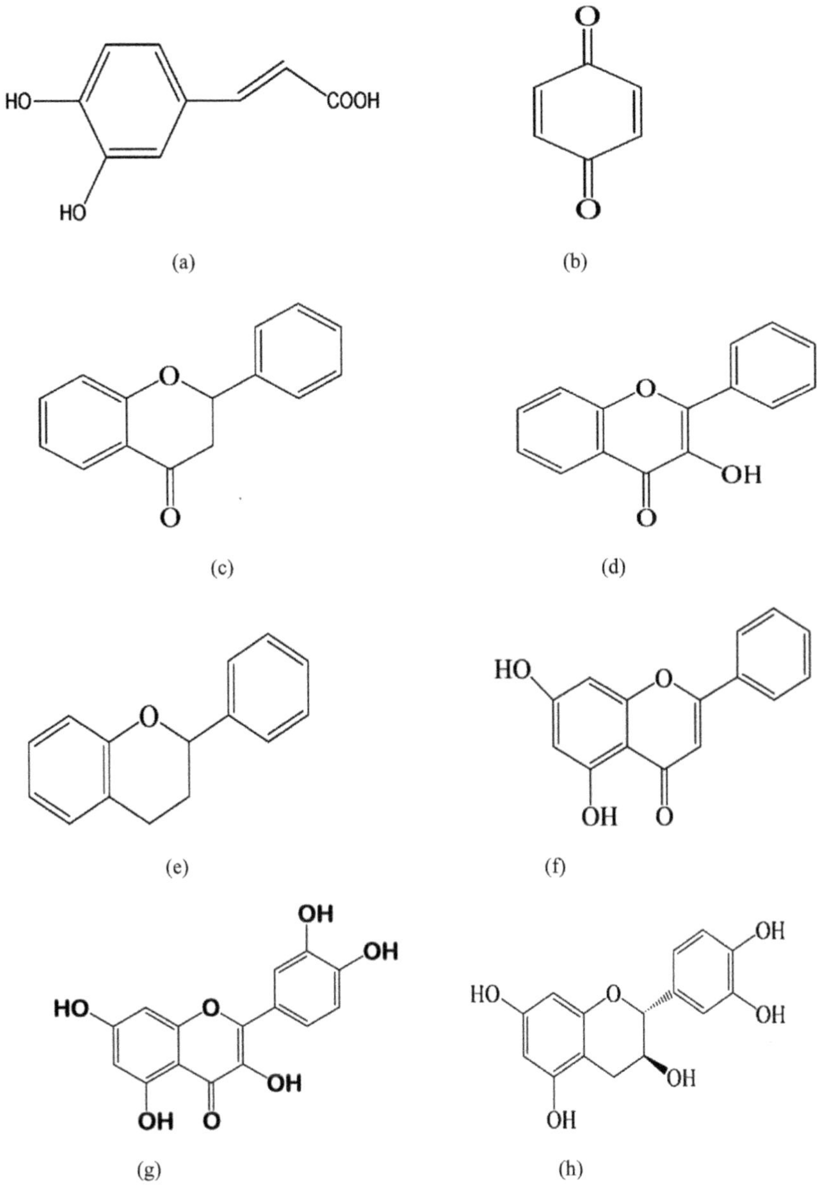

(a)

(b)

(c)

(d)

(e)

(f)

(g)

(h)

FIGURE 8.1 *(Continued)*

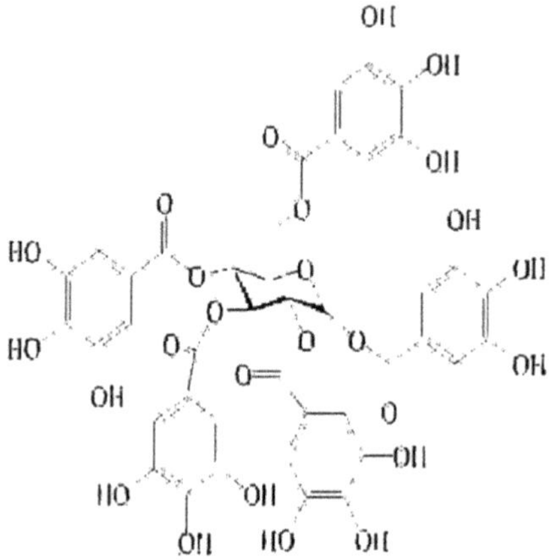

(i)

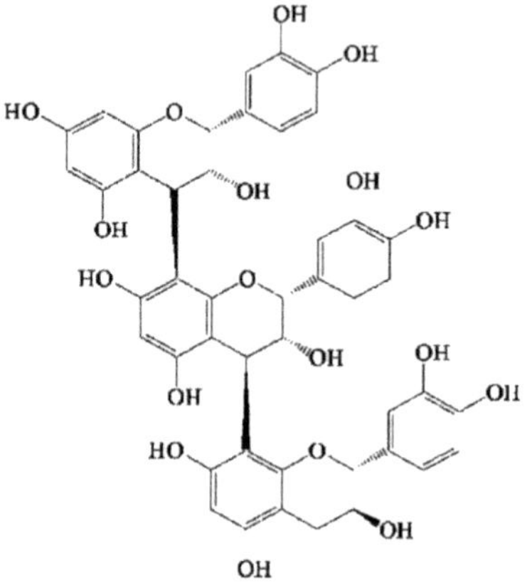

(j)

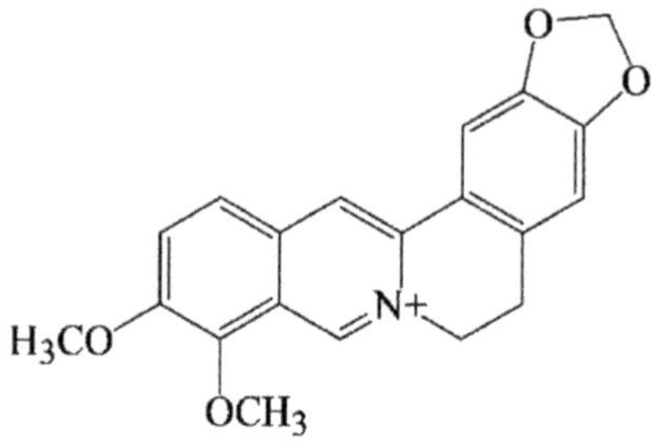

(k)

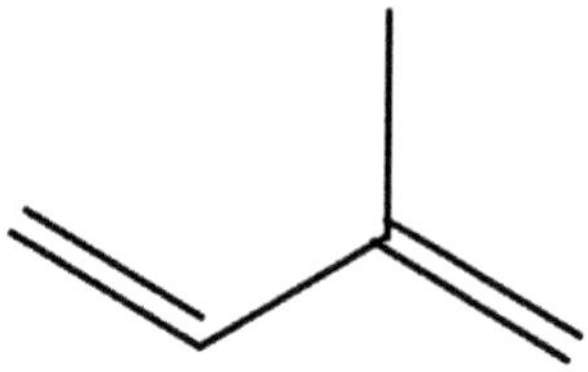

(L)

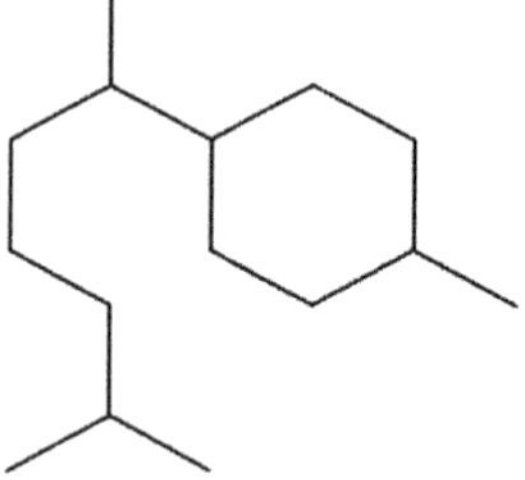

(m)

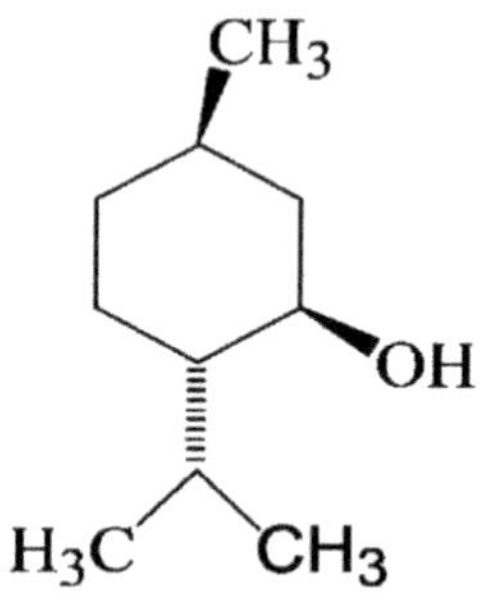

(n)

FIGURE 8.1   *(Continued)*

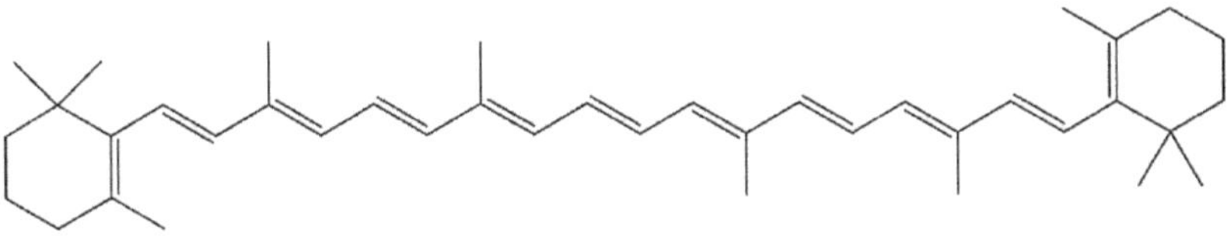

(o)

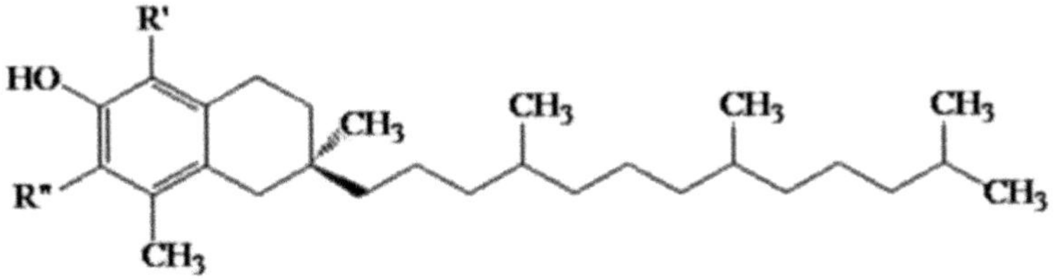

(p)

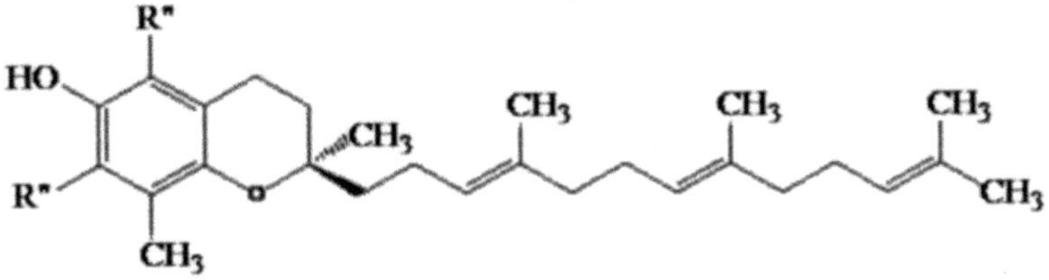

(q)

FIGURE 8.1 Structure of some compounds: (a) caffeic acid (poly phenols); (b) quinone; (c) flavone; (d) flavonol; (e) flavonoid; (f) chrysin; (g) catechins (flavanol); (h) iso-flavone (alkaloid); (i) hydrolyzable tannins (pentagalloyl glucose or pgg); (j) condensed tannins (procyanidin); (k) berberine; (l) isoprene (m) sesquiterpene; (n) menthol (terpenoid); (o) β-carotene; (p) tocopherol; and (q) tocotrienol.

Catechins are one of the main compounds of flavonoids class and the C_3-ring is in reduced form in this class. The flavonoids, due to their existence in green teas, have been investigated widely. Later it was found that the antimicrobial properties of teas encompass fusion of catechins compounds.[42] In vitro studies reveal that *Shigella, Vibrio cholera, Streptococcus mutans,* some other bacteria strains and other microorganisms are inhibited by these compounds.[16,122,123,145,147]

Inhibitory effects are demonstrated by flavonoid compounds found in several viruses. In contrary to HIV, the flavonoids such as chrysin shown in Figure 8.1f[33] have been valued in various renewed studies. It has been found that derivatives of flavone are inhibitor against respiratory syncytial virus (RSV).[13,82] In vitro studies on cell monolayers, about the mode of action of quercetin shown in Figure 8.1g, naringin, hesperetin, and cate-chins shown in Figure 8.1h have been delivered in concise form by Kaul,[82] who found that all these flavonoid compounds are operative in several ways except naringin, which was not inhibitory against para-influenza virus type 3, poliovirus type 1, herpes simplex virus type 1 (HSV-1). According to Kaul, the differences in the activity of these compounds are critically due to small structural differences. Furthermore, the low toxic potential is another piercing benefit of many plant derivatives.[32]

On structural basis, flavonoids have shown more response against microorganisms, those have OH-group on the β-ring instead of two hydroxyl groups[22] and the cellular membrane seems to be microbial targeted according to this conclusion. Therefore, based on membrane structure, this form of flavonoids may be more interrupting to lipophilic compounds. Nevertheless, contrasting effects have been originated according to several authors, that is, microbial activity was increased by increasing hydroxylation.[127] However, analogues results have been found as compared with former conclusions as for simple phenolics (see above). Therefore, it can be concluded that degree of hydroxylation and toxicity to microorganisms have not evidently been projected.[17,49,90]

8.2.4 TANNINS

Tannins belong to phenolic family polymeric in nature, having molecular weights ranging 500–3000.[63] Tannins generally are present in bark, leaves, roots, fruits, and wood in virtually every single plant.[128] Hydrolyzable and condensed tannins (Fig. 8.1i and Fig. 8.1j) are two categories in which they are classified. Generally, gallic acid as multiple esters with glucose molecules are the base of hydrolyzable tannins; and monomers of flavo-noids are source of supplementary abundant tannins.

Condensation of derivatives of flavonoids is leading to tannins forma-tion, which is relocated to woody tissues of plant. Polymerization of quinone units is another way of tannins formation.[49] Meanwhile, it has been recommended that variety of diseases can be inhibited by depletion of beverages encompassing tannins, particularly red wines and green teas.[133]

Tannins have been allocated to several physiological activities of humans, such as host-mediated tumor activity, activity against infection in broad array, and stimulation of phagocytic cells.[63]

Analogues to quinones, their mode of action against microbes reflects their aptitude to inactivate the microbial adhesions, cell envelop transport proteins, inactivation of enzymes, etc. Direct inactivation of microorganisms is evident in germ tubes of *Crinipellis perniciosa*[18] that morphologically was changed by small amount of tannins. Insect proceeding in plants[131] and disrupt digestive actions in ruminal animals are impeded by tennins.[19] Antimicrobial assets of tannins have also been reviewed by Scalbert,[128] who acknowledged hindered activities of tannins by listing 31 studies. Overall, these studies recommended that tannins can be lethal to filamentous fungi, bacteria, and yeasts.[32,49]

8.2.5 ALKALOIDS

Alkaloids are class of heterocyclic nitrogenous compounds. Among different alkaloids morphine, one isolated in 1805 from the opium poppy (*Papaver somniferum*) was the example of first medically useful psychoactive alkaloid.[43] Heroin and codeine are derivatives of morphine. Most of the alkaloids are extracted from members of Ranunculaceae family[80] or buttercup family,[11] which show antimicrobial activities.

Solamargine, a glyco-alkaloid along with other alkaloids, isolated from the berries of *Solanum khasianum* has been found effective against different infections including HIV[101,134] and intestinal infections associated with AIDS. A significant important member of alkaloid group is berberine (shown in Fig. 8.1k). It is effective against plasmodia[110] and trypanosomes.[45]

8.2.6 ISOPRENE AND TERPENOIDS

Essential oils (mixture of metabolically active volatile compounds) are secondary metabolites; and isoprene structures as shown in Figure 8.1L are the base of these oils. Generally, they are termed as terpenes, such as $C_{10}H_{16}$, a monoterpene along with hemiterpenes (C_5) and sesquiterpenes (C_{15}) shown in Figure 8.1m. They also occur as di-, tri-, and tetraterpenes (C_{20}, C_{30}, and C_{40}, respectively). Terpenoids are formed with the addition of oxygen atom in their basic structure. Generally, they are derived from acetate. Structurally, they share the beginning with fatty acid but in contrast to fatty acid they are cyclized and have extensive branching. For example, camphor and methanol shown in Figure 8.1n as monoterpenes, artemisin, and farnesol as sesquiterpenes are examples of common terpenes.[32,61]

It has been found that terpenoids or terpenes have inhibiting effects on fungi,[12,60,89] viruses,[46,62,115] and bacteria.[5,7,14,59] It was found that more than 60% derivatives of essential oils are effective against wide range of microorganisms including fungi and bacteria. However, the mechanism of antimicrobial action of terpenes is not fully defined but is speculated due to the lipophilic nature of terpenes, and it may involve in membrane disruption.[32]

8.2.7 CAROTENOIDS

Carotenoids are categorized as subclass of terpenes and plant pigments generally found in bright orange, red, and yellow color, mostly occurring in vegetables such as spinach, tomatoes, pink grape fruit, oranges, parsley, red palm oil, etc. Many plants possess bright colors due to high content of carotenoids. Bright colors in animals by plants are due to carotenoids such as shellfish and flamingos. Carotenoids are responsible for yellow color of egg yolks shielded unsaturated fats in yolk. Carotenoids are classified in two main different types based on molecular structure:

- First type is carotenes, which chemically consist of 40-C tetraterpenes, without any addition of specific chemical group such as keto or hydroxyl group. Well-known carotenes include beta-carotene shown in Figure 8.1o and lycopene[34] categorized as type one.
- Xanthophyll is other type of carotenoids, which have chemical compounds termed as alcoholic carotenoids and keto-carotenoids. Cryo-xanthin, zea xanthin, and astaxanthin fall in second category. Almost more than 600 carotenoids exist naturally. These phytonutrients are considered as forerunners of vitamin A, while the vitamin A activity has been shown only by fewer than 10% carotenoids. Generally, the vitamin A activity is owed by alpha, beta, and epsilon carotenes among other carotenes. Most active is beta carotene. It is reported that 50–54% antioxidant activity is due to beta carotene and alpha carotene, while 42–50% antioxidant activity is shown by epsilon carotene.

8.2.8   TOCOLS (TOCOTRIENOLS AND TOCOPHEROLS)

Tocopherols and tocotrienols belong to class of compounds known as tocols. Tocopherols shown in Figure 8.1p belong to a conventional class of Vitamin E, while tocotrienols shown in Figure 8.1q are known as precursor of vitamin E (tocopherols) with numerous health benefits. Both these biochemicals act as strong antioxidants based on their structural importance. Palm oil and cereal grains are rich source of natural tocotrienols and tocopherols. Studies reveal that growth of breast cancer cells has been restrained by tocotrienols while the tocopherols enable to do that. Studies on tocopherols and tocotrienols revealed that their biological properties are quite different to each other. Based on cholesterol lowering effects, tocotrienols were most studied.[111]

8.3   ANTIOXIDANT POTENTIAL OF BIORESOURCE NATURAL PRODUCTS

An antioxidant can be defined as "a molecule which can stop the oxidation chain reaction by scavenging the free radicals and delay or inhibit the oxidation of other molecules." Oxidation is the process in which electrons are transferred from one substance to another oxidative agent. Being the first line of defense against free radical damages, natural antioxidant compounds are able of stabilizing/deactivating free radicals before their damaging effects to cellular membranes.[24,107,116]

In recent era in the development of new drugs, natural substances are under observations and seeking great attention in many developing countries for replacement of synthetic compounds. For example, butylated hydroxytolune (BHT) and butylated hydroxyanisol (BHA) are being commonly used as synthetic antioxidants but the toxicity concerns about these antioxidants have limited their uses.[76] Therefore in recent years, plant-based natural antioxidants are broadly active due to health concerns.[150] It has been found that the use of natural antioxidants have lower risk of specific diseases such as stroke, heart diseases, and cancer due to their active in increasing the plasma antioxidant capacity.[24]

Plants and other natural products are main source of natural antioxidants.[70,72,73,107] Wide variety of natural occurring antioxidants are present in plants with different composition, chemical structure, physical properties, and biological activities.[35,58,107] It has also been found that the content of effective phytochemicals is plant part (root, stem, leaf, and flower), plant growth stage as well as specific season.[69] Normally the roots and fruits are rich source of phytochemicals due to source sink movement of chemicals for storage or as defense against adverse environmental conditions. Based on the phytochemical constituents, different plant parts are effective against different diseases.[58]

Some species of plants have been investigated for their antioxidant properties based on different plant parts.[24,28] Measuring of the antioxidant activity of plant extracts/botanicals by using different methods helps in screening of plants. Methods are 2,2-diphenyl-1-picryl hydrazyl method (DPPH), nitric oxide method, 2,2-azinobis-3-ethylbinzothiazoline-6-sulfonic acid method (ABTS), β-carotene bleaching test (BCBT), oxygen radical absorbance capacity (ORAC) method, N,N-dimethyl-p-Phenylenediamine (DMPD) assay, and thiobarbituric acid reactive species (TBARS) assay, etc.[24,48,69,70,72,73,75]

In plants, natural occurring antioxidants categories are: (1) Nonenzymatic mechanisms, phenolic defense compounds (flavonoids, vitamin E, phenolic acid, and other phenols), ascorbic acid, cartenoids, nitrogenous compounds (amino acid, alkaloids, and amines), and chlorophyll derivatives; (2) Enzymatic peptide defense mechanisms (superoxide dismutase, catalases, peroxidases, ascorbate peroxidases, glutathione peroxidase/reductase, and other enzymatic proteins). Both categories of natural antioxidants have been playing important role in human health.[113]

The major endogenous enzymatic antioxidants related to the neutralization of reactive oxygen species (ROS) are: peroxidases (POD), superoxide dismutase (SOD), glutathione reductase (GRx), and glutathione peroxidase (GPx) and catalase (CAT). These antioxidant enzymes metabolize the oxidative toxic agents. These require micronutrients like copper, iron, selenium, manganese, and zinc as cofactors for optimum catalytic activity.[116,118,132]

The endogenous nonenzymatic antioxidants (also called metabolic nonenzymatic antioxidants) are produced by metabolism in the body and are: lipoid acid, glutathione, L-arginine, ubiquinone (coenzyme Q10), uric acid, bilirubin, melatonin, transferrin, etc. Alpha lipoic acid (ALA) and glutathione are members of more general group called thiols, while transferrin, together with albumin, lactoferrin, ceruloplasmin, metallothionein, ferritin, and myoglobin, categorized in a group of metal binding proteins. Lipoic acid is a sulphur-containing molecule that is able to show antifree radical potential in both aqueous and lipid media. It also improves its antioxidant effect when is bounded to pro-oxidant metals.

Glutathione, another important antioxidant, can directly suppress the oxidative effects of ROS like lipid peroxides, and thus plays main role in xenobiotic metabolism. The study has shown that vitamin C and gluta-thione jointly act as antioxidant and thus reduce the free radicals and have a sparing effect upon each other.[77] Natural antioxidants, in pure or concen-trated form, such as CQ-10,L-arginine, and melatonin, have recently been used as supplements for the treatment and prevention of various diseases especially, chronic and degenerative diseases.[118]

Plants have been studied for their antioxidant properties and their use as health supplements (exogenous use). Another name has been proposed for them as nutrient-derived antioxidants or dietary antioxidants. All members of this group are nonenzymatic in nature, including polyphenols, some essential fatty acids (omega-3 and omega-6 fatty acids), tocols including vitamin E, ascorbic acid (vitamin C), carotenoids (β-carotene, lycopene), flavonoids, as well as trace metals (selenium, manganese, zinc), etc.[34,77,116]

Lot of research has been performed to study the antioxidant activity of phytochemicals as shown in Table 8.1. The antioxidant activities in different extracts of *Piper cubeba* of family Piperaceae[28] has been studied in different fractions of the extract of plant *Stachys schtschegleevi* of family Lamiaceae.[2,68] Natural antioxidants can be found in many different kind of edible plants, mainly in plant-derived foods such as vegetables and fruits. Especially the vegetables and fruits, which are bright in color, are high in antioxidants.[24,58]

Among fruits, berries are prime rich source of antioxidants. Blackber-ries, grapes, raspberries, strawberries, cranberries, and blueberries are all rich in flavonoids. Oranges, grape fruits, and other citrus varieties are other important popular sources of antioxidative vitamins. However, not all the antioxidants are found in one type of fruit and vegetable.[152] The tea plant *Camellia sinensis* has been studied extensively for its antioxidant proper-ties. Leaves of green tea are rich source of flavonoids. Green tea, which is rich in phenolic compounds and flavonoids, is being used in traditional Chinese medicine.[42]

Edible herbs like parsley, coriander, and dill are also considered as excellent source of antioxidant. In view of the information available in literature, most spices are rich in antioxidants. Therefore, antioxidants cannot only be obtained naturally but also from supplements,[124,128] although

TABLE 8.1 Antioxidant Bioactivity of Secondary Metabolites.

Medicine plants	Parts studied	Bioactive compounds	Antioxidant assays	References
Agaricus brasiliensis (Mushroom)	Whole mushroom	Linoleic and palmitic acids	FRAP and DPPH radical scavenging assays	47, 100
Apios Americana (Ground Nut)	Nut	Isoflvone: genistein-7-*O*-gentiobioside	DPPH radical scavenging assay	142
Berberis thunbergii (Rose Glow/Red berry)	Roots	Flavonoidan alkaloids	DPPH radical scavenging activity method	152
Bupleurum longiradiatum (Bupleuri)	Roots	Thymol, butylidene phythalide, and 5-indolol heptanol	DPPH radical scavenging and inhibition of lipid peroxidation methods	138
Chamaemelum nobile (Chamomile)	Flower head and leafy stem	Polyphenols including phenolic acids and flavonoids	DPPH freeradicals scavenging assay, reducing power and inhibition of lipid peroxidation.	56
Citrus aurantium L. (Bitter orange)	Flower	5-Hydroxy-6,7,3',4' tetramethoxyflvone and limonexic acid	DPPH radical scavenging activity assay	126, 154
Combretum zeyheri (fruited bush willow)	Leaf	Ursolic acid, 2α,3β-dihydroxyurs-12-en-28-oic acid, maslinic acid, 6-β-hydroxymaslinic and Triterpenoids	DPPH and TEAC Assays	99, 121
Coriandrum sativum L. (Coriander)	Leaves, seeds and oil.	Linalool, geranyl acetate, mono and sesquiterpenes camphor menthol and camphor	DPPH assay	10, 124
Crocus sativus (Saffron)	Whole plant	Lauric acid, crocin, and hexadecanoic acid	DPPH assay and β-carotene-linoleic acid assay.	3, 95, 155
Curcuma domestica Valeton (Turmeric)	Leaf	Curcumin, β-pinene. camphene, eugenol. β-sitosterol.	DPPH free radical scavenging assay	58, 65

TABLE 8.1 (Continued)

Medicine plants	Parts studied	Bioactive compounds	Antioxidant assays	References
Cuscuta reflexa Roxb. (Dodder)	Stem	Flavonoids, Dulcitol, bergenin, Coumarins, glycosides and lactone	DPPH Free radical scavenging Assay	58, 149
Daucus carota L (Carrot)	Root	Carotenes, carotenoids, glycosides, flavonoids, anthocyanin	ABTS and DPPH radical scavenging assay	25, 58
Emblica officinalis Gaertn (Amla)	Fruit	Polyphenols (Ellagic acid, gallic acid, tannins) and vitamin C	HRS, DPPH and FRAP	58, 96
Falcaria vulgaris (Sickle weed)	Flowers, stem, and leaf	α-Terpinyl acetate, α-pinene & β-caryophyllene	DPPH assay	136
Foeniculum vulgare Mill. (Fennel)	Leaf and seed essential oil and whole seeds	Volatile oil, fenchone, anethole, limonene, anisaldehyde and estragole.	Superoxide anion scavenging activity, DPPH assay and scavenging of hydrogen peroxide	9, 26, 58, 109
Glycyrrhiza glabra L. (Licorice)	Root	Glycyrrhizin, flavonoids, liquiritin, isoliquiritin, rhamnoliquiritin, 2-methylisoflavones.	DPPH Free radical scavenging assay	29, 58, 144
Heracleum Thomsonii (Clarke)	Aerial parts	Neryl acetate, coumarins	DPPH assay	57
Ipomoea hederacea (Morning-glory, Ivy leaf)	Seeds and leaves	γ-Tocopherol and β-tocopherol, saponins, alkaloids, caffeic acid and carbohydrates	TEAC, FRAP, DPPH and TRAP antioxidant assays	139, 153, 157
Litchi chinensis (Lychee)	Fruit	Phenolic acids including 3,4 dihydroxyl benzoate,2-(2-Hydroxyl-5 (methoxycarbonyl) phenoxy) benzoic acid, kaempferol, isolariciresinol, stigmasterol, methyl shikimate and ethyl shikimate	ORAC and DPPH Radical scavenging assay	79, 91

TABLE 8.1 *(Continued)*

Medicine plants	Parts studied	Bioactive compounds	Antioxidant assays	References
Mangifera indica L. (Mango)	Root, leaf and fruit.	Polyphenols including phenolic acids, flavonoids, Cyanogenetic glycosides, vitamin A & C, mangiferin, β-sitosterol, quercetin, ellagic acid gallic acid.	DPPH and ABTS assays	58. 98, 120
Matricaria chamomilla (Chamomil)	Flower head and leafy stem	Phenolic compounds	β-carotenelinoleic acid and DPPH assay	1, 55
Mentha longifolia (Wild Mint)	Aerial parts and essential oil	Borneol, Germacrene D, β-carophylene, piperitenone and piperitenoneoxide,	DPPH assay, bleaching β-carotene, and inhibition of linoleic acid peroxidation assays	75
Momordica charantia L. (Bitter Melon)	Root, leaf, seed andfruit.	Triterpene glycosides and stearic acid	DPPH & RAP assays	44, 58
Nepeta Persica and *Cataria* (Catnip)	Flowers, stem, leaf, roots and aerial parts	α-pinene, β-pinene and essential oil	DPPH assay and β-carotene bleaching test	78, 97, 137
Nigella sativa Linn (Cumin/Kalonji)	Seed	Phenols, flavonoids, essential oils, proteins, alkaloids and saponins.	DPPH and 2,2'-azino-bis (3-ethylbenzthiazoline-6-sulphonic acid assays, ABTS test	51, 119
Ocimun sanctum (Tulsi)	Leaf and seeds	Volatile oil, terpenoids, eugenol, thymol, estragole, Linalool	Reducing power assay and DPPH	58, 85, 135
Origanum dictamnus (Hop marjoram)	Aerial parts	α-tocopherol, rosmarinic acid & methyl ester	Antioxidant activity by the Rancimat method	27, 52
Oroxylum indicum (Indian Trumpet tree/Midnight horror)	Bark	Flavonoids and phenolics	ABTS, DPPH and FRAP Assays.	81, 103

TABLE 8.1   (Continued)

Medicine plants	Parts studied	Bioactive compounds	Antioxidant assays	References
Piper krukoffi (Pepper)	Leaves and aerial parts	Lignans(-)-kusunokin, alkaloids, flavonoids, phenylpropanoids, lignansand terpenes	DPPH ABTS and TEAC Assay	83, 104
Psoralea corylifolia L. (Babchi/Bawchi)	Seed	Essential oil, fixed oil, resin, bakuchiol (monoterpene phenol)	Superoxide scavenging assay, HRS and DPPH assay	58, 87
Punica granatum L. (Pomegranate)	Peel leaves, seeds and flowers.	Polyphenols, lavonoids, anthocyanins and hydrolysable tannins	TEAC, TAC, ABTS radical scavenging activity and DPPH assay	39, 106
Quercus (resinosa, laeta, grisea, obtusata) (Oak)	Leaf	Tannins, flavonoids, alkaloids, saponins	DPPH and HRS Assay.	125
Rosmarinus officinalis (Rosmary)	Essential oil and whole plant	1,8-cineole, camphor, α-pinene, limonene, camphene and linalool	DPPH assay and inhibition of peroxidation in linoleic acid system.	70, 105, 151
Santalum album L. (Chandan)	bark	Volatile oil, santaol,α-santalol, β-santalol, β-sitosterol	Nitric oxide free radical scavenging, ABTS free radical scavenging, (TAC), and DPPH free radical scavenging assays.	58, 102
Satureja montana (Savory)	Leaves, flowers and stem	Carvacrol, triterpenes and phenolic compounds	DPPH assay	20, 21, 37, 64
Solanum nigrum L. (Black night shade)	Leaf	Polyphenolics compounds, flavonoids, steroids	Hydrogen peroxide radical scavenging assay, Reducing power assay, nitric oxide radical scavenging and ß-carotene linoleate.	58, 143
Swertia chirayita (Chirata)	Whole plant	Xanthones, mangiferin, swertinin, chiratin, arginine	TPC, FRAP, HRS, Nitric oxide radical scavenging and DPPH	4, 58, 66

TABLE 8.1 (Continued)

Medicine plants	Parts studied	Bioactive compounds	Antioxidant assays	References
Thymus linearis and *Thymus serpyllum* (Wild thyme)	Whole plant	Terpenes and terpenoids including monoterpene hydrocarbons, oxygenated monoterpenes, sesquiterpene and hydrocarbons	DPPH free radical scavenging assay and bleaching of β-carotene in linoleic acid system	67
Vitis vinifera (Grape wine)	Leaf and fruits	Phenolic acids (trans-caffoyltartaric and trans-coumaroyltartaric acids) quercetin-3-*O*-galactoside quercetin-3-*O*-glucoside,myricetin-3-*O*-glucoside, and kaempferol-3-*O*-glucoside	DPPH radical scavenging and Assay of HRS.	6, 41, 114
Waldheimia glabra (Smooth ground daisy)	Whole plant	Phenolics	DPPH free radical scavenging assay	50
Withania somnifera Dunal (Winter cheery)	Root, leaf, and seed	Steroidal lactone, withanolides, glycine, withanine	TAC, DPPH radical scavenging assay, H RS superoxide radical scavenging, and hydrogen peroxide scavenging assay	23, 58

the natural sources are undoubtedly preferred. The common rosemary (*Rosmarinus officinalis* L.) contains a lot of compounds with antioxidant activity.[105] The research group of this chapter has shown that besides its antioxidant properties, *Rosmarinus officinalis* has also some antiproliferative and antibacterial activities.[15,70,151]

Some tropical fruits have also been studied for their antioxidant properties, such as mangosteen (*Garcinia mangostana*), guava (*Psidium guajava*), papaya (*Carica papaya* L.), dragon fruit (*Hylocereus undatus*), and star fruit (*Averrhoa carambola*), as well as banana (*Musa sapientum*) and water apple (*Syzygium aqueum*) that are well known for their antioxidant properties.[92] Pomegranate (*Punica granatum* L.) has been widely used in medicine for a number of therapeutic reasons due to its high antioxidant capacity.[39,106]

Similarly, the cultivation of mango (*Mangifera indica* Linn.) was started approximately 4000 years ago and its production and consumption is directly proportion to growth rate of population.[120] Peel and kernel are the byproducts of mango processing, and have exhibited great antioxidant activity.[58,98] It was found that kernel subsidizes the fruit about 17–22%.[140] The natural phenolic antioxidant potential of mango seed kernel is the result of acid hydrolysis.[120]

Cuscuts reflexa, belonging to Convolvolaceae family, possesses bitter sharp taste. The plant extract is rich source of phenolic compounds[94] and has exhibited strong antioxidant potential[149] and antimicrobial activities.[8] *Withania somnifera* is vital medicinal plant being used as medicine since 3000 years.[58] Study highlighted its phytochemical potential and demonstrated its antioxidant potential by using different antioxidant assays.[23]

Similarly, *Embilca officinals* Gaertn (Indian goose berry) commonly known as amla, is extensively distributed in China, India, Indonesia, and Thailand[93] in varied environmental conditions. It has great importance in medicinal plants.[86] Numerous diseases such as cold, fever, and stomach ache, etc. were being cured by using amla fruit with combination of different part of plants. Review article of Luqman and Kumar[96] demonstrated the antioxidant property of *Embilca officinals* by applying assays such as DPPH, HRS, and FRAP.

Human are utilizing the *Glycyrrhiza* species (Licorice) since 4000 years to cure different diseases especially cough. More than 30 species are included in the genus *Glycyrrhiza* and 15 out of 30 have been studied regarding their medicinal use. *Glycyrrhiza glabra* is being used in the preparation of varying medicines.[29] Key component of *G. glabra* as shown Table 8.1 is glycyrrhizic acid (triterpene glucoside), which is 50 times more sweet as compared with sugar.[144] Root extract of licorice is good to cure liver diseases and its antioxidant activities were studied by performing DPPH Assay.[29,58] In Indo-Pak subcontinent, *Salanum nigrum* L (commonly called Black nightshade) has broad uses in medicines.[117] The different parts of *Solanum nigrum* such as stem fruit and leaves against different assays are rich source of antioxidative compounds.[143]

Moreover, the plant genus *Nepeta* consists of about 250 species, present widely in North America, Europe, and Asia. About 67 species of Nepata are native to Iran and 39 out of them are only limited to Iran. Studies demonstrate that the chemical composition of essential oil of *N. persica*[78,84,137] has the potential against varying diseases. The free radical scavenging, antioxidant, antibacterial, antifungal activities of *Nepeta persica* and the results showed that the essential oil of *N. persica* was effective.[97]

Santalum album L. is a natural source of variety of sesquiterpenoid alcohols and essential oil. Different antioxidant assays performed on extracts of *S. album* concluded that due to presence of antioxidant components, it has great potential as a renewable resource.[58,102]

Study revealed antioxidative potential of *O. dictamnus* plant as showed in Table 8.1.[52] It is an aromatic plant and is used as a spice in foods or as hot drink after boiling in water. It is deliberated harmless. *O. dictamnus* is a self-growing plant, found in dry soil and rocky areas and dearth tolerant.[52,58]

Swertia chirata (Gentianaceae family) is a persistent herb, commonly termed as chirata. Xanthones and arginine are its key bioactive compounds found.[58,66] Literature has highlighted the antioxidant potential of *S. chirata* extracts and successive health remunerations.[4]

Crocus sativus L. (Saffron plant) is a proliferated vegetative plant and is being consumed as great source of colorants, food additives, and ingredient of traditional medicines.[40] Saffron derived from dry stigmas of *C. sativus* is the most costly spice used in industry, in textile dye as well as in gastronomic assistant.[95] Studies reveal the cardioprotective effects[53] and antioxidant properties of crocin, which is a component of *C. sativus*.[156] It has also been demonstrated the antioxidant potential of *Crocus* species. Furthermore, *Matricaria chamomilla* L. is also a medicinal plant.[3] In vitro, studies revealed the antioxidant and antimicrobial potential of methanolic extracts as well as the essential oil of *M. chamomilla*.[1,55]

8.4   SUMMARY

A primordial time, plants have been used as efficient resource to heal different diseases. They have hold on couple of compounds, which directly strike the targeted area or diseases. Scientists have investigated different species of plants by accomplishing different experiments and have disclosed numerous natural chemical compounds such as polyphenols, flavonoids, alkaloids, tocols, quinones, terpenes, carotenoids, tannins and have examined their mode of action and defense mechanism. Plants exhibit excellent pharmacological properties due to presence of vitamins and antioxidant compounds against excessive free radicals as a result of different biological activities. Antioxidants safe guard our intercellular system. Some selected plants have been discussed in this review. Isolation and identification of phytochemicals has been done by applying different techniques and different suitable combination of bioactive compounds and has been tested by performing multiple assays. Research still continues to evaluate antioxidative potential of bioresources.

The appropriately designed drugs undoubtedly will bring significant advances in traditional and modern healthcare system. Here point of ponder is, despite of numerous advances in technology and various studies of bioactive constituents of bioresources, still several vigorous and productive compounds are not available in supplements and medicines in market.

KEYWORDS

- alcoloids
- antioxidant
- carotenoids
- flavones
- flavonoids

REFERENCES

1. Abdoul-Latif, F. M.; Mohamed, N.; Edou, P.; Ali, A. A.; Djama, S. O.; Obame, L. C.; Bassole, I. H. N.; Dicko, M. H. Antimicrobial and Antioxidant Activities of Essential

Oil and Methanol Extract of *Matricaria chamomilla* L. from Djibouti. *J. Med. Plants Res.* **2011**, *5* (9), 1512–1517.

2. Abichandani, M.; Nahar, L.; Nigam, P.; Chitnis, R.; Nazemiyeh, H.; Delazar, A.; Sarker, S. D. Antibacterial and Free-Radical-Scavenging Properties of *Stachys schtschegleevii* (Lamiaceae). *Arch. Biol. Sci.* **2010**, *62*, 941–945.

3. Acar, G.; Dogan, N. M.; Duru, M. E.; Kıvrak, I. Phenolic Profiles, Antimicrobial And Antioxidant Activity of the Various Extracts of *Crocus* Species in Anatolia. *Afr. J. Microbiol. Res.* **2010**, *4 (*11), 1154–1161.

4. Ahirwal, L.; Singh, S.; Dubey, M. K.; Bharti, V.; Mehta, A. Investigation of Antioxidant Potential of Methanolic Extract of *Swertia chirata* Buch. Ham. *Eur. J. Med. Plants* **2014**, *4* (11), 1345–1355.

5. Ahmed, A. A.; Mahmoud, A. A.; Williams, H. J.; Scott, A. I.; Reibenspies, J. H.; Mabry, T. J. New Sesquiterpene a-methylene Lactones from the Egyptian Plant *Jasonia candicans*. *J. Natural Prod.* **1993**, *56*, 1276–1280.

6. Ali, K.; Maltese, F.; Choi, Y. H.; Verpoorte, R. Metabolic Constituents of Grape Vine and Grape-derived Products. *Phytochem. Rev.* **2010**, *9* (3), 352–378.

7. Amaral, J. A.; Ekins, A.; Richards, S. R.; Knowles, R. Effect of Selected Monoterpenes on Methane Oxidation, Denitrification, and Aerobic Metabolism by Bacteria in Pure Culture. *Appl. Environ. Microbiol.* **1998**, *64*, 520–525.

8. Amol, P.; Vikas, P.; Kundan, C.; Vijay, P.; Rajesh C. In Vitro Free Radicals Scavenging Activity of Stems of *Cuscuta reflexa*. *J. Pharm. Res.* **2009**, *2* (1), 58–61.

9. Anwar, F.; Ali, M.; Hussain, A. I.; Shahid, M. Antioxidant and Antimicrobial Activities of Essential Oil and Extracts of Fennel (*Foeniculum vulgare* Mill.) Seeds from Pakistan Antioxidant and Antimicrobial Activities of Essential Oil and Extracts of Fennel. *Flavor and Fragrance J.* **2009**, *24*, 170–176.

10. Anwar, F.; Sulman, M.; Hussain, A. I.; Saari, N.; Iqbal, S.; Rashid, U. Physicochemical Composition of Hydro-distilled Essential Oil from Coriander *(Coriandrum sativum* L.) Seeds Cultivated in Pakistan. *J. Med. Plants Res.* **2011**, *5* (15), 3537–3544.

11. Atta-ur-Rahman.; Choudhary, M. I. Diterpenoid and Steroidal Alkaloids. *Nat. Prod. Comm.* **1995**, *12*, 361–379.

12. Ayafor, J. F.; Tchuendem, M. H. K.; Nyasse, B. Novel Bioactive Diteroenoids from *Aframomum aulacocarpos*. *J. Nat. Prod.* **1994**, *57*, 917–923.

13. Barnard, D. L.; Huffman, J. H.; Meyerson, L. R; Sidwell, R. W. Mode of Inhibition of Respiratory Syncytial Virus by a Plant Flavonoid. *Chemotherapy* **1993**, *39*, 212–217.

14. Barre, J. T.; Bowden, B. F.; Coll, J. C.; Jesus, J.; Fuente, V. E.; G. C. Janairo, G. C.; Ragasa C. Y. Bioactive Triterpene from *Lantana camara*. *Phytochemistry* **1997**, *45,* 321–324.

15. Bernardes, W. A.; Lucarini, R.; Tozatti, M. G.; Flauzino, L. G.; Turatti, I. C.; Andrade e Silva, M. L.; Martins, C. H.; da Silva Filho, A. A.; Cunha,W. R.. Antibacterial activity of the Essential oil from Rosmarinus Officinalis and its Major Components against Oral Pathogens. *J. Biosci.* **2010**, *65* (9–10), 588–593.

16. Borris, R. P. Natural Products Research: Perspectives from a Major Pharmaceutical Company. *J. Ethnopharmacol.* **1996**, *51*, 29–38.

17. Brigitte, A.; Graf, G.; Paul, E.; Milbury, K.; Blumberg, J. B. Flavonols, Flavones, Flavanones, and Human Health: Epidemiological Evidence. *J. Med. Food* **2005**, *8* (3), 281–290.

18. Brownlee, H. E.; McEuen, A. R.; Hedger, J.; Scott, I. M. Antifungal Effects of Cocoa Tannin on the Witches' Broom Pathogen *Crinipelli sperniciosa. Physiol. Mol. Plant Pathol.* **1990**, *36*, 39–48.

19. Butler, L. G. *Effects of Condensed Tannin on Animal Nutrition.* In *Chemistry and Significance of Condensed Tannins.* Springer: Boston, MA, 1989; pp 391–402.

20. Cetkovic, G. S.; Mandic, A. I.; Canadanovic-Brunet, J. M.; Djilas, S. M.; Tumbas, V. T. HPLC Screening of Phenolic Compounds in Winter Savory (*Satureja Montana L.)* Extracts. *J. Liquid Chromatogr. Related Technol.* **2007**, *30*, 293–306.

21. Cetkovic, G. S.; Canadanovic-Brunet, J. M.; Djilas, S. M.; Tumbas, V. T.; Sinisa L.; Markov, S. L.; Cvetkovic, D. D. Antioxidant Potential, Lipid Peroxidation Inhibition and Antimicrobial Activities of *Satureja Montana* L. subsp. *Kitaibelii* Extracts. *Int. J. Mol. Sci.* **2007**, *8* (10), 1013–1027.

22. Chabot, S.; Bel-Rhlid, R.; Chenevert, R.; Piche, Y. Hyphal Growth Promotion In Vitro of the VA Mycorrhizal Fungus, *Gigaspora margarita* Becker and Hall, by the Activity of Structurally Specific Flavonoid Compounds under CO 2-enriched Conditions. *New Phytol.* **1992**, *122*, 461–467.

23. Chaudhuri, D.; Ghate, N. B.; Sarkar, R.; Mandal, N. Phytochemical Analysis and Evaluation of Antioxidant and Free Radical Scavenging Activity of *Withania somnifera* Root. *Asian J. Pharm. Clin. Res.* **2012**, *5* (4), 193–199.

24. Chanda, S.; Dave, R. A Review of In Vitro Models for Antioxidant Activity Evaluation and Some Medicinal Plants Possessing Antioxidant Properties. *Afr. J. Microbiol. Res.* **2009**, *3* (13), 981–996.

25. Chatatikun, M.; Chiabchalard, A. Phyytochemical Screening and Free Radical Scavenging Activities of Orange Baby Carrot and Carrot (*Daucus carota* Linn.) Root Crude Extracts. *J. Chem. Pharm. Res.* **2013**, *5* (4), 97–102.

26. Chatterjee, S.; Goswami, N.; Bhatnagar, P. Estimation of Phenolic Components and In Vitro Antioxidant Activity of Fennel (*Foeniculum vulgare*) and Ajwain (*Trachyspermum ammi*) Seeds. *Adv. Biores.* **2012**, *3* (2), 109–118.

27. Chatzopoulou, A.; Karioti, A.; Gousiadou, C.; Vivancos, V. L.; Kyriazopoulos, P. Golegou, S. Depsides and Other Polar Constituents from *Origanum dictamnus L.* and their In Vitro Antimicrobial Activity in Clinical Strains. *J. Agric. Food Chem.* **2010**, *58* (10), 6064–6068.

28. Chitnis, R.; Abichandani M.; Nigam, P.; Nahar, L.; Sarker, S. D. Antioxidant and Antibacterial Activity of the Extracts of *Piper cubeba. ARS Pharm.* **2007**, *48* (4), 343–350.

29. Chopra, P. K. P. G.; Saraf, B. D.; Inam, F.; Deo, S. S. Antimicrobial and Antioxidant Activities of Methanol Extract Roots of *glycyrrhiza glabra* and HPLC Analysis. *Int. J. Pharm. Pharm. Sci.* **2013**, *5* (2), 157–160.

30. Chu, W. L.; Radhakrishnan, A. K. A Review of Research on Bioactive Molecules: Achievements and the Way Forward. *Int. e-J. Sci. Med. Educ.* **2008**, *1*, 21–24.

31. Colla, G.; Rouphael, Y. Biostimulants in Horticulture. *Scientia Horticulturae Sci.* **2015**, *196*, 1–2.

32. Cowan, M. M. Plant Products as Antimicrobial Agents. *Clin. Microbiol. Rev.* **1999**, *12*, 564–582.

33. Critchfield, J. W.; Butera, S. T.; Folks, T. M. Inhibition of HIV Activation in Latently Infected Cells by Flavonoid Compounds. *AIDS Res. Human Retroviruse* **1996**, *12* (1), 39–46.

34. Dahan, K.; Fennal, M.; Kumar, N. B. Lycopene in the Prevention of Prostate Cancer. *J. Soc. Integrative Oncol.* **2008**, *6*, 29–36.

35. Dai, J., Mumper, R. J. Review: Plant Phenolic: Extraction, Analysis and Their Antioxidant and Anticancer Properties. *Molecules* **2010**, *15*, 7313–7352.

36. Dixon, R. A.; Dey, P. M.; Lamb, C. J. Phytoalexins: Enzymology and Molecular Biology. *Adv. Enzymol.* **1983**, *55*, 1–69.

37. Dorman, H. J. D.; Bachmayer, O.; Kosar, M.; Hiltunen, R. Antioxidant Properties of Aqueous Extracts from Selected *Lamiaceae* Species Grown in Turkey. *J. Agric. Food Chem.* **2004**, *52*, 762–770.

38. Duke, J. A. *Handbook of Medicinal Herbs.* CRC Press, Inc.: Boca Raton, FL, 1985, p. 532.

39. Elfalleh, W.; Hannachi, H.; Tlili, N.; Yahia, Y.; Nasri, N.; Ferchichi, A. Total Phenolic Contents and Antioxidant Activities of Pomegranate Peel, Seed, Leaf and Flower. *J. Med. Plants Res.* **2012**, *6*, 4724–4730.

40. Escribano, J.; Rios, I.; Fernandez, A. Isolation and Cytotoxic Properties of a Novel Glycoconjugate from Corms of Saffron Plant (*Crocus sativus* L.). *Biochimica ET Biophysica Acta* **1999**, *1426* (1), 217–222.

41. Fernandes, F.; Ramalhosa, E.; Pires, P. *Vitis vinifera* Leaves Towards Bioactivity. *Ind. Crops Prod.* **2013**, *43* (1), 434–440.

42. Ferrara, L.; Montesano, D.; Senatore, A. The Distribution of Minerals and Flavonoids in the Tea Plant (*Camellia sinensis*). *Farmaco.* **2001**, *56*, 397–401.

43. Fessenden, R. J.; Fessenden, J. S. *Organic chemistry*, 2nd ed.; Willard Grant Press: Boston, MS, 1982; p 313.

44. Fidrianny, I.; Darmawati, A.; Sukrasno. Antioxidant Capacities From Different Polarities Extracts Of Cucurbitaceae Leaves Using Frap, DPPH Assays And Correlation With Phenolic, Flavonoid, Carotenoid Content. *Int. J. Pharm. Pharm. Sci.* **2014**, *6* (2), 0975–1491.

45. Freiburghaus, F.; Kaminsky, R.; Nkunya, M. H. H.; Brun, R. Evaluation of African Medicinal Plants for their In Vitro Trypanocidal Activity. *J. Ethnopharmacol.* **1996**, *55*, 1–11.

46. Fujioka, T.; Kashiwada, Y. Anti-AIDS Agents: Betulinic Acid And Platanic Acid As Anti-HIV Principles from *Syzigium claviflorum*, and the Anti-HIV Activity of Structurally Related Triterpenoids. *J. Natural Prod.* **1994**, *57*, 243–247.

47. Gan, C. H.; Nurul Amira, B.; Asmah, R. Antioxidant Analysis Of Different Types Of Edible Mushrooms (*Agaricus bisporous* and *Agaricus brasiliensis*). *Int. Food Res. J.* **2013**, *20* (3), 1095–1102.

48. Gazzani, G.; Papetti, A.; Massolini, G.; Daglia, M. Antioxidant and Prooxidant Activity of Soluble Components of Some Common Diet Vegetables and Effect of Thermal Treatment. *J. Agric. Food Chem.* **1998**, *46*, 4118–4122.

49. Geissman, T. A. Flavonoid Compounds, Tannins, Lignins and Related Compounds. *In Pyrrole Pigments, Isoprenoid Compounds And Phenolic Plant Constituents*; Florkin, M., Stotz, E. H., Eds; Elsevier: New York, 1963; p. 265.

50. Giorgi, A.; Panseri, S.; Mattara, M. Secondary Metabolites and Antioxidant Capacities of *Waldheimia glabra* (Decne.) Regel from Nepal. *J. Sci. Food Agric.* **2013,** *93* (5), 1026–1034.

51. Goga, A.; Hasic, S.; Becirovic, S.; Cavar, S. Phenolic Compounds and Antioxidant Activity of Extracts of *Nigella sativa* L. *Bull. Chem. Technol. Bosnia and Herzegovina* **2012,** *15,* 19–22.

52. Gortzi, O.; Lalas, S.; Chinou, I.; Tsaknis, J. Evaluation of the Antimicrobial and Antioxidant Activities of *Origanum dictamnus* Extracts Before and After Encapsulation in Liposomes. *Molecules* **2007,** *12,* 932–945.

53. Goyal, S. N.; Arora, S.; Sharma, A. K.; Joshi, S.; Ray, R.; Bhatia, J.; Kumari, S.; Arya, D. S. Preventive Effect of Crocin of *Crocus sativus* on Hemodynamic, Biochemical, Histopathological and Ultra Structural Alterations in Isoproterenol-induced Cardio Toxicity in Rats. *Phytomedicine* **2010,** *17,* 227–232.

54. Granger, M.; Samson, E.; Sauvage, S.; Majumdar, A.; Nigam, P.; Nahar, L.; Celik, S.; Sarker, S. D. Bioactivity of Extracts of *Centaurea Polyclada* Dc. (Asteraceae). *Arch. Biol. Sc. Belgrade* **2009,** *61* (3), 447–452.

55. Guimaraes, R.; Barros, L.; Duenas, M. Infusion and Decoction of Wild German Chamomile: Bioactivity and Characterization of Organic Acids and Phenolic Compounds. *Food Chem.* **2013,** *136* (2), 947–954.

56. Guimaraes, R.; Barros, L.; Duenas, M. Nutrients, Phytochemicals and Bioactivity of Wild Roman Chamomile: A Comparison Between the Herb and Its Preparations. *Food Chem.* **2013,** *136* (2), 718–725.

57. Guleria, S.; Sainia, R.; Jaitaka, V.; Kaula.; Lalb, B.; Rahic, P. Composition and Antimicrobial Activity of the Essential Oil of He*racleum thomsonii* (Clarke) from the Cold Desert of the Western Himalayas. *Nat. Prod. Res.* **2011,** *25* (13), 1250–1260.

58. Gupta, V. K.; Sharma, S. K. A Review of Plants as Natural Antioxidants. *Nat. Prod. Radiance* **2006,** *5* (4), 326–334.

59. Habtemariam, S.; Gray, A. L.; Waterman, P. G. A New Antibacterial Sesquiterpenes from *Premna oligotricha. J. Nat. Prod.* **1993,** *56,* 140–143.

60. Harrigan, G. G.; Ahmad, A.; Baj, N.; Glass, T. E.; Gunatilaka, A. A. L.; Kingston, D. G. I. Bioactive and Other Sesquiterpenoids from *Porellacordeana. J. Nat. Prod.* **1993,** *56,* 921–925.

61. Harris, R. S. Vitamins K. In: *Pyrrole Pigments, Isoprenoid Compounds and Phenolic Plant Constituents*; Florkin, M., Stotz, E., Eds; Elsevier: New York, 1963; pp 192–198.

62. Hasegawa, H.; Matsumiya, S.; Uchiyama, M.; Kurokawa, T.; Inouye, Y.; Kasai, R.; Ishibashi, S.; Yamasaki, K. Inhibitory Effect of Some Triterpenoid Saponins on Glucose Transport in Tumor Cells and Its Application to In Vitro Cytotoxic and Antiviral Activities. *Planta Medica* **1994,** *6,* 240–243.

63. Haslam, E. Natural Polyphenols (vegetable tannins) as Drugs: Possible Modes of Action. *J. Nat. Prod.* **1996,** *59,* 205–215.

64. Hassanein, H. D.; Said-al Ahl H. A. H; Abdelmohsen, M. M. Antioxidant Polyphenolic Constituents of *Satureja Montana* L. Growing in Egypt. *Int. J. Pharm. Pharm. Sci.* **2014,** *6* (4), 578–581.

65. Hincapie, C. A.; Monsalve, Z.; Seigler, D. S.; Alarcon, J.; Cespedes, C. L. Antioxidant Activity of *Blechnum chilense* (Kaulf.) Mett., *Curcuma domestica* Valeton and

Tagetes verticillata Lag. & Rodriguez. *Boletin Latinoamericano y del Caribe de Plantas Medicinales y Aromáticas* **2011**, *10* (4), 315–324.

66. Hussain, A. I.; Anwar, F.; Bhatti, H. N.; Rashid, U. Phytochemical and In Vitro Anthelmintic Screening of *Butea frondosa* and *Swertia chirata* from Pakistan. *J. Chem. Soc. Pakistan* **2006**, *28* (1), 84–92.

67. Hussain, A. I.; Anwar, F.; Chatha, S.; Latif, S.; Sherazi, S. T. H.; Ahmad, A.; Worthington, J.; Sarker, S. D. Chemical Composition and Bioactivity Studies of the Essential Oils from Two *Thymus* Species from the Pakistani Flora. *LWT-Food Sci. Technol.* **2012**, *3*, 1–8.

68. Hussain, A. I.; Anwar, F.; Nigam, P. S.; Sarker, S. D. Antimicrobial and Antioxidant Activities of the Essential Oils of Selected Species of the Lamiaceae. *Annals Nutr. Metab.* **2009**, *55*, 86–86.

69. Hussain, A. I.; Anwar, F.; Sherazi, S. T. H.; Przybylski, R.; Chemical Composition, Antioxidant and Antimicrobial Activities of Basil (*Ocimum basilicum*) Essential Oils Depends on Seasonal Variations. *Food Chem.* **2008**, *108* (3), 986–995.

70. Hussain, A. I.; Anwar, F.; Chatha, S.; Jabbar, A.; Mahboob, S.; Nigam, P. *Rosmarinus officinalis* Essential Oil: Antiproliferative, Antioxidant and Antibacterial Activities. *Brazilian J. Microbiol.* **2010**, *41*, 1070–1078.

71. Hussain, A. I.; Anwar, F.; Nigam, P.; Ashraf, M.; Gilani, A. H. Seasonal Variation in Content, Chemical Composition and Antimicrobial and Cytotoxic Activities of Essential Oils from Four Mentha Species. *J. Sci. Food Agric.* **2010**, *90*, 1827–1836.

72. Hussain, A. I.; Chatha, S. A. S.; Noor, S.; Arshad, M. U.; Ali, Z.; Rathore, H. A.; Sattar, M. Z. A. Effect of Extraction Techniques and Solvent Systems for the Extraction of Antioxidant Components from Peas (*Arachis hypogaea* L.) hulls. *Food Analytical Methods* **2012**, *5,* 890896.

73. Hussain, A. I.; Rathore, H. A.; Sattar, M. Z. A.; Chatha, S. A. S.; Ahmad, F.; Ahmad, A.; Johns, E. J. Phenolic Profile and Antioxidant Activity of Various Extracts from *Citrullus clocnthis (L.)* from the Pakistan Flora. *Ind. Crops Prod.* **2013**, *45*, 416–422.

74. Hussain, I. A.; Anwar F.; Nigam, P.; Sarker, S. D.; Moore J. E.; Rao, J. R.; Mazumdar, A. Antibacterial Activity of Some *Lamiaceae* Essential Oils Using Resazurin as an Indicator of Cell Growth. *LWT-Food Sci. Technol.* **2011**, *44*, 1199–1206.

75. Iqbal, T.; Hussain, A. I.; Chatha, S. I. S.; Naqvi, S. A. R.; Bokhari, T. H. Antioxidant Activity and Volatile and Phenolic Profiles of Essential Oil and Different Extracts of Wild Mint (*Mentha longifolia*) from the Pakistani Flora. *J. Analytical Methods Chem.* **2013**, *1*, 1–6.

76. Ito, N.; Fukushima, S.; Tsuda, H. Carcinogenicity and Modification of the Carcinogenic Response by BHA and BHT and Other Anti-oxidants. *CRC Crit. Rev. Toxicol.* **1985**, *15* (2), 109−150.

77. Jacob, R. A. The Integrated Antioxidant System. *Nutr. Res.* **1995**, *15* (5), 755–766.

78. Javidnia, K.; Miri, R.; Safavi, F.; Azarpira, A.; Shafiee, A. Composition of the Essential Oil of *Nepeta persica Boiss* from Iran. *Flavor Fragrance J.* **2002**, *17*, 20–22.

79. Jiang, G.; Lin, S.; Wen, L. Identification of a Novel Phenolic Compound in Litchi (*Litchi chinensis* Sonn.) Pericarp and Bioactivity Evaluation. *Food Chem.* **2013**, *136* (2), 563–568.

80. Jones, S. B.; Jr.; Luchsinger, A. E. *Plant Systematics*. McGraw Hill Book Co.: New York, N.Y., 1986; p 214.

81. Joshi, N.; Shukla, A.; Nailwal, T. K. A Review of Taxonomic and Phytomedicine Properties of *Oroxylum indicum (L.)* Vent: A Wonderful Gift of Nature. *J. Med. Plant Res.* **2014,** *8* (38), 1148–1155.

82. Kaul, T. N.; Middletown, E.; Jr.; Ogra, P. L. Antiviral Effect of Flavonoids on Human Viruses. *J. Med. Virol.* **1985,** *15*, 71–79.

83. da Silva, J. K.; Andrade, E. H.; Kato, M. J.; Carreira, L. M.; Guimaraes, E. F.; Maria, L. Antioxidant Capacity and Larvicidal and Antifungal Activities of Essential Oils and Extracts from *Piper krukoffii. Nat. Prod. Comm.* **2011,** *6* (9), 1361–1366.

84. Khajeh, M.; Yamini, Y.; Shariati, S. H. Comparison of Essential Oils Compositions of *Nepeta persica* Obtained by Supercritical Carbon Dioxide Extraction and Steam Distillation Methods. *Food Bioprod. Proc.* **2010,** *8* (8), 227–232.

85. Khan, A.; Ahmad, A.; Akhtar, F.; Yousaf, S.; Xess, I.; Khan, L. *Ocimum Sanctum* Essential Oil and Its Active Principles Exert Their Antifungal by Disrupting Ergosterol Biosynthesis and Membrane Integrity. *Res. Microbiol.* **2010,** *161* (10), 816–823.

86. Khan, K. H. Role of *Emblica officinalis* in Medicine–A Review. *Botanical Res. Int.* **2009,** *2* (4), 218–228.

87. Kiran, B.; Raveesha, K. A. In vitro Evaluation of Antioxidant Potentiality of Seeds of *Psoralea corylifolia* L. *World Appl. Sci. J.* **2010,** *8* (8), 985–990.

88. Krishnaiah, D.; Sarbatly, R.; Bono, A. Phytochemical Antioxidants for Health and Medicine: A Move Towards Nature. *Biotechnol. Mol. Biol. Rev.* **2007,** *1*, 97–104.

89. Kubo, I.; Muroi, H. Himejima, M. Combination Effects of Antifungal Nagilactones Against *Candida albicans* and Two Other Fungi with Phenyl Propanoids. *J. Nat. Prod.* **1993,** *56*, 220–226.

90. Kumar, S.; Pandey, A. K. Review of 88 Chemistry and Biological Activities of Flavonoids. *Scientific World J.* **2013,** *1*, 1–16.

91. Li, W.; Liang, H.; Zhang, M. W.; Zhang, R. F.; Deng, Y. Y.; Wei, Z. C.; Zhang, Y.; Tang, X. J. Phenolic Profiles and Antioxidant Activity of Litchi (*Litchi Chinensis* Sonn.) Fruit Pericarp from Different Commercially Available Cultivars. *Molecules* **2012,** *17* (12), 14954–14967.

92. Lim, Y.; Lim, T.; Tee, J. Antioxidant Properties of Several Tropical Fruits: A Comparative Study. *Food Chem.* **2007,** *103*, 1003–1008.

93. Liu, X.; Cui, C.; Zhao, M.; Wang, J.; Luo, W.; Yang, B.; Jiang, Y. Identification of Phenolics in the Fruit of Emblica (*Phyllanthusemblica* L.) and their Antioxidant Activities. *Food Chem.* **2008,** *109*, 909–915.

94. Loffler, C.; Sahm, A.; Wray, V., Soluble Phenolic Constituents from *Cuscuta reflexa* and *Cuscuta platyloba. Biochem. Syst. Ecol.* **1995,** *23* (2), 121–128.

95. Lozano, P.; Castellar, M. R.; Simancas, M. J.; Iborra, J. L. Quantitative high-performance Liquid Chromatographic Method To Analyses Commercial Saffron (*Crocus sativus* L.) Products. *J. Chromatogr. A.* **1999,** *830* (2), 477–483.

96. Luqman, S.; Kumar, R. Correlation Between Scavenging Property and Antioxidant Activity in the Extracts of *Emblica officinalis* Gaertn, syn. *Phyllanthus emblica* L. Fruit. *Annals Phytomed.* **2012,** *1* (1), 54–61.

97. Mahboubi, M.; Kazempour, N.; Ghazian, F.; Taghizadeh, M. Chemical Composition, Antioxidant and Antimicrobial Activity of *Nepeta persica Boiss.* Essential oil. *Herba Polonica* **2011,** *57* (1), 63–71.

98. Maisuthisakul, P.; Gordon, M. H. Antioxidant and Tyrosinase Inhibitory Activity of Mango Seed Kernel by Product. *Food Chem.* **2009,** *117* (2), 332–341.

99. Masoko, P.; Eloff, J. N. Screening of Twenty-four South African *combretum* and Six *terminalia* Species (combretaceae) for Antioxidant Activities. *Afr. J. Tradit. Complement. Altern. Med.* **2007,** *4* (2), 231–239.

100. Mazzutti, I. S.; Ferreira, S.; Riehl, C. Supercritical FlUid Extraction of *Agaricus Brasiliensis*: Antioxidant and Antimicrobial Activities. *J. Supercrit. Fluids* **2012,** *70,* 48–56.

101. McMahon, J. B.; Currens, M. J.; Gulakowski, R. J.; Buckheit, R. W. J.; Lackman-Smith, C.; Hallock, Y. F.; Boyd, M. R. Michellamine-B, A Novel Plant Alkaloid, Inhibits Human Immunodeficiency Virus-induced Cell Killing by at Least Two Distinct Mechanisms. *Antimicrob. Agents Chemother.* **1995,** *39,* 484–488.

102. Misra, B. B.; Dey, S. Phytochemical Analyses and Evaluation of Antioxidant Efficacy of In Vitro Callus Extract of East Indian Sandalwood Tree (*Santalum album* L.). *J. Pharm. Phytochem.* **2012,** *1* (3), 1–16.

103. Moirangthem, D.; Talukdar, N.; Bora, U. Differential Effects of *Oroxylum indicum* Bark Extracts: Antioxidant, Antimicrobial, Cytotoxic and Apoptotic Study. *Cytotechnology* **2013,** *65* (1), 83–95.

104. Morandim-Giannetti, A. A.; Pin, A. R.; Pietro, N. A. S.; Oliveira, H. C.; Mendes-Giannini, M. J. S.; Alecio, A. C.; Kato, M. J.; Oliveira, J. E.; Furlan, M. Composition and Antifungal Activity Against *Candidaalbicans, Candida parapsilosis, Candida krusei* and*Cryptococcus neoformans* of Essential Oils from Leaves of *Piper* and *Peperomia* Species. *J. Med. Plants Res.* **2010,** *4* (17), 1810–1814.

105. Moreno, S.; Scheyer, T.; Romano, C. S; Vojnov, A. A. Antioxidant and Antimicrobial Activities of Rosemary Extracts Linked to their Polyphenolic Composition. *Free Radical Res.* **2006,** *40* (2), 223–231.

106. Murthy, K. N. C.; Jayaprakasha, G. K.; Singh, R. P. Studies on Antioxidant Activity of Pomegranate (*Punica granatum*) Peel Extract Using In Vivo Models. *J. Agric. Food Chem.* **2002,** *50* (17), 4791–4795.

107. Naik, S. R. Antioxidants and Their Role in Biological Functions: An Overview. *Indian Drugs* **2003,** *40* (9), 501–508.

108. Nostro, A.; Germano, M. P.; Dangelo, V.; Marino, A.; Cannatelli, M. A. Extraction Methods and Bioautography for Evaluation of Medicinal Plant Antimicrobial Activity. *Lett. Appl. Microbiol.* **2000,** *30,* 379–384.

109. Oktay, M.; Gulcin, I.; Kufrevioglu, O. I. Determination of In Vitro Antioxidant Activity of Fennel (*Foeniculum vulgare*) Seed Extracts. *LWT-Food Sci. Techol.* **2003,** *36,* 263–271.

110. Omulokoli, E.; Khan, B.; Chhabra, S. C. Antiplasmodial Activity Of Four Kenyan Medicinal Plants. *J. Ethnopharmacol.* **1997,** *56,* 133–137.

111. Osuntokun O. T.; Olajubu F. A. Comparative Study of Phytochemical and Proximate Analysis of Seven Nigerian Medicinal Plants. *Appl. Sci. Res. J.* **2014,** *2* (1), 10–26.

112. Pan, M. H.; Ghai, G.; Ho, C. T. Food Bioactives, Apoptosis, and Cancer. *Mol. Nutr. Food Res.* **2008,** *52,* 43–52.

113. Panda, S. K. Assay Guided Comparison for Enzymatic and Non-enzymatic Antioxidant Activities with Special Reference to Medicinal Plants. *INTECH* **2012,** *14,* 382–400.

114. Parekh, J.; Chanda, S. *In-vitro* Antimicrobial Activities of Extracts of *Launaea procumbens* Roxb. (Labiateae), *Vitis vinifera* L. (Vitaceae) and *Cyperus rotundus* L. (Cyperaceae). *Afr. J. Biomed. Res.* **2006**, *9*, 89–93.

115. Pengsuparp, T.; Cai, L.; Fong, H. H. S.; A. D.; Kinghorn, A. D.; Pezzuto, J. M.; Wani, M. C.; Wall, M. E. Pentacyclic Triterpenes Derived from *Maprounea africana* are Potent Inhibitors of HIV-1 Reverse Transcriptase. *J. Nat. Prod.* **1994**, *57*, 415–418.

116. Percival, M. Antioxidants. *Clin. Nutr. Insights* **1998**, *31* (10), 1–4.

117. Pereez, R. M.; Perez, J. A.; Garcia, L. M. D.; Sossa, H. M. Neuro Pharmacological Activity of *Solanum nigrum* Fruit. *J. Ethnopharmacol.* **1998**, *62*, 43–48.

118. Pham-Huy, L. A.; He H.; Pham-Huy, C. Free Radicals, Antioxidants in Disease and Health. *Int. J. Biomed. Sci.* **2008**, *4* (2), 89–96.

119. Piras, A.; Rosa, A.; Marongiu, B. Chemical Composition and In Vitro Bioactivity of the Volatile Oils of *Nigella sativa* L. Extracted by Supercritical Carbon Dioxide. *Ind. Crops Prod.* **2013**, *46*, 317–323.

120. Pitchaon, M. Antioxidant Capacity of Extracts and Fractions from Mango (*Mangifera indica* Linn.) Seed Kernels. *Int. Food Res. J.* **2011**, *18*, 523–528.

121. Runyoro, D.; Srivastava, S.; Darokar, M. Anticandidiasis Agents from Tanzanian Plant *Combretum zeyheri*. *Med. Chem. Res.* **2013**, *22* (3), 1258–1262.

122. Sakanaka, S.; Kim, M.; Taniguchi, M.; Yamamoto, T. Antibacterial Substances in Japanese Green Tea Extract Against *Streptococcus mutans*, A Cariogenic Bacterium. *Agric. Biol. Chem.* **1989**, *53*, 2307–2311.

123. Sakanaka, S.; Shimura, N.; Aizawa, M.; Kim, M.; Yamamoto T. Preventive Effect of Green Tea Polyphenols Against Dental Caries in Conventional Rats. *Biosci. Biotechnol. Biochem.* **1992**, *56*, 592–594.

124. Samojlik, I.; Lakic, N.; Mimica-Dukic, N.; Dakovic-Svajcer, K.; Bozin, B. Antioxidant and Hepatoprotective Potential of Essential Oils of Coriander (*Coriandrum sativum* L.) and Caraway (*Carum carvi* L.) (Apiaceae). *J. Agric. Food Chem.* **2010**, *58* (15), 8848–8853.

125. Sanchez-Burgos, J. A.; Larrosa, M. M.; Gallegos-Infante, J. A.; Gonzalez-Laredo, R. F.; Medina-Torres, L.; Rocha, N. E. Antioxidant, Antimicrobial, Antitopoisomerase and Gastroprotective Effect of Herbal Infusions from Four *Quercus species*. *Ind. Crops Prod.* **2013**, *42*, 57–62.

126. Sarrou, E.; Chatzopoulou, P.; Dimassi-Theriou, K.; Therios, I. Volatile Constituents and Antioxidant Activity of Peel, Flowers and Leaf Oils of *Citrus aurantium* L. Growing in Greece. *Molecules* **2013**, *18*, 10639–10647.

127. Sato, M.; Fujiwara, S.; Tsuchiya, H.; Fujii, T.; Iinuma, M.; Tosa, H.; Ohkawa, Y. Flavones with Antibacterial Activity Against Cariogenic Bacteria. *J. Ethnopharmacol.* **1996**, *54*, 171–176.

128. Scalbert, A. Antimicrobial Properties of Tannins. *Phytochemistry* **1991**, *30*, 3875–3883.

129. Scalbert, A.; Johnson, I. T.; Saltmarsh, M. Polyphenols: Antioxidants and Beyond. *Am. J. Clin. Nutr.* **2005**, *8*, 215–217.

130. Schmidt, H. *Phenol Oxidase (Ed 14.18.), A Marker Enzyme for Defense Cells: Progress tn Histochemistry and Cytochemistry*. Gustav Fischer: New York, 1988, p. 23.

131. Schultz, J. C. Tannin-insect Interactions. In: *Chemistry and Significance of Condensed Tannins*; Hemingway, R. W., Karchesy, J. J., Eds; Plenum Press: New York, 1988, p. 553.

132. Sen, S.; Chakraborty, R.; Sridhar, C.; Reddy, Y. S. R. De, B. Free Radicals, Antioxidants, Diseases and Phytomedicines: Current Status and Future Prospect. *Int. J. Pharm. Sci. Rev. Res.* **2010,** *3* (1), 91–100.

133. Serafini, M.; Ghiselli, A.; Ferro-Luzzi, A. Red Wine, Tea and Anti-oxidants. *Lancet* **1994,** *8,* 344–626.

134. Sethi, M. L. Inhibition of Reverse Transcriptase Activity by Benzo Phenanthridine Alkaloids. *J. Nat. Prod.* **1979,** *42,* 187–196.

135. Sethi, S.; Prakash, O.; Chandra, M.; Punetha, H.; Pant, A. K. Antifungal Activity of Essential Oils of Some *Ocimum* Species Collected from Different Locations of Uttarakhand. *Ind. J. Nat. Prod. Res.* **2013,** *4* (4), 392–397.

136. Shafaghat, A. Free Radical Scavenging and Antibacterial Activities, and Gc/Ms Analysis of Essential Oils from Different Parts of *Falcaria vulgaris* from Two Regions. *Nat. Prod. Comm.* **2010,** *5* (6), 981–984.

137. Shafaghat, A.; Oji, K. Nepeta Lactone Content and Antibacterial Activity of the Essential Oils from Different Parts of *Nepeta persica. Nat. Prod. Comm.* **2010,** *5* (4), 625–628.

138. Shi, B.; Liu, W.; Wei, S. P.; Wu, W. J. Chemical Composition, Antibacterial and Antioxidant Activity of the Essential Oil of *Bupleurum longiratum. Nat. Prod. Comm.* **2011,** *5* (7), 1139–1142.

139. Singh, B.; Singh, S. Chemical Investigation of Seed of *Ipomoea hederacea* and its Biological Activity. *J. Chem. Pharm. Res.* **2012,** *4* (2), 1441–1448.

140. Soong, Y.; Barlow, P. J. Antioxidant Activity and Phenolic Content of Selected Fruit Seeds. *Food Chem.* **2004,** *88,* 411–417.

141. Stern, J. L.; Hagerman, A. E.; Steinberg, P. D.; Mason, P. K. Phlorotannin-protein Interactions. *J. Chem. Ecol.* **1996,** *22,* 1887–1899.

142. Takashima, M.; Nara, K.; Niki, E. Evaluation of Biological Activities of a Groundnut (*Apios Americana* Medik) Extract Containing a Novel Isoflvone. *Food Chem.* **2013,** *138* (1), 298–305.

143. Thangaraj, S. M. S.; Thangaraj, I. M. K. S.; Palaniswamy, M. Antioxidant and Antimicrobial Activity of Different Parts of *Solanum nigrum Linn. J. Pharm. Res.* **2012,** *5* (4), 2082–2086.

144. Tian, M.; Yan, H.; Row, K. H. Extraction of Glycyrrhizic Acid and Glabridin from Licorice. *Int. J. Mol. Sci.* **2008,** *9,* 571–577.

145. Tsuchiya, H.; Sato, M.; Iinuma, M.; Yokoyama, J. Inhibition of the Growth of Cariogenic Bacteria In Vitro by Plant Flavanones. *Experientia* **1994,** *50,* 846–849.

146. Tsuchiya, H.; Sato, M.; Miyazaki, T.; Fujiwara, S.; Tanigaki, S.; M. Ohyama, M.; Tanaka, T.; Iinuma, M. Comparative Study on the Antibacterial Activity of Phytochemical Flavanones Against Methicillin-resistant *Staphylococcus aureus. J. Ethnopharmacol.* **1996,** *50,* 27–34.

147. Vijaya, K., Ananthan, S.; Nalini, R. Antibacterial Effect of Theaflavin, Polyphenon 60 (*Camellia sinensis*) and *Euphorbia hirta* on *Shigella* spp.—A Cell Culture Study. *J. Ethnopharmacol.* **1995,** *49,* 115–118.

148. Wild, R., Ed. *The Complete Book of Natural and Medicinal Cures.* Rodale Press, Inc.: Emmaus, PA, 1994, p. 314.

149. Yadav, S. B.; Tripathi, V.; Singh, R. K.; Pandey, H. P. Antioxidant Activity of *Cuscuta reflexa* Stems. *Ind. J. Pharm. Sci.* **2000,** *62* (6), 477–480.

150. Yen, G. C.; Chang, Y. C.; Su, S. W. Anti-oxidant Activity and Active Compounds Of Rice Koji Fermented with *Aspegillus candidus*. *Food Chem.* **2003**, *83*, 49–54.

151. Zaouali, Y.; Bouzaine, T.; Boussaid, M. Essential Oils Composition in Two Rosmarinus Officinalis L. Varie and Incidence for Antimicrobial and Antioxidant Activities. *Food Chem. Toxicol.* **2010**, *28* (11), 3144–3152.

152. Zhang, C. R.; Schutzki, R.; Nair, M. Antioxidant and Anti-inflammatory Compounds in the Popular Landscape Plant *Berberis thunbergii* var. *atropurpurea*. *Nat. Prod. Comm.* **2013**, *8* (2), 165–168.

153. Zhang, L.; Tu, Z.; Yuan, T.; Wang, H.; Xie, X.; Fu, Z. Antioxidants and α-glucosidase Inhibitors from Ipomoea Batatas Leaves Identified by Bioassay-guided Approach and Structure-activity Relationships. *Food Chem.* **2016**, *208*, 61–67.

154. Zhao, H. Y.; Yang, L.; Wei, J. Bioactivity Evaluations of Ingredients Extracted from the Flowers of *Citrus aurantium* L. var. *amara* Engl. *Food Chem.* **2012**, *135* (4), 2175–2181.

155. Zheng, C. J.; Li, L.; Ma, W. H.; Han, T.; Qin, L. P. Chemical Constituents and Bioactivities of the Liposoluble Fraction from Different Medicinal Parts of *Crocus sativus*. *Pharm. Biol.* **2011**, *49* (7), 756–763.

156. Zheng, Y. Q.; Liu, J. X.; Wang, J. N.; Xu, L. Effects of Crocin on Reperfusion-induced Oxidative/Nitrative Injury to Cerebral Micro Vessels after Global Cerebral Ischemia. *Brain Res.* **2007**, *23*, 86–94.

157. Zia-Ul-Haq, M.; Ahmad, S.; Calani, L.; Mazzeo, T.; Rio, D. D.; Pellegrini, N.; Feo, V. D. Compositional Study and Antioxidant Potential of *Ipomoea hederacea* Jacq. and *Lepidium sativum* L. Seeds. *Molecules* **2012**, *17*, 10306–10321.

CHAPTER 9

POTENTIAL SIGNIFICANCE OF PROTEASES: SOURCES AND APPLICATIONS

MARWA WAHEED, MUHAMMAD BILAL HUSSAIN,
SADIA HASSAN, MOHAMMAD ALI SHARIATI, and
OLUWAFEMI ADELEKE OJO

ABSTRACT

Food from plant origin is a rich source of beneficial components, such as phytochemicals, antioxidants, plant hormones, and enzymes. This study is based on the view point of various sources of proteases specifically plant-derived proteases and their potential applications. Proteases play an important role in different industries, particularly in food, pharmaceuticals, digestive diseases, and detergent industries, diagnostic and fine chemical industries. Interest has been growing in plant proteases that are natural, eco-friendly and have commercial importance as compared with animal or microbial proteases. It is clear that these enzymes, as compared to other sources can break more fat, protein, and carbohydrate in the broader range of pH and temperature. Protease as coagulating enzyme in dairy industry is considered a significant source for the preparation of cheese. In addition to this, an overview on the application of proteases in meat tenderizing, bread making, beverage (as clarifying agent), and leather industry is also addressed briefly.

9.1 INTRODUCTION

Plants are affluent source of nutritional components with health benefits for consumers. They are loaded with many secondary metabolites. Thus, plants are said to be the potential source of innovative drugs to control diseases in humans.[8] Enzymes, hormones, and phytochemicals are the beneficial ingredients, which are derived from plants, cost effective and easily available throughout the year. Vitamins, nitrogenous, and phenolic compounds have elevated antioxidant activity. They reveal actions against tumor formation, bacterial growth reduction, antiinflammatory, anticancer, antiatherosclerotic, antiviral, and antimutagenic actions. Thus, phytochemicals as plant derived components are helpful for disease protection.[10] Extracts and oils of plants have found their application in natural therapies, preservation of whole and processed food, pharmaceuticals, and alternative medicine.[39] Therefore, medicinal plants are of interest to scientific communities and pharmaceutical industries due to their antimicrobial prospective.[39]

Essential oils from plants, like aromatic herbal plants (i.e., fennel, lavender, peppermint, and thyme) as well as volatile compounds (i.e., phenylpropanoids and monoterpenes) had been recounted because of their action on viruses, yeast, fungi, Gram-positive, and Gram-negative bacteria.[60] Essential oil mixture components act as exopolysaccharide synthesis, bacterial adherence, as well as bacterial growth inhibitors.[14] Plant extracts having high ratio of protein hydrolyzing enzymes are utilized in usual medicine for an extended period of time.[58] Natural sources, such as pineapple, *Aspergillus* fungi, and papaya are used as diet relating enzyme supplements.[25] Majority of dietary supplement enzymes are derived from fungi and are also refined or purified particularly from the same source. The enzymes extracted from plant source are regulated as food by Food and Drug Administration (FDA), and due to this reason, these do not need any kind of prescription.[28]

This chapter presents an overview of various sources of plant derived proteases and their potential applications in meat tenderizing, bread making, beverage (as clarifying agent), and leather industry.

9.2　ACTIVE COMPONENTS OF PLANTS

9.2.1　PHYTOCHEMICALS

Biologically active compounds offer good health benefits for humans as evident from micro- and macronutrients.[72] They act as plant disease inhibitors as well as impart sensory characteristics to plants. Phytochemicals are considered as secondary metabolites, which exhibit significant biological activities, for example, antimicrobial, antioxidant, decrease in platelet aggregation, immune system stimulation, anticancer activity, and detoxification of enzyme modulation. Phytochemicals not only protect plants against physical damage but are also beneficial for the treatment of human diseases. They are defined as antioxidants because of redox-active molecules. Removal of nitrogenous species, reactive oxygen, and free radicals can be done by antioxidants.[87]

Phytochemicals are biochemicals found in nature, which give texture, smell, flavor, and color to the plants and can help to avert diseases. They are natural products, which activate catechins in tea, limonoids in citrus fruit, lycopene, lignans, and glucosinolates in cruciferous vegetables, and so on. Recently, phytochemicals with anticancer activities are been given more attention by the researchers.[66]

Flavonoids are active plant compounds, which demonstrate their anti-cancerous property through their action as antioxidants. They are sourced from soy foods in the form of isoflavones, turmeric rich in curcumin compound and green tea rich in epigallocatechin gallate, as well as from citrus fruits. Lycopene and carotenoids are also suggested as imperative substances for human health. Lycopene is believed to have an antioxidant activity and is a quencher of reactive oxygen species (ROS) because of its unsaturated temperament.

β-carotene is the most active antioxidant among the carotenes. It is among the most general type of carotenoids and can be sourced from yellow, green, and orange fruits and vegetables.[23] Tannin is another phytochemical, also known as proanthocyanidins. It causes detoxification of carcinogens and is a scavenger of free radicals that are harmful. Ellagic acid prevents cancer and is utilized in medicine. It is found in red raspberry seeds, cranberries, pomegranates, strawberries, pecans, and walnuts.[84]

9.2.2 HORMONES

There are three important types of plant hormones, for example, auxin, gibberellins, and abscisic acid (ABA). Auxin (indole acetic acid) promotes growth and stem elongation in plants. Other advantages are adventitious and lateral roots formation, leaf loss inhibition, cell division, increased production of ethylene, and which enforce lateral bud's dormancy, which is produced by other immature parts and shoot apical meristems.

It is responsible for phototropism. Auxin causes cell elongation on the unlighted side of shoot by making cell walls softer and expansion of the cell cytoplasm.[33]

Gibberellins hormone controls the movement of soluble sugars from starch in the cereals and transfer hereditary dwarfs of rice, pea, and corn in mature plants phenotypically.[80] ABA sustains in the control of stomata closing and adaptation to different stresses. They cause commencement of undeveloped structure, for example, seeds and winter buds in higher plants species and deciduous trees, respectively. Amalgamation of storage proteins and seed maturation is also an important aspect of ABA. It is appropriate to mention that ethylene demonstrates inhibitory responses related to the ripening of fruits.[75]

9.2.3 ENZYMES

Enzymes are protein in nature that accelerate the biochemical reactions as per life processes, which are essential, including respiration, digestion, maintenance of tissues, and metabolism. They have an ability to remove or add atoms to a molecule, join together smaller molecules, and split a large molecule into smaller ones. In other words, they are highly specific biological catalysts. Enzymes have the ability to work under more or less mild conditions and they become superlative or ideal catalyst to be used in food technology.[15] Enzymes catalyze many biochemical reactions needed to sustain the life and present in all living organisms. They are metabolized and broken down after intake similar to other proteins of diet. Enzymes are believed to be naturally safe and considered as nontoxic.[17]

Enzymes are important and are also crucial for the production of food that we eat, the clothes we wear, and even production of fuel for automobiles. The number of plant-based enzymes that are industrially employed is growing fast but the number is still small. Enzymes can be extracted from any living organism. Today, the enzyme industry is the result of a quick improvement, observed mainly in the previous four decades. Natural enzymes have been utilized in food products like sour dough, beer, as well as manufacturing of goods, for example, linen, indigo, and leather from prehistoric times.[82]

Enzyme applications are vary from texture to flavor in the food industry. In the enzyme optimization, numerous developments have taken place. Therefore, by using the recombinant production of protein, well organized mono-component enzymes are provided without any possible harmful side effect.[73] Recently, a lot of work have been conducted on the transglutaminase application as an agent that stabilizes the formation and texture of yoghurt, noodles, and sausages.[56]

9.2.4 SOURCES OF ENZYMES

9.2.4.1 MAMMALS AND MARINE ANIMALS

Due to Halal and Haram issue, animal-based enzymes are prohibited for utilization in food industries. They are rennet, proteases, pepsin, trypsin, and chymotrypsin. Fish-processing wastes are produced globally in large amount. Bones, frames, viscera, head, and waste water from fish processing constitute around 70% of the total mass of aquatic animals. Among different byproducts, fish viscera are potential sources of enzymes.[48]

In recent era, characterization and recovery of enzymes from marine animals had been accomplished, that further led to the occurrence of enzymes during food processing. The retrieval of proteinases from by-products of fish has great importance because of the cost effectiveness that could be able to introduce new industrial applications. Removal of enzymes from fish-processing waste and their utilization in food industry can extensively decrease the contamination problem of local society. Protein hydrolyzing enzymes are mostly created by marine animal's digestive glands.

Marine animals have beneficial digestive proteinases that accelerate the removal of proteins and are similar to the proteinases from microorganisms, animals, and plants. Marine animals had been adapted to diverse ecological condition, and these adaptations together with inter- and intra-species hereditary variation are related to certain exclusive properties of their proteases.[86]

Usually, during the hydrolysis of proteins, pepsin is considered as an important acid protease that is utilized further in the extraction of collagen, removal of gelatin as a rennet replacement, and in the treatment of digestibility.[90] Pepsin had been investigated and isolated from different mammals including bovine, goat, rat and rabbit, human, and Japanese monkey. Similarly, pepsin can also be taken out from waste products of fish, especially intestines, which demonstrate their share as 5% of the total fish weight.[89]

Animal (bovine and calf) rennet contains pepsin and chymosin, usually in a ratio of 9:1. It is then commercially used in the manufacturing of cheese. The high cost of animal-derived proteases forced the industry to discover the new promising substitutes, such as fish pepsin. Pepsin from cod has an ability to coagulate the milk than chymosin from calf, and cheese manufactured from cod pepsin exhibited good flavor characteristics after being tested by the sensory board. Studies have found that pepsin from cod has lesser temperature coefficient for milk coagulation than that of calf rennet.[91]

9.2.4.2 MICROBES

Enzymes from microbial sources are known to be superior, particularly for their usages in industries. Microbial enzymes were previously used in different profitable processes.[57] Preferred microorganisms, together with yeasts, fungi, and bacteria have been internationally considered in enzyme preparations, which are synthesized biologically and considered economically feasible.[37] Microorganisms discharge huge concentration of enzymes in the culture media, which belong to the genus *Bacillus*. Various species of *Bacillus* produce protease during the exponential growth period. While other species of it produce extracellular proteases through stationary growth phases.[16] Proteases are highly significant enzymes in pharmaceutical, food, detergent, leather, silver recovery, and waste management systems.[36]

Fungus protease elaborates a wide range of enzymes as compared to bacteria. *Aspergillus oryzae* produces neutral, alkaline, and acid proteases. Fungal proteases are vigorous above large pH variety and exhibit broad substrate specificity. Proteases from fungi have the best possible pH involving 4 and 4.5, and are constant from pH 2.5 to 6.0. They are also used in the modification of food protein. Thermophilic bacteria are good source of thermostable industrialized enzymes and moreover, they show an advanced grade of confrontation to protein denaturing factors for example, pH, detergents, and organic solvents when compared with corresponding mesophilic enzymes.[83] Proteases have been utilized in food processing for centuries and any proof of the discovery of their activities had been misplaced in the mist of time. Rennet obtained from the stomach of unweaned calves had been used conventionally in the production of cheese.[64]

Alkaline microbial proteases are crucial, owing to their stability and activity in severe environmental conditions: high pH, presence of surfactants, and temperature ranging from 50°C to 90°C.[6] Another type of microbial proteases is serine alkaline proteases that are common in bacteria and fungi and owned their name as serine because of the presence of serine essential amino acid at the active site. These are endoproteases and have specificity in substrate selection with the molecular weight between 20 and 50 D.[3]

Lipid digesting enzymes (lipases) are extensively present in nature, but these enzymes from microbial sources are important because of their easy availability with wider stability and low cost of production than animal and plant lipases. They might be originating from molds, bacteria, or fungi and most of them are produced extracellularly. Generally, lipases do not require cofactors nor do they catalyze the side reactions.[9]

Most industrial microbial lipases result from bacteria and fungi. Some lipase producing microorganisms, as well as their utilization in various food processes were testified. *A. oryzae* used for the synthesis of saturated fatty acids, *Bacillus subtilis* for bread making, and *Candida rugosa* used for cheese production. *Rhizomucor meihei* surfactants are considered to be utilized for baking, dairy industry, and noodles whereas, *Geotrichum* sp. had food applications in bread making.[68]

9.2.4.3 PLANTS

Plant enzymes are recognized as a group of biochemical accelerators that catalyze a huge number of chemical reactions and are profitably used in food, detergents, diagnostic industries, fine chemical industries, and pharmaceutical industries.[73] These are the most popular choices for enzymes. The most common among the plant enzymes are amylase, cellulase, bromelain, peroxidase, papain, and protease, which are present in starch, cellulosic material, pineapple, horse radish, meat, and papaya, respectively. Enzymes from plant sources are generally acid resistant and initiate their action in the abdomen. They are used in food processing industries, baking, oil extraction, detergent, cosmetics, textile, and leather industries.[47]

Amylases are starch hydrolyzing plant enzymes. Starch is a substrate that can be obtained for commercial utilization from the roots and tubers of plants, for example, potato, cassava, pith of sago palm, arrowroot, and seeds of plants, such as wheat, corn, rice, or sorghum. Corn is the most important commercialized source of thickener or starch, from where starch compounds can be extracted by wet milling procedure. Amylase enzymes are currently being used for different purposes. The most important share of starch-processing industry can be justified by the transformation of starch into dextrin, syrup, and sugar. Hydrolysates of starch compounds are being utilized in fermentation as a source of carbon along with sweetness source in a variety of synthetic beverages and food products.[76]

Cellulose digesting enzymes from plant sources were used extensively in beverage and food industries. Additionally, they also have extensive series of applications in food biotechnology. Cellulases ensure their significant utilization as a part of complex enzyme system, for example, cellulases, xylanases, and pectinases, which are being used for clarification of fruit and vegetable juices. Consumption of cellulases could reduce the dependency on fossil fuels via offering a renewable and appropriate source of energy in the form of glucose and their utilization could also solve modern waste disposal problems.[43]

Proteases were recognized from the sap or latex of various families of plants, for example, apocynaceae, asteraceae, moraceae, euphorbiaceae, and caricaceae. Most of the plant proteases have been categorized as cysteine protease and aspartic proteases. Numerous industrialized processes include protein breakdown by proteases and some of the proteases are obtained from plants. Protein modification by enzymatic proteolysis can improve functional properties of protein isolates and also used in the production of special food, which are destined for children, the old people, or sportsmen. Medicinal seeds were also reported for their proteolytic activity and these seeds are used to treat different diseases. Proteases have been used in the processing of rennet or chymosin for cheese making where, chymosin breaks the specific peptide bond.[79]

9.3 PROPERTIES AND APPLICATIONS OF DIFFERENT PLANT-BASED ENZYMES

9.3.1 PEROXIDASE

It is the most common enzyme having a place in the oxidoreductase class of enzyme and usually acts as a catalyst during the reaction between hydrogen peroxide (H_2O_2) and various substrates through the release of O_2 from H_2O_2.[92] Rich sources of peroxidase enzymes are plants, especially sprouts and roots of higher plants that include beet root, horse radish, and potato tuber. They are generally used in pharmaceutical industry and food industry, as well as biosensor construction.[52] Some new applications of peroxidases consist of treatment of wastewater, production of different aromatic chemicals, and elimination of peroxide from industrial wastes and foodstuffs.[2] Plant peroxidases have potentials for biopulping, bioremediation of chlorinated phenols and cresols, biobleaching in paper manufacturing industry and for degradation of fabric dye.[29]

9.3.2 BROMELAIN

Bromelain is a protein hydrolyzing enzyme and is present in the stems and immature fruit of pineapple as a crude and aqueous extract. Stem bromelain had been utilized in different processes of food industries, for example, baking processes, tenderization of meat, beer clarification, prevention from fruit juice browning, and production of hydrolysates of protein. Likewise, bromelain has been used for softening and pretanning of skin in leather industries. It is used for the treatment of acute inflammation and sports injuries and is also easily available for the pharmacies of USA and Europe as well as for food stores concerned with public health.[18] This plant enzyme is also used for treating few malignant diseases, inflammatory, and blood coagulation related diseases.[81]

9.3.3 PAPAIN

It is an endolytic cysteine protease enzyme which can be extracted or removed from latex of papaya (*Carica papaya* L.) plant through cutting of the peel of the young green fruit. The greener is the fruit, more active is the papain and it is a crucial proteolytic enzyme that can be used in a lot of dynamic natural processes.[7] Papain is used extensively in meat tenderizers and considered as a common constituent in meat industry and brewery.[46]

It has been demonstrated from previous research studies that papain is used to improve the stretchability and meltability of Nabulsi cheese, as well as gives fibrous structure in pastries, pizza, and kunafa.[1] Even as a protein digesting enzyme, papain is used for fighting against digestive disorders, dyspepsia, and disturbance of the gastrointestinal tract.[40] These enzymes represent anti-inflammatory, antibacterial, and antifungal properties.[26]

9.3.4 PROTEASE

Peptide bonds hydrolyze by protease enzymes. An exopeptidase is any peptidase that catalyzes the cleavage of the terminal peptide bond; the process releases a single amino acid or dipeptide from the peptide chain. There are five classes of endoproteases such as aspartic, serine, threonine, metallo, and cysteine. Protease sources are leguminous flour, flowers of *Cynara cardunculus*, papaya, moringa flowers, ginger, and pineapple.[67] The action of proteolytic enzymes is essential in the cell division, blood clotting cascade, digestion of food proteins, signal transduction, and apoptosis.[71]

These are extensively utilized in food, pharmaceutical, leather tanning, and detergent industries.[35] The massive diversity of proteases along with their specific action had engrossed the consideration to take improvement of their biotechnological applications, as well as physiological applications.[41] Proteases have been considered as eco-friendly because the enzymes produced for industrial application are suggested as nonpathogenic, nontoxic, and further recognized as safe.[36]

9.4   SIGNIFICANCE OF PLANT-BASED PROTEASE

Plant derived enzymes are most popular choice and among these, protease are the major constituent of these enzymes.[65] Proteases, like cardosins from *Cynara* sp., ficin from *Ficus* sp., and papain from *C. papaya* are sometimes constituents of sap, roots, seeds, latex, and fruits, but largely of their flowers or leave. Proteases from plant sources take part in numerous biological processes, such as hydrolysis of chloroplast proteins that are damaged by light, flower senescence, and storage protein mobilization. These enzymes are generally acid resistant and instigate their action in the abdomen.

Natural sources, such as pineapple and papaya are used as diet relating enzyme supplements.[25] A plant-based alternate would be of great importance in view of its essentially unlimited availability. Proteases from plants have been utilized in clotting of milk for cheese manufacturing. These milk clotting agents are suggested as the substitute of calf rennet owing to the high cost, infrequent availability, spiritual factors, and prohibition on the recombinant calf rennet in various nations. The universal increase in cheese consumption along with ethical concerns and shortage of rennet has established a general significance for milk clotting plant proteases.[4]

Therefore, the new enzyme search from other sources is still continued to make it lucrative and appropriate to industry.[88]

Several factors (vegetarianism and religious beliefs) are important for getting attention to the use of microbial coagulants and coagulants extracted from plants.[21] The use of plant enzymes in cheese making are beneficial because they are natural, cheap, easy to prepare, allow undemanding process, and used to produce cheese for ecological markets. They suggest a better opportunity for people, whose eating habits, health, biotechnological level, law, and financial situation restrain the use of animal and microbial enzymes. However, many plant extracts have proven unsuitable for cheese making due to their large proteolytic action. Commercially, plant-based proteolytic enzymes are important because of their action in broad array of temperature, pH, and specificity of substrate. The use of plant coagulants is associated with many advantages as they are natural, cheap, easy to prepare, allow straightforward process, and are used to produce industrial products for ecological markets.[85]

9.5 NATURAL SOURCES OF PROTEASE

9.5.1 PAPAYA

Reproductive ability, fast growth, high-photosynthetic rate of plant leaves, low construction cost of hollow twigs, and high yield of seeds and fruits illustrate many benefits of *C. papaya* L., the pioneer plant of tropical climate. Adult papaya plants can have the ability to sustain dozens of fruits, flowering, and vegetative growth at different developmental stages altogether. Owing to the nutritional value, high yield, perennial fruit production, and functional properties of this fruit plant are undeniable.[19]

Papaya has a high nutritional value, particularly when the fruitlet reaches at its ripening stage. Protein, fat, and crude fiber are more in unripe fruit of papaya than the ripe fruit whereas, carbohydrates, energy, and carotenes are comparatively more than in unripe green fruit. The fruits have low calories and rich concentration of minerals and vitamins. Papaya fruit is significant because of the rich amount of fiber, vitamin C, folate, vitamin A, calcium, riboflavin, niacin, iron, potassium, and thiamine. Likewise, papaya has the highest position among per serving value of potassium, carotenoids, ascorbic acid, and fiber content than other fruits.

The ascorbic acid content of papaya is 108 mg/100 g of fresh fruit as compared to orange, that is, 67 mg/100 g fruit.[49]

Owing to the nutritional significance, serotonin content, flavor compounds, and good digestive properties, papaya is valued globally.[31] Serotonin content (0.99 mg/100 mg) in papaya enables the circulatory system to facilitate the activity of impulse and to lessen the risk of blood clotting inside the vessels. Papaya with fermentation has auspicious value of nutraceuticals that play an important role as antioxidants for the patients having deficiency of antioxidants.

Ripened or mature fruit with dry skin is a dietary source for the consumption of chickens. Papain and chymopapain are two major enzymes, which have the ability to diminish the risk of drugs.[55] It has been revealed that proteases extracted from the fruit plant latex vary greatly from the nonfruit parts, as well as from the newly cut or wounded fruits of the papaya plant. A chain of proteins with low-molecular weight are being obtained from freshly wounded fruitlets, and continuous cut or injury of fruits activate more than a few enzymes comprising caricain, chymopapain, and papain.[12]

Papaya proteases are used commonly in the process of cell isolation because they have been proved less destructive with more proficiency for certain tissues as compared with other proteases. Therefore, papaya leaves used for traditional wrapping of meat, act as tenderizer, is not a surprising fact. Other usages have been reported, such as milk coagulation, chewing-gum manufacturing, beer clarification, cosmetic industry, and in the pharmaceutical industry as a medicine for digestive system.[24] Papain-based gel has stated its benefits for being utilized in chemo-mechanical removal of dental caries.[53,61]

Some extracts of papaya have been utilized traditionally for the treatment of poison or infection by intestinal nematodes that are reported dangerous for livestock farming and humans as well. In another study, scientists appraised the efficiency of this routine method, beside the nematodes, due to the presence of papain enzyme that causes damage to nematode cuticles.[40] Pectin in papaya fruit acts as a gelling agent in confectionary in a manner that it upturns the intestinal viscosity that causes the reduction in the absorption of cholesterol from food or bile and afterwards reduces the serum cholesterol concentration. Intestinal microorganisms cut down pectin and give short chain fatty acids as by-products that are considered beneficial for intestinal health.[78]

9.5.2 MORINGA

Globally, *Moringa oleifera* can be used for food and has some advantageous properties. It is a rich source of nutrients, amino acids, antioxidants, and also has anticancer and antiinflammatory properties.[77] Moringa could be an enormously important source of food, as it is an extremely good source of sulfur containing amino acids like cystine and methionine, minerals (particularly iron), and vitamins A, B, and C. It can be used in instant noodles, spices, bread, fortifying sauces, juices, and milk.[54]

Leaves of moringa are economical source of proteins, vitamins, and minerals for developing countries. Moringa leaves contain vitamin C more than seven times than oranges, potassium more than three times than bananas, iron more than three times than that of spinach, protein more than two times than milk, and calcium equal to four times of milk.[42] The leaves and pods of moringa have high quantity of Na, Zn, Cu, P, Ca, Fe, K, Mg, and Mn. Nearly all plant parts are used ethnically for their medicinal properties, flavor nutritional value, and taste and also as a seed and vegetable. Epidemiological studies have showed that leaves of moringa reveal antiulcer, anticonvulsant, antiinflammatory, antiatherosclerotic, and antitumor activities.[27] Moringa leaves are also added to the feed of animals, which endure healthy livestock. Moringa powder can be used in aqua cultural systems as a food of fish and also for animals as a supplement of protein. Lactating mothers and pregnant women use the leaves' powder to increase the nourishment of their children, particularly in the countries facing malnutrition.[30]

Moringa seeds have considerable amount of edible oil that can be used as cooking oil for the purpose of frying and as salad oil for salad dressing. The fatty acid composition of solvent or enzyme extracted oil from moringa seeds exhibited 70.0% in enzyme extract, and 67.9% oleic acid in solvent extract. Other conspicuous fatty acid compounds in moringa oil are acid palmitic (7.8%), behenic (6.2%), and stearic (7.6%). Because of the high ratio of monounsaturated to saturated fatty acids, seed oil from moringa will be considered as a proper substitute for monounsaturated oil. Moringa is entitled as "miracle plant," this is the only fact that all the parts of moringa plant can be utilized as feed, food, medicine, livestock, biogas, fodder, bio pesticide, and water purification agent.[34]

Numerous scientists have isolated proteases from various plant leaves, such as *M. oleifera*, *Coriandrum sativum*, *Murraya koenigii*, and *Nicotiana tobaccum*. Protein concentration was analyzed after partial purification of these plant leaves and concentration of protein was more in *M. oleifera*.[74] In another study, the milk coagulating detected and caseinolytic activities in flowers of *M. oliefera* using substrate skim milk and azocasein, respectively.[63] Aqueous extract of *M. oliefera* roots and leaves showed casienolytic activities.[70] According to some researchers, precipitated protein fraction from flowers of *M. oleifera* is potentially a helpful tool in the processes of cheese manufacturing, as more breakdown of α- and β-caseins had not occurred.[20]

9.5.3 PINEAPPLE

Protein digesting enzymes of pineapple plant are known as bromelain. Economic importance of pineapple protease is associated with the replacement of trypsin and pepsin in pancreatic treatment, manufacturing of pharmaceuticals, and their results on the digestive system. Bromelain is also used in the cure of surgical trauma, heart diseases, and rheumatoid arthritis. It also provides an antiinflammatory effect and reduces edema.[45] Moreover, bromelain has a wide-ranging applications in food industry, for example, beer clarification, cheese manufacturing, preparation of dietetic and infant foods, softening of meat, in the production of detergents, and also in textile industries.[32]

9.5.4 GINGER

Protease from ginger can be used in coagulation of milk that gives a traditional Chinese milk product and being suggested as a source of rennet substitute that could be further utilized in dairy products. Ginger rhizomes exist in Southeast Asia, Africa, India, West Indies, and tropical regions of Asia. Rhizome contains protein hydrolyzing enzymes, which can be divided through a diethylaminoethyl-cellulose (DEAE-cellulose) column, into two segments, which have their molecular weight of around 22.5 kDa.[38]

9.5.5 PANEERBOOTI (VEGETABLE RENNET)

Withania coagulans (paneerbooti) is generally called as "vegetable rennet" or "Indian cheese maker," as leaves and fruits of the plant are utilized for clotting of milk. Various studies have reported that it is possible to prepare soft cheese, using rennet preparations from *W. coagulans* as protease source.[59] Cheddar cheeses are manufactured using rennet preparation from this source had a good texture and flavor but gave a perceptible bitter taste, which nevertheless could be reduced to some extent, by prolonging the ripening period. Milk clotting activity of fruits is accredited to the husk berries and pulp containing Withanin enzyme with coagulating property. Other main components except the milk clotting enzyme in berries are, free amino acids, alkaloids, and essential oils.[13]

9.6 INDUSTRIAL APPLICATIONS OF PROTEASES

Study of proteases is extremely valid by their main function in numerous disciplines of industry. The universal market of industrial enzymes was considered to reach in 2010, at a value of $3.3 billion and proteases are related to the biggest section of this market.[69]

9.6.1 DAIRY INDUSTRY

Application of proteases in the dairy industry includes cheese manufacturing. Worldwide unavailability of calf rennet because of the increasing demand of cheese manufacturing has increased the exploration for alternate milk coagulants from plant and microbial sources. In cheese making, the key role of proteases is to break or hydrolyze the particular peptide bond that generates para-κ-casein and macropeptides. In West African, Southern European, and Mediterranean countries, plant coagulants are used for cheese making. West African countries are using *Calotropis procera* commonly known as (sodom apple) for the manufacturing of traditional cheese.[71]

9.6.2 MEAT TENDERIZATION

Softness of meat is considered as the most important aspect of its quality. Meat toughness depends mainly on the activity of endogenous proteolytic enzymes, length of sarcomere, and also the intramuscular connective tissues.[44] There are five exogenous enzymes (ficin, *Bacillus subtilis* protease, papain, *A. oryzae* protease, and bromelain), which have been suggested recently as "generally recognized as safe" (GRAS) by United States Department of Agriculture (USDA). Plant proteases are better than enzymes from bacterial sources due to some safety issues like, harmful health effects and pathogenecity. However, proper enzyme (protein hydrolyzing) assay needs to be done for the reason that a disproportionate quantity will cause putrefaction of meat and meat products.[22]

9.6.3 BAKING INDUSTRY

In baking industry, wheat flour is a key constituent of baking processes. Wheat flour contains a protein called gluten, which is insoluble and defines the quality of dough used for baking. By partial proteolysis, exo-, and endoproteinases obtained from *A. oryzae* have been utilized to transform wheat gluten. Addition of proteases reduce the dough mixing time that gives loaf a puffy volume, whereas bacterial proteases are being used to increase the strength and extensibility of dough.[51]

9.6.4 BEVERAGE INDUSTRY

In brewery, microbial proteases have also been considered as important. Alcoholic drinks (beer) have complex protein compounds that are soluble poorly at low temperature and gives unclear or turbid appearance when it is cold. Utilization of protease enzymes for the breakdown of proteins that cause the turbid appearance of drinks can be a solution of this problem.[62]

9.6.5 PHARMACEUTICAL INDUSTRY

Proteases contribute in the life cycle of infectious microorganisms and this property has bound them to become a target or aim for the development of medical treatments for deadly diseases like cancer of all types and acquired immune deficiency syndrome (AIDS). From microorganisms, proteases are used primarily in the treatment of several conditions, for example, cardiovascular diseases, swelling, and necrotic injuries. These enzymes can also be used as an immune stimulatory agent. Protease enzymes are utilized widely in pharmacological industry for the provision and preparation of medicines, for example, wound healing ointment and in detergents of contact lenses.[58]

9.6.6   DETERGENT INDUSTRY

Proteases have also been found widespread for their utilization in washing detergents or cleaners. Because of the thermostability, as well as stability of proteases at higher pH and the decreasing of toxic wastes ensured, these enzymes are a perfect contender for the application in laundry. At home and in industrial institutions, alkaline proteases are being added in various trademarks of detergents for their use. Protein digesting enzymes have been supplemented in laundry cleaners ever since 50 years, which enable the release of protein material in pigments like milk and blood. Protease enzyme not only removes bloodstain but also other constituents comprising of protein contents from the body secretion plus proteinaceous food for example, egg, meat, fish, and milk. The best of enzymes in detergent industry must be stable and active in cleaning solution and also have adequate stability in temperature.[11]

9.6.7   LEATHER INDUSTRY

There is a new development of plant proteases due to their utilization for dehairing and bating of hides in the leather industry.[5] In leather industry, removal of hair, and unwanted adhering to subcutaneous layer by chemicals causes a problem. In leather industry, tanners are doubtful in the utilization of enzyme due to some limitations in their utilization at marketable level because of some reasons of enzyme stability factors at numerous environmental conditions including temperature, pH along with the cost of production.[50]

9.7   SUMMARY

This review is focused primarily on the worldwide features of protease enzymes and gives a special importance to plant proteases, as well as their industrial applications. They play a crucial role in food, agriculture, detergent, pharmaceutical and leather industries. Currently, the worldwide expected sale value of industrial enzymes is more than US \$3 billion and from which, protein digesting enzymes (proteases) constitute around 60% of the total sale. Plant proteases play a vital role in numerous industries, primarily in feed, food, leather processing, chemical, detergent industries, and pharmaceutical industries. Innovation and advancement in the field of food biochemistry will offer an industrious status for proteases and will go on assisting their usages to bring a defensible position for the improvement of human life.

ACKNOWLEDGMENTS

The authors are highly obliged to the Library Department, Government College University Faisalabad (GCUF) and IT Department, Higher Education Commission (HEC, Islamabad) for access to journals, books, and valuable database.

KEYWORDS

- eco-friendly
- enzymes
- hormones
- pharmaceuticals
- phytochemicals
- protease

REFERENCES

1. Abu-Alruz, K.; Mazahreh, A. S.; Quasem, J. M.; Hejazin, R. K.; El-Qudah, J. M. Effect of Proteases on Meltability and Stretchability of Nabulsi Cheese. *Am. J. Agric. Biol. Sci.* **2009,** *4,* 173–178.
2. Agostini, E.; Tigier, H. A.; Ascota, M. A Peroxidase Isoenzyme Secreted by Turnip (*Brassica napus*) Hairy-Root Culture Inactivation by Hydrogen Peroxide and Application in Diagnosis Kits. *J. Biotechnol. Appl. Biochem.* **2002,** *35,* 1–7.
3. Akel, H.; Al-Quadan, F.; Yousef, T. K. Characterization of a Purified Thermostable Protease from Hyper Thermophilic *Bacillus* Strain. *Eur. J. Sci. Res.* **2009,** *31,* 280–288.

4. Akhtaruzzaman, M.; Mozumder, N. H. M.; Jamal, R.; Rahman, A.; Rahman, T. Isolation and Characterization of Protease Enzyme from Leguminous Seeds. *Agric. Sci. Res. J.* **2012,** *2,* 434–444.

5. Akram, M.; Shafaat, S.; Bukhari, D. A.; Rehman, A. Characterization of a Thermostable Alkaline Protease from *Staphylococcus aureus* S-2 Isolated from Chicken Waste. *Pak. J. Zool.* **2014,** *46,* 1125–1132.

6. Alqueres, S. M. C.; Almeida, R. V.; Clementino, M. M.; Vieira, R. P.; Almeida, W. I.; Cardoso, A. M.; Martins, O. B. Exploring the Biotechnological Applications in the Archaeal Domain. *Braz. J. Microbiol.* **2007,** *38,* 398–405.

7. Amri, E.; Mamboya, F. Papain, a Plant Enzyme of Biological Importance: A Review. *Am. J. Biochem. Biotechnol.* **2012,** *8* (2), 99–104.

8. Antony, M. L.; Singh, S. V. Molecular Mechanisms and Targets of Cancer Chemoprevention by Garlic-Derived Bioactive Compound Diallyl Trisulfide. *Ind. J. Exp. Biol.* **2011,** *49,* 805–816.

9. Aravindan, R.; Anbumathi, P.; Viruthagiri, T. Lipase Applications in Food Industry. *Ind. J. Biotechnol.* **2007,** *6,* 141–158.

10. Arts, I. C.; Hollman, P. C. Polyphenols and Disease Risk in Epidemiologic Studies. *Am. J. Clin. Nutr.* **2005,** *81,* 317–325.

11. Aurachalam, C.; Saritha, K. Protease Enzyme: An Eco-Friendly Alternative for Leather Industry. *Ind. J. Sci. Technol.* **2009,** *2,* 29–32.

12. Azarkan, M.; Dibiani, R.; Baulard, C.; Baeyens-Volant, D. Effects of Mechanical Wouding on Carica papaya Cysteine Endopeptidases Accumulation and Activity. *Int. J. Biol. Macromol.* **2006,** *38,* 216–224.

13. Aziz, M. A.; Adnan, M.; Begum, S.; Azizullah, A.; Nazir, R.; Iram, S. A Review on the Elemental Contents of Pakistani Medicinal Plants: Implications for Folk Medicines. *J. Ethnopharmacol.* **2016,** *188,* 177–192.

14. Bakkali, F.; Averback, S.; Averback, D.; Idaomar, M. Biological Effects of Essential Oils: A Review. *Food Chem. Toxicol.* **2008,** *46,* 446–475.

15. Batra, N.; Walia, M. Production and Characterization of Alkaline Protease from Bacteria Strains Isolated from Cotton Field. *Afr. J. Microbiol. Res.* **2014,** *8,* 702–709.

16. Beg, Q. K.; Gupta, R. Purification and Characterization of an Oxidation Stable Thiol Dependent Serine Alkaline Protease from *Bacillus mojavensis. Enzyme Microbiol. Biotechnol.* **2006,** *32,* 294–304.

17. Bornscheuer, U.T. Enzymes in Lipid Modification. *Annu. Rev. Food Sci. Technol.* **2018,** *9* (1), 87–97.

18. Brien, S.; Lewith, G.; Walker, A.; Hicks, S. M.; Middleton, D. Bromelain as a Treatment for Osteoarthritis: A Review of Clinical Studies. *Evid. Complemen. Alternat. Med.* **2004,** *1* (3), 251–257.

19. Brown, J. E.; Bauman, J. M.; Lawrie, J. F.; Rocha, O. J.; Moore, R.C. The Structure of Morphological and Genetic Diversity in Natural Populations of *Carica papaya* (Caricaceae) in Costa Rica. *Biotropica* **2012,** *44* (2), 179–188.

20. Bruno, M. A.; Lazza, C. M.; Errasti, M. E.; Lopez, L. M. I.; Caffini, N. O.; Pardo, M. F. Milk Clotting and Proteolytic Activity of an Enzyme Preparation from *Bromelia hieronymi* Fruits. *Food Sci. Technol.* **2010,** *43,* 695–701.

21. Chazarra, S.; Sidrach, L.; Lopez-Molina, D.; Rodriguez-Lopez, J. N. Characterization of the Milk Clotting Properties of Extracts from Artichoke (*Cynarascolymus* L.) Flowers. *Int. Dairy J.* **2007**, *17*, 1393–1400.
22. Chen, Q. H.; He, G. Q.; Jiao, Y. C.; Ni, H. Effects of Elastase from a Bacillus Strain on the Tenderization of Beef Meat. *Food Chem.* **2006**, *98*, 624–629.
23. Cheng, H. M.; Koutsidis, G.; Lodge, J. K.; Ashor, A.; Siervo, M.; Lara, J. Tomato and Lycopene Supplementation and Cardiovascular Risk Factors: A Systematic Review and Meta-Analysis. *Atherosclerosis* **2017**, *257*, 100–108.
24. Choudhary, D.; Roy, S.; Chakrabarti, C.; Biswas, S.; Dattagupta, J. K. Production and Recovery of Recombinant Propapain with High Yield. *J. Phytochem.* **2009**, *70* (1), 465–472.
25. Chu, W. H. Optimization of Extracellular Alkaline Protease Production from Species of *Bacillus*. *J. Ind. Microbiol. Biotechnol.* **2007**, *34*, 241–245.
26. Chukwuemeka, N. O.; Anthoni, A. B. Antifungal Effects of Pawpaw Seed Extracts and Papain on Postharvest *Carica papaya* L. Fruit Rot. *Afr. J. Agric. Res.* **2010**, *5*, 1531–1535.
27. Chumark, P.; Khunawat, P.; Sanvarinda, Y.; Phornchirasilp, S.; Phivthongngam, P. N. L.; Ratanachamnong, P.; Srisawat, S.; Pongrapeeporn, K. S. The In Vitro and *Ex Vivo* Antioxidant Properties, Hypolipidaemic and Anti-Atherosclerotic Activities of the Water Extract of *Moringa oleifera* Lam. Leaves. *J. Ethnopharmacol.* **2008**, *116*, 439–446.
28. Duarte, A. R.; Duarte. D. M. R.; Moreira, K. A.; Cavalcanti, M. T. H.; Lima-Filho, J. L.; Figueiredo, A. L. *Jacaratia corumbensis* O. Kuntze a New Vegetable Source of Milk-Clotting Enzymes. *Braz. Arch. Biol. Technol.* **2009**, *52*, 1–9.
29. Dutt, D.; Tyagi, C. H. Agnihotri, S.; Kumar, A. Bio-Soda Pulping of Ligino Cellulosic Residue of Plmarosa Grass: An Attempt to Energy Conversion. *Ind. J. Chem. Technol.* **2010**, *17*, 60–70.
30. Fahey, J. W. *Moringa oleifera:* A Review of the Medical Evidence for its Nutritional, Therapeutic, and Prophylactic Properties, Part 1. *Trees Life J.* **2005**, *1* (5), 1–24.
31. Fernandes, F. A. N.; Rodrigues, S. Optimization of Osmotic Dehydration of Papaya Followed by Air-Drying. *Food Res. Int.* **2006**, *39* (4), 492–498.
32. Ferreira, J. F.; Santana, J. C. C.; Tambourgi, E. B. The Effect of pH on Bromelain Partition from *Ananas comosus* by PEG4000/Phosphate ATPS. *Braz. Arch. Biol. Technol.* **2011**, *54*, 125–132.
33. Friml, J. Auxin Transport—Shaping the Plant. *Curr. Opin. Plant Biol.* **2003**, *6*, 1–6.
34. Gopalakrishnan, L.; Doriya, K.; Kumar, D. S. *Moringa oleifera:* A Review on nutritive Importance and its Medicinal Application. *Food Sci. Human Wellness* **2016**, *5* (2), 49–56.
35. Gupta, A.; Khare, S. K. Enhanced production and characterization of a solvent stable protease from solvent tolerant *Pseudomonas aeruginosa*. *Enzyme Microbial Technology,* **2007**, *42*, 11–16.
36. Gupta, R.; Beg, Q. K.; Lorenz, P. Bacterial alkaline proteases: molecular approaches and industrial applications. *Applied Microbiology and Biotechnology,* **2002**, *59*, 15–32.
37. Gurung, N.; Ray, S.; Bose, S.; Rai, V. A Broader View: Microbial Enzymes and Their Relevance in Industries, Medicine, and Beyond. *Bio. Med. Res. Int.* **2013**, pp. 18; http://dx.doi.org/10.1155/2013/329121

38. Hashim, M.; Iqbal, M. F.; Xiaohong, C.; Mingsheng, D. Impact of Processing Conditions on the Milk Clotting Activity of Crude Protease Extracted from Chinese Ginger. *Int. Conf. Food Eng. Biotechnol.* **2011,** *9,* 327–328.
39. Hoet, S.; Opperdoes, F.; Brun, R.; Quetin- Leclerq, J. Natural Products Active Against African Trypanosomes: A Step Towards New Drug. *Nat. Prod. Rep.* **2004,** *21* (1), 353–364.
40. Huet, J.; Looze, Y.; Bartik, K.; Raussens, V.; Wintjens, R. Structural Characterization of the Papaya Cysteine Proteinases at Low pH. *Biochem. Biophys. Res. Commun.* **2006,** *341,* 620–626.
41. Kalpana, D. M.; Rasheedha, B. A.; Gnanaprabhal, G. R.; Pradeep, B. V.; Palaniswamy, M. Purification, Characterization of Alkaline Protease Enzyme from Native Isolate *Aspergillus niger* and its Compatibility with Commercial Detergents. *Ind. J. Sci. Technol.* **2008,** *1,* 1–6.
42. Kamal, M. *Moringa oleifera* Lam—The Miracle Tree. *Trees Life J.* **2008,** *4* (2), 215–230.
43. Kapdan, L. K.; Kaegi, F. Biohydrogen Production from Waste Materials. *Enzyme Microbiol. Technol.* **2006,** *38,* 569–582.
44. Kemp, C. M.; Parr, T. Advances in Apoptotic Mediated Proteolysis in Meat Tenderization. *Meat Sci.* **2012,** *92,* 252–259.
45. Ketnawa, S.; Chaiwut, P.; Rawdkuen, S. Pineapple Wastes: A Potential Source for Bromelain Extraction. *Food Bioprod. Proc.* **2012,** *90,* 385–391.
46. Khanna, N.; Panda, P. C. The Effect of Papain on Tenderization and functional properties of Spending Hen Meat Cuts. *Ind. J. Anim. Res.* **2007,** *41,* 55–58.
47. Kirk, O.; Borchert, T. V.; Fuglsang, C. C. Industrial Enzyme Applications. *Curr. Opin. Biotechnol.* **2002,** *13* (4), 345–351.
48. Klomklao, S.; Benjakul, S.; Visessanguan, W. Comparative Studies on Proteolytic Activity of Spleen Extracts from Three Tuna species Commonly used in Thailand. *J. Food Biochem.* **2004,** *28,* 355–372.
49. Krishna, K. L.; Paridhavi, M.; Patel, J. A. Review on Nutritional, Medicinal and Pharmacological Properties of Papaya (*Carica papaya*). *Nat. Prod. Radiance* **2008,** *7* (4), 364–373.
50. Kumar, R.; Vats, R. Protease Production by Bacillus subtilis Immobilized on Different Matrices. *NY Sci. J.* **2010,** *3* (7), 20–24.
51. Kumar, S.; Sharma, N. S.; Saharan, M. R.; Singh, R. Extracellular Acid Protease from *Rhizopus oryzae:* Purification and Characterization. *Proc. Biochem.* **2005,** *40,* 1701–1705.
52. Lavery, C. B.; Macinnis, M. C.; Macdonald, M. J.; Williams, J. B.; Spencer, C. A.; Burke, A. A.; Irwin, D. J.; D'Cunha, G. B. Purification of Peroxidase from Horseradish (*Armoracia rusticana*) Roots. *J. Agric. Food Chem.* **2010,** *58,* 8471–8476.
53. Lopes, M. C.; Mascarini, R. C.; De Silva, B. M.; Florio. F. M.; Basting, R. T. Effect of a Papain-Based Gel for Chemo Mechanical Caries Removal on Dentin Shear Bond Strength. *J. Dent. Child.* **2007,** *74,* 93–97.
54. Manzoor, M.; Anwar, F.; Iqbal, T.; Bhnager, M. I. Physicochemical Characterization of *Moringa concanensis* Seeds and Seed Oil. *J. Am. Oil Chem. Soc.* **2007,** *84,* 413–419.
55. Marotta, F.; Pavasuthipaisit, K. Relationship Between Aging and Susceptibility of Erythrocytes to Oxidative Damage: In View of Nutraceutical Interventions. *Rejuv. Res.* **2006,** *9* (2), 227–230.

56. Mirzaei, M. Microbial Transglutaminase Application in Food Industry. *Digestion (Seguro, Kumazawa, Kuraishi. 1996)*, **2011**, *4*, 7–10.
57. Nigam, P.S. Microbial Enzymes with Special Characteristics for Biotechnological Applications. *Biomolecules* **2013**, *3* (1), 597–611.
58. Otsuki, N.; Dang, N. H.; Kumagai, E.; Kondo, A.; Iwata, S.; Morimoto, C. Aqueous Extract of *Carica papaya* Leaves Exhibits Antitumor Activity and Immunomodulatory Effects. *J. Ethnopharmacol.* **2010**, *127*, 760–772.
59. Pezeshki, A.; Hesari, J.; Ahmadi Zonoz, A.; Ghambarzadeh, B. Influence of *Withania coagulans* Protease as a Vegetable Rennet on Proteolysis of Iranian UF White Cheese. *J. Agric. Sci. Technol.* **2011**, *13*, 567–576.
60. Pitman, S. K.; Drew, R. H.; Perfect, J. R. Addressing Current Medical needs in Invasive Fungal Infection Prevention and Treatment with New Antifungal Agents, Strategies and Formulations. *Expert Opin. Emerg. Drugs* **2011**, *16*, 559–586.
61. Piva, E.; Ogliari, F. A.; Moraes, R. R. D.; Cora, F.; Henn, S. Papain-Based Gel for Biochemical Caries Removal: Influence on Microtensile Bond Strength to Dentin. *Braz. Oral Res.* **2008**, *22*, 364–370.
62. Piwolo, M. Industrial Uses of Enzymes. *Biotechnol. VI* **2009**, *6*, 135–138.
63. Pontual, E. V.; Carvalho, B. E. A.; Bezerra, R. S.; Coelho, L. C. B.; Napoleao, T. H.; Paiva, P. M. G. Caseinolytic and Milk Clotting Activities from *Moringa oleifera* Flowers. *Food Chem.* **2012**, *135*, 1848–1854.
64. Rahman, R. N.; Lee, P. G.; Basri, M.; Salleh, A. B. An Organic Solvent Tolerant Protease from *Pseudomonas aeruginosa* Strain K: Nutritional Factors Affecting Protease Production. *Enzyme Microb. Technol.* **2005**, *36*, 749–757.
65. Rai, S. K.; Roy, J. K.; Mukherjee, A. K. Characterization of a Detergent-Stable Alkaline Protease from a Novel Thermophilic Strain *Paenibacillus tezpurensis* sp. *Appl. Microbiol. Biotechnol.* **2010**, *85*, 1437–1450.
66. Ramaa, C. S.; Shirode, A. R.; Mundada, A. S.; Kadam, V. J. Nutraceuticals-an emerging era in the treatment and prevention of cardiovascular diseases. *Curr. Pharm. Biotechnol.* **2006**, *7*, 15–23.
67. Rawlings, N. D.; Barrett, A. J.; Bateman, A. MEROPS: The Peptidase Database. *Nucl. Acids Res.* **2010**, *38*, 227–233.
68. Sanchez, M.; Prim, N.; Rendez-Gil, F.; Pastor, F. I. J.; Diaz, P. Engineering of Baker's Yeasts, *Escherichia coli* and *Bacillus* Hosts for the Production of *B. subtilis* Lipase. *J. Biotechnol. Bioeng.* **2002**, *78* (1), 339–345.
69. Sarrouh, B.; Santos, T. M.; Miyoshi, A.; Dias, R.; Azevedo, V. Up-to-Date Insight on Industrial Enzymes Applications and Global Market. *J. Bioproc. Biotechnol.* **2012**, *3*, 124–137.
70. Satish, A.; Sairam, S.; Ahmed, F.; Urooj, A. *Moringa oleifera* Lam.: Protease Activity Against Blood Coagulation Cascade. *Pharm. Res.* **2012**, *4*, 44–49.
71. Sawant, R.; Nagendran, S. Protease: An Enzyme with Multiple Industrial Applications. *World J. Pharm. Pharm. Sci.* **2014**, *3* (6), 568–579.
72. Saxena, M.; Saxena, J.; Nema, R.; Singh, D.; Gupta, A. Phytochemistry of Medicinal Plants. *J. Pharmacogn. Phytochem.* **2013**, *1* (6), 168–182.
73. Seifzadeh, S.; Sajedi, R. H.; Sariri, R. Isolation and Characterization of Thermophilic Alkaline Proteases Resistant to Sodium Dodecyl Sulfate and Ethylene Diaminetetra Acetic Acid from *Bacillus* sp. *Iran. J. Biotechnol.* **2008**, *6*, 214–221.

74. Sharmila, S.; Rebecca, L. J.; Das, M. P.; Saduzzaman, M. D. Isolation and Partial Purification of Protease from Plant Leaves. *J. Chem. Pharm. Res.* **2012,** *4* (8), 3808–3812.

75. Shinozaki, K.; Yamaguchi-Shinozaki, K.; Seki, M. Regulatory Network of Gene Expression in the Drought and Cold Stress Responses. *Curr. Opin. Plant Biol.* **2003,** *6,* 410–417.

76. Souza, P. M. D. Application of Microbial α-Amylase in Industry—A Review. *Braz. J. Microbiol.* **2010,** *41* (4), 850–861.

77. Sreelatha, S.; Padma, P. R. Antioxidant Activity and Total Phenolic Contents of *Moringa oleifera* Leaves in Two Stages of Maturity. *Plant Food. Hum. Nutr.* **2009,** *64,* 303–311.

78. Srivastava, P.; Malviya, R. Sources of Pectin, Extraction and its Applications in Pharmaceutical Industry—An Overview. *Ind. J. Nat. Prod. Res.* **2011,** *2* (1), 10–18.

79. Synowiecki, J. Some applications of thermophiles and their enzymes for protein processing. *Afr. J. Biotechnol.* **2010,** *9,* 7020–7025.

80. Thomas, S. G.; Hedden, P. Gibberellin Metabolism and Signal Transduction. *Ann. Plant Rev.* **2018,** *24,* 147–184.

81. Tochi, B. N.; Zhang, W.; Ying, X.; Wenbin, Z. Therapeutic Application of Pineapple Protease (Bromelain): A Review. *Pak. J. Nutr.* **2008,** *7* (4), 513–520.

82. vanBeilen, J. B.; Li, Z. Enzyme Technology: An Overview. *Curr. Opin. Biotechnol.* **2002,** *13* (4), 338–344.

83. Van den Burg, B. Extremophiles as Source for Novel Enzymes. *Curr. Opin. Microbiol.* **2003,** *6,* 213–218.

84. Vattem, D. A.; Shetty, K. Biological Function of Ellagic Acid: A Review. *J. Food Biochem.* **2005,** *29* (3), 234–266.

85. Vidyalakshmi, A.; Selvi, S. E. Protease Activity of Floral Extracts of *Jasminum grandiflorum* L., a Wound Healing Herb. *J. Med. Plant Stud.* **2013,** *4,* 11–15.

86. Wong, T. Y.; Preston, L. A.; Schiller, N. L. Alginate Lyase: Review of Major Sources and Enzyme Characteristics, Structure-Function Analysis, Biological Roles, and Applications. *Ann. Rev. Microbiol.* **2000,** *54* (1), 289–340.

87. Wood, J. G.; Rogina, B.; Lavu, S.; Howitz, K.; Helfand, S. L.; Tatar, M.; Sinclair, D. Sirtuin Activators Mimic Caloric Restriction and Delay Ageing in Metazoans. *Nature* **2004,** *430,* 686–689.

88. Yadav, S. K.; Shikha, D. B.; Darmwal, N. S. Oxidant and Solvent Stable Alkaline Protease from *Aspergillus flavus* and its Characterization. *Afr. J. Biotechnol.* **2011,** *10* (43), 8630–8640.

89. Yoruk, R.; Marshall, M. R. Physicochemical Properties and Function of Plant Polyphenol Oxidase: A Review. *J. Food Biochem.* **2003,** *27* (5), 361–422.

90. Zhang, Y.; Liu, W.; Li, G.; Shi, B.; Miao, Y.; Wu, X. Isolation and Partial Characterization of Pepsin Soluble Collagen from the Skin of Grass Carp (*Ctenopharyngodon idella*). *J. Food Chem.* **2007,** *103,* 906–912.

91. Zhao, L.; Budge, S. M.; Ghaly, A. E.; Brooks, M. S.; Dave, D. Extraction, purification and characterization of fish pepsin: a critical review. *J. Food Proc. Technol.* **2011,** *2* (6), 1–14.

92. Zia, M. A.; Kousar, M.; Ahmed, I.; Iqbal, H. M. N.; Abbas, R. Z. Comparative Study of Peroxidase Purification from Apple and Orange Seeds. *Afr. J. Biotechnol.* **2011,** *10* (33), 6300–6303.

CHAPTER 10

MICROALGAL PIGMENTS AS NATURAL COLOR: SCOPE AND APPLICATIONS

K. G. SREEKALA, MALAIRAJ SATHUVAN, JAVEE ANAND,
KARUPPAN RAMAMOORTHY, VENGATESH BABU, and
S. NAGARAJ

ABSTRACT

The microalgal resources have become excellent sources of bioactive and functional compounds for our benefits. Many species are being cultivated for the production of biochemicals pertaining to the fields of therapeutics, cosmetics, nutraceuticals, pharmaceuticals, food and feed, aquaculture, poultry, energy production, and so on. The microalgae possess an array of widely distributed natural pigments belonging to three classes—chlorophylls, carotenoids, and phycobilins—out of which carotenoids consist of large number of pigments. The microalgal pigments are employed as natural colorants in food, feed, cosmetics, confectionary, poultry, and aquaculture, and in fluorescent cell imaging techniques. Apart from this, many possess anti-inflammatory, antioxidant, anticancer, anti-angiogenic, anti-obesity, antidiabetic, immune stimulating and neuroprotective bioactivities. All these applications point out vast potential of microalgae for our welfare.

10.1 INTRODUCTION

Pigments are photosynthetic colored substance that can harvest solar light energy in microalgae. These are broadly classified into chlorophylls (higher plants and photosynthetic algae), carotenoids (algae), and phycobilins (cyanobacteria and some rhodophyceae). They have been utilized as supplements in many bioproducts in our life, and are hence significant compounds.[33] The naturally occurring carotenoids are classified into: (1) carotenes of lycopene, β-carotene, and α-carotene, which are linear or cyclized hydrocarbons, and (2) xanthophylls such as astaxanthin, canthaxanthin, lutein, and zeaxanthin, which are the oxygenated derivatives of carotenes. Lycopene is the common precursor from which astaxanthin and lutein are synthesized through two divergent pathways with β-carotene and α-carotene as intermediates, respectively.[1] The research on microalgae-based nutrients has exhibited promising results, but the commercial products currently available are still a few to count. Two types of products are obtained from microalgae, targeting the food market:

- The first one includes the highly nutritious (vitamin B12, C, D2, etc.) dried algae (especially the microalgae species *Chlorella* and *Spirulina*).
- The second one is the significant products that are obtained from the microalgae and can be incorporated in food or feed to enhance their nutritional quality.

These products include pigments like astaxanthin, antioxidants like β-carotene, proteins like phycocyanin, and fatty acids such as omega-3, DHA, or EPA.[60] Algal pigments have a high market value that indicates their potential for commercial success in a not too distant future. Algal pigments are responsible for harvesting light, fixing CO_2, protecting algae cells from over illumination damage, and also algal culture. The three main types of pigments (Fig. 10.1) impart different colors to the microalgae: (1) the carotenoids, carotenes impart an orange color; (2) xanthophylls give yellowish shade; (3) the phycobilins impart red or blue coloration; and (4) the chlorophylls green coloration.

The pigment fractions of algae find application as nutritional supplements E160a, E306, E307, E308 for pharmaceutical, veterinary, and medical purposes (anti-inflammatory, anti-oxidative, cancer preventive effects), and in cosmetic industry and food technology (Tables 10.1 and 10.3). In addition to this, Table 10.2 gives a glimpse of the use of β-carotene and lutein in poultry feeding for yellow–orange coloration of egg yolk and in aquaculture.[27]

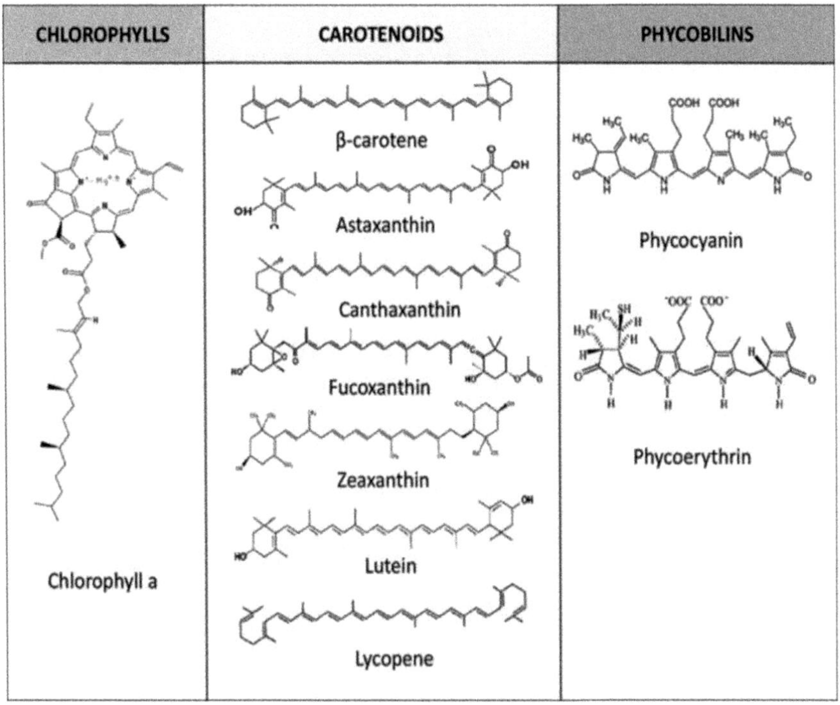

CHLOROPHYLLS	CAROTENOIDS	PHYCOBILINS

FIGURE 10.1 Chemical structure of some common and commercially significant pigments found in microalgae.

TABLE 10.1 Microalgal Pigments Used as Antioxidants, and their Sources.

Algal source	Pigments used as antioxidant	Reference
Chaetoceros calcitransis	Fucoxanthin	15
Coccomyxa onubensis	Lutein	59
Dunaliella salina	β-carotene	29
Haematococcus pluvialis	Astaxanthin	3
Nannochloropsis gaditana	β-carotene	34
Nostoc muscorum	Phycobilin	51
Spirulina platensis	Phycoerythrocyanins	48
T. arborum	Zeaxanthin	10

The cyanobacteria is in the limelight recently because of abundant source of novel chemicals of industrial, biotechnological, and medicinal

TABLE 10.2 Microalgal Pigments used as Natural Colorant, and their Sources.

Algal source	Pigments used as colorant	Potential application	Reference
Chlorella zofingiensis	Lutein	Aquaculture and poultry farming Pharmaceuticals, cosmetics, and food.	33, 61
Dunaliella salina	β-carotene	Food stuffs.	29, 44
Haematococcus pluvialis	Astaxanthin	To replace synthetic colors like Tartrazine (E102), Sunset Yellow (E110), E rythrosine (E127), Allura Red (E129).	29
		Aquaculture pigmentation of salmonids, trout, ornamental, and teleost fish; Poultry industry. Food colorant	14, 35, 44
Nannochloropsis gaditana	β-carotene	Poultry feed to color egg yolk and skin.	34
Nannochloropsis salina	Canthaxanthin	Aquaculture of salmonids. Sausage coloration. Tanning pills in Canada	27
Spirulina platensis	Chlorophyll	Foods & Beverages	27
	C-Phycocyanin Phycoerythro- cyanin Allophycocyanin	Confectionary and dairy products.	47
T. arborum	Zeaxanthin	Human consumption, cosmetics, and pharmaceuticals; fish, poultry, and eggs.	10

significance such as pigments, polysaccharides, fatty acids, proteins, and other compounds, which exhibit an array of bioactivities.[19] The high structural diversity of pigments imparts better pharmacological properties that are related to their potentials to modulate scavenge reactive oxygen species (ROS), thus modulating oxidation reactions. Significant in vitro and in vivo antioxidant activities have been reported for this source. There is an increasing demand for antioxidants for use in food, beauty products, and other goods that may be vulnerable to oxidation, and therefore the quest for nature-derived alternatives is on the go. As a result, microalgal biotechnology has been developed as a significant method for commercial production of such biocompounds.[45]

In this review, the recent applications of pigments from microalgae are brought together, with the aim of bringing the commercial significance of natural pigments into limelight.

TABLE 10.3 Microalgal Pigments used as Against Various Cancer Cell Lines.

Algal source	Pigments used	Cell lines under study	Reference
Botryococcus braunii	Zeaxanthin	Non-small cell lung cancer (NSCLC) cells.	27
Chaetoceros calcitransis	Fucoxanthin	Colon cancer Human chronic myeloid leukemia cell line (K562).	17, 40
Dunaliella salina	Zeaxanthin	Colon cancer	27
Dunaliella tertiolecta	Fucoxanthin	MCF-7 human mammary cancer cells; Colon cancer.	8
Haematococcus pluvialis	Astaxanthin	A549 & H1703 Colon cancer	26
Nannochloropsis gaditana	Zeaxanthin	MCF-7 and MDA-MB-231 cells	27
Nostoc muscorum Oscillatoria sp.	Phycobilin	HeLa cells and K-562 cells	13, 20, 51
Spirulina platensis	Chlorophyllin	HCT116 human colon cancer cells	27
	Phycoerythro-cyanin	Ehrlich ascites carcinoma cell (EACC), Human hepatocellular cancer cell line (HepG2), Lung cancer cell line (A549).	48
T. arborum	Cryptoxanthin	Human hepatocarcinoma cell line HepG2.	28

10.2 NATURAL PIGMENTS

10.2.1 CHLOROPHYLLS

Chlorophylls are most prominent algal pigments, and absorb light mainly in the blue zone and in the red zone to a minor extent, while moderately absorbing green regions of the solar spectrum. This accounts for the typical green color of algae that possess chlorophyll as main pigment. The well-known representative is chlorophyll-a that occurs in all microalgal species. Chlorophylls are found to have many applications in foods: these pigments are approved as food additive (E140) for coloration.

Many popular world cuisines use chlorophyll as a green colorant in a lot of dishes and drinks, such as pasta or absinthe. In addition, wasabi consisting of "simple" horseradish colored with *Spirulina* chlorophyll is also available. Also a study reports the benefit of chlorophyll derivatives (chlorophyllin) against colon cancer cells, if provided as dietary supplements. The response of HCT116 human colon cancer cells, to treatment with chlorophyll derivative (chlorophyllin), demonstrated its effectiveness as chemopreventive agent. Chlorophyll-a is also used in products such as deodorants and in formulations to fight bad breath, because it has an excellent deodorant quality.[27]

10.2.2 CAROTENOIDS

Carotenoids are pigments that are not synthesized by human body and require to be obtained from diet. Carotenoids possess many health benefits, creating interest in the production of drugs and functional foods containing carotenoids. To meet the increasing worldwide demand for carotenoids, its chemical synthesis is employed. The synthetic β-carotene contains 100% trans-isomers, while natural carotenoid is a combination of trans- and cis-isomers in which the latter exhibits better therapeutic nature. This type of a mixture cannot be synthesized chemically. This puts forth the importance of obtaining carotenoids naturally for commercial applications.

Recently, some microorganisms like bacteria, fungi, algae, etc. have been utilized for carotenoids production.[19] Carotenoids in plants and algae capture excess light energy, protecting chlorophyll from photo-damage; and include β-carotene, xanthophylls, astaxanthin, canthaxanthin, zeaxanthin, and echinenone. While in humans, they serve as antioxidants.[30]

Carotenoids have been used extensively in the food industry as dyes and as feed additives for live-stock, poultry, and fish farming. The importance of carotenoids is not limited only to their well-known coloring properties but also to their antioxidant capacity, nutritional quality as provitamin-A, and beneficial health characteristics. Different studies have demonstrated their ability to prevent a variety of ailments like cataract, age-related macular degeneration, atherosclerosis, and selected cancers. All these beneficial features have encouraged the use of carotenoids as food additives and nutraceuticals.

Besides, the stringent regulation in the use of synthetic molecules and the social demand for safer food have stimulated the industry to use carotenoids from natural sources such as microalgae. As a result, microalgae-derived carotenoids are experiencing strong market demand, in spite of its price that is almost double compared with synthetic carotenoids. Considering these facts, carotenoids are one of the co-products that have been considered to make biofuel production from microalgae economically feasible.[34,56,57]

Carotenoids are available commercially as natural colorants, such as: β-carotene, astaxanthin, and lutein. Many of microalgal species have carotenoids in them; and the main ones used commonly in commercial production and culturing are: *D. salina, H. pluvialis* (for astaxanthin), *C. zofingiensis, C. vulgaris, S. platensis* (β-carotene), and *Botryococcus braunii* (lutein).[33] Animal experiments have shown that the administration of *S. platensis* can lower plasma cholesterol levels and blood pressure.[44]

Dunaliella salina and *Haematococcus pluvialis* under stress conditions accumulate appreciable amounts of β-carotene and astaxanthin, respectively. These carotenoids are already in use as food colorants, vitamin supplements, food additives, and cosmetics ingredients.[2] β-carotene exhibits a color range including yellow, orange, and red. Therefore, it is a better substitute for unsafe synthetic colorants, such as: tartrazine (E102), sunset yellow FCF (E110), erythrosine (E127), and allura red (E129). Further, β-carotene has an equivalent antioxidant activity as that of the synthetic antioxidants beta-hydroxy acid (BHA) and Butylated hydroxy-toluene (BHT), etc.[29]

The photosynthetic efficiency of microalgae is higher as compared with terrestrial plants, due to high content of chlorophylls and carotenoids. The photosynthetic pigments, extracted from *Chlorella vulgaris* used for wastewater purification, can be used to replace toxic synthetic dyes to produce photocurrent. Microalgal pigments, which are nontoxic and less expensive, can be utilized as a sensitizer on the nano photoelectrode of dye-sensitized solar cell (DSSC) to achieve the conversion of light into electricity. These DSSCs based on natural pigments are less efficient, cheap, plentiful, biodegradable, and can be prepared easily than those based on synthetic dyes like ruthenium polypyridyl complexes.[36]

Several species of microalgae contain valuable polyunsaturated fatty acids or fatty acids that resemble common vegetable oils, which can be put to various applications. The presence of substantial amounts of carotenoids and tocopherol in algal lipids helps in improving the stability of vegetable oils.[30]

The carotenoids have been associated with relevant bioactivities such as lowering risk of coronary heart diseases and cancer prevention. Several studies have demonstrated the antiproliferative activity of carotenoids, such as zeaxanthin, fucoxanthin, violaxanthin, or lutein isolated from microalgae against different cancer cells. The antiproliferative effect of violaxanthin-enriched fraction obtained from *Dunaliella tertiolecta* has been demonstrated on MCF-7 human mammary cancer cells. Besides the cytoprotective attributes of carotenoids, evidence also suggests that these compounds may inhibit cancer cell proliferation and, in some cases, induce cell death.[8] *Nannochloropsis gaditana* is a microalga characterized by its ability to accumulate 65–70% lipids (dry weight), and carotenoids such as astaxanthin, β-carotene, canthaxanthin, neoxanthin, zeaxanthin, and violaxanthin.[34]

Astaxanthin is used in the pigmentation of salmonids, ornamental fish, and poultry.[35] Many of nervous, endocrine, and dietary processes influence the coloration of teleost fishes that in turn leads to information transfer within color-based visual signals. The significance of carotenoid pigments as dietary requirement, their antioxidant property, and the quality of altering coloration is very well mentioned in teleost fish and many other taxa. Also, the change in color brought about by consuming carotenoid can have effects on color-based behaviors in ornamental fishes.[14]

Carotenoids are group of diverse lipophilic pigments >600 members that play vital role in light harvesting and photo-protection in plants and microorganisms. However, given their diverse and ubiquitous nature, these have been used as important tools for identifying the presence of certain microalgal groups in different aquatic systems. These microalgal groups in various aquatic habitats have been found to display a fixed pattern of carotenoids during specific growth stage, which is often useful for their identification.

The constituent pigments of these groups are considered excellent chemotaxonomic bio markers due to their specificity. High performance liquid chromatography (HPLC) characterization of such pigments can lead to wealth of information about the taxonomic composition and prevailing physiological conditions. Often, these studies give an indication on the influence of climatic and anthropogenic activities on phytoplankton response on a large geographical area. Several research groups have recorded their observations on prevailing phytoplankton populations in specific areas based on such pigment profiles.[41]

The microalgae cultivation is more advantageous as compared with the traditional plant cultivation as it involves less time-consuming processes. Even microalgae can be cultured on waste materials. Due to their antioxidant property, carotenoids like astaxanthin and canthaxanthin promote the functioning and maintenance of photosynthetic machinery. Carotenoids maintain cell membrane fluidity in high or low temperature, and highlight conditions, thereby promoting membrane integrity essential for cell survival.[17]

10.2.2.1   ASTAXANTHIN

Astaxanthin is a keto-carotenoid, fat-soluble, orange to red pigment, found in certain microalgae, microorganisms, and marine animals. It has a potential antioxidant property responsible for reducing various human and animal diseases, thus gaining nutraceutical and pharmaceutical importance.[3] It is used to replace synthetic astaxanthin used in production of feed for breeding of fish and crustaceans in aviculture and aquaculture. Astaxanthin obtained naturally from yeasts and microalgae is in huge demand due to rising concern about using chemical substances in foods.[31] In the United States, it is a Food and Drug Administration-approved colorant in fish feed for pigmentation of salmon in aquaculture. Astaxanthin has been found to protect liver cell mitochondria against lipid peroxidation in rats and humans.[44,62]

Astaxanthin is a xanthophyll carotenoid, possessing many bioactivities such as prevention of cancers, CVDs, metabolic syndrome, diabetes, hepatic ailments, Parkinson, Alzheimer, etc. It was found to exhibit moderate antioxidant effect on white blood cells in human and animal models, noted as a decline in superoxide and hydrogen peroxide. Evidence also suggest that astaxanthin in combination with vitamin-C can improve phagocytic capacity of neutrophil, decline ROS, IL-1β, and TNF-α release.

These studies suggest that an adjuvant therapy may be developed for many ailments including diabetes mellitus.[18]

Some studies have reported that astaxanthin ($C_{40}H_{52}O_4$) possesses antioxidant activity 100- to 500-fold higher than other antioxidants, for example, α-tocopherol and β-carotene. Astaxanthin do not exhibit provitamin-A activity but a panoply of important pharmacological properties that may be involved in treating some diseases like cancer, hypertension, diabetes, and neurological disease, among others.[11] The green microalgae *C. zofingiensis* is a natural source of astaxanthin, which is major carotenoid having role against excessive oxidative damage.[23]

Astaxanthin from green microalga *Haematococcus pluvialis* was found to inhibit LDL peroxidation preventing lipoproteins from being converted into proatherogenic particle.[5] Production of the astaxanthin from *H. pluvialis* is now the most hopeful application of microalgal biotechnology and has been used in the development of nutraceuticals and cosmetics.[25] The diverse therapeutic applications of astaxanthin include: combating immune failure, anti-inflammation, carpal tunnel syndrome, muscle soreness and cancer.[24]

Its other benefits include protection of skin and eyes against UV radiation, healing of gastric ulcer caused by *Helicobacter pylori*, and ROS-mediated neurodegenerative processes. This led to the assumption that combination of fish oil and astaxanthin intake balances antioxidative processes in activated neutrophils in a much better way, improving anti-inflammatory responses.[6] Astaxanthin has photo protective role, protects cell lines (viz., U937 cells), and biomembrane systems against oxidative stress, and it can span the cellular membrane bilayer and penetrate even the blood–brain barrier.[55]

Formerly also known as hematochrome, astaxanthin is even discussed as a "fountain of youth" by ameliorating age-related diseases. Cosmetic industry uses this pigment in sunscreen creams due to its photo-protection activity, which is higher than that of vitamin E, and due to its waterproofness.[27] It is used for antitumor therapies and treatment of neural damage related with age-related macular degeneration, Parkinson and Alzheimer diseases. Besides, it is considered as a natural super food aimed at enhancing the athletic performance by increasing stamina and reducing the time of muscle recovery.[42] Strenuous exercise induces oxidative stress, inflammation, lipid peroxidation, and muscle damage. Therefore, astaxanthin supplementation may be important in sports nutrition. Natural astaxanthin has been documented to increase muscular strength and endurance in five out of six human clinical studies as well as four supporting animal trials.[7]

Treatment of colon cancer in mouse model with astaxanthin was shown to generate normal expression of NF-κB, MMP9, IL-6, TNF-α, COX-2, and inhibit proliferation and induce apoptosis. Also, dietary astaxanthin significantly suppressed the formation of colonic mucosal ulcers and dysplastic crypts in an animal model. DMH-induced colon cancer was markedly reduced by astaxanthin. It has a good chemopreventive effect on antioxidant status, lipid peroxidation, cell proliferation, and total number of aberrant crypt foci (ACF), and eventually reduced the histological lesions in a rat model.[28]

Astaxanthin treatment was found to inhibit cell survival and multiplication of NSCLC, A549, and H1703 cells. A combination of mitomycin C (MMC) and astaxanthin could induce cell death in NSCLC cells.[26]

The hypoglycemic, antioxidant, and anti-apoptotic effects of astaxanthin aid in treating complications of diabetes mellitus, as well as improve cognition by protecting neurons against inflammation injury.[63]

Astaxanthin is commonly used in the pigmentation of ornamental fish, salmonids, and in poultry industry.[35,43] In addition to coloration requested by the customer, the pigment acts on the immune system performance of these fishes, thereby positively impacting their fertility.[27] In aquaculture, it is widely used to pigment the flesh of fish such as salmon or trout with a characteristic pink color, and as a nutritional component that improves survival in juvenile stages of fish.

However, its use in the food industry is limited because of its highly conjugated structure and unsaturated nature. During food processing it becomes unstable and susceptible to degradation, especially when removed from its biological matrix and exposed to chemical changes induced by the conditions of the industry. Carotenoid degradation not only leads to loss of its functional properties, but also decreased nutritional value of food, reduced coloration, and loss of organoleptic characteristics.[12]

10.2.2.2 FUCOXANTHIN

It is a microalgae-derived carotenoid that has no major market value, but is available commercially as an anti-obesity, anticancer, and anti-inflammatory supplement. There are reports that it is nontoxic at lower doses, and is safe for human health.[17] Numerous studies have supported the antiproliferative ability of carotenoids, such as fucoxanthin, zeaxanthin, violaxanthin, and lutein isolated from microalgae, against different cancer cells.[8]

Recently, there has been growing demand for fucoxanthin obtained from brown algae and diatoms. Reports indicate its potential to prevent cell growth and initiate apoptosis in human tumor cells, and antioxidant, anti-inflammatory, antidiabetic, and anti-obesity properties.[2] *Chaetoceros calcitrans*, a representative marine species from the class Bacillariophyceae, accounts for the base of marine food chain. The major light harvesting complexes in this diatom is fucoxanthin, which makes up to the major part of the total production of carotenoids in nature. Previous studies revealed fucoxanthin to be an effective free radical scavenger that enhances the expression of HO-1 and NQO-1 in the cells.[15]

The carotenoid fucoxanthin has a distinctive structure with epoxy, carbonyl, hydroxyl, and carboxyl groups in a hydrocarbon chain with multiple double bonds, and has photosynthetic and protective functions. The antioxidant property of fucoxanthin has cancer chemoprevention potential, besides which there is a pro-oxidant activity on tumors, thereby inducing death in cancer cells. The other mechanisms involve regulation of cell death, cell cycle arrest, and metastasis. Fucoxanthin was found to exhibit antiproliferative effect in some carcinomas including breast cancer (MCF-7) cells. Fucoxanthin and fucoxanthinol (metabolite of fucoxanthin) induced apoptosis and inhibited NF-kB pathway in MCF-7 and MDA-MB-231 cells, respectively.[40]

The characteristic golden brown color of diatoms is due to the presence of photosynthetic pigment fucoxanthin. *Amphora* sp. is being reported for the first time for a light dependent biosynthesis of silver nanoparticles (SNPs) using its aqueous extract and $AgNO_3$. The light sensitive photosynthetic pigment fucoxanthin has been demonstrated to reduce Ag^+ ion to Ag^0. Furthermore, the biologically synthesized metal nanoparticles showed antibacterial activity against both gram-positive and gram-negative bacteria. In the view of above remarkable properties, this study primarily and conclusively demonstrated the feasibility of green biosynthesis of SNPs using fresh cell extract of a ubiquitous diatom, *Amphora* sp. Owing to its microscopic size, ubiquitous nature, and modest nutritional requirement, the diatom *Amphora* may prove to be biotechnologically important in future for green technology.[22]

The anti-obesity activity of fucoxanthin or its derivatives attributed the ability to inhibit lipogenic enzymes while induce lipolytic enzymes, with the assistance from leptin and adiponectin hormones. Fucoxanthin also suppresses insulin resistance, liver, and visceral body fat accumulation.

Thus, consumption of fucoxanthin or its derivatives holds hope for the treatment of type-2 diabetes, metabolic syndrome, and CVDs.[38]

Fucoxanthin has been well known to lower body fat in obese subjects. The anti-obesity effect of fucoxanthin was partly due to the stimulated energy expenditure via uncoupling protein. Fucoxanthin and its metabolite fucoxanthinol both attenuated lipid accumulation in an adipocyte differentiation system. Both compounds downregulated adipogenic genes like PPARγ in a dose-dependent way. Both carotenoids showed strong inhibition against pancreatic lipase activity. In addition, they both suppressed triacyl glycerol absorption after oral infusion with oil. Interestingly, fucoxanthin only reduced body fat in obese models, while it did not affect body fat of normal animals, suggesting a specific effect targeting on adiposity in the development of obesity. Besides, both original and metabolic forms of fucoxanthin possess potent anti-obesity effect. In view of these findings, fucoxanthin can be promising compound for obesity management.[21]

10.2.2.3   ZEAXANTHIN

Zeaxanthin is an orange–yellow carotenoid pigment (xanthophyll), found in the microalgal species, *Botryococcus braunii*, *Dunaliella salina*, *Nannochloropsis oculata*, and *Nannochloropsis gaditana*, and is commercially used as food additive E 161 h, animal feed, pharmaceuticals for colon cancer, and eye health.[27] There are number of reports on the antioxidant capacity of *Dunaliella*, *Botryococcus*, *Chlorella*, *Nostoc*, *Arthrospira*, *Phaeodactylum*, *Polysiphonia*, *Scytosiphon*, and *Synechocystis*.

Zeaxanthin is an important carotenoid for the prevention of age-related macular degeneration and cataract, and for ophthalmic protection. *D. salina* and *D. tertiolecta* have minor amounts of this pigment.[2] Lutein and zeaxanthin are also conquering the nutraceutical market because of their role in eye health, and their strong antioxidant property can decrease the risk of nearly 60 chronic diseases. These xanthophyll pigments are not toxic and are safe for human consumption.[17]

Microalgae such as *C. zofingiensis*, *C. vulgaris*, *C. protothecoides*, *S. almeriensis*, and *H. pluvialis* produce good amount of zeaxanthin. Zeaxanthin and lutein occur in retina and helps to maintain vision. The supplementation of zeaxanthin and lutein can efficiently lower the possibilities of cataract, macular degeneration, and atherosclerosis occurrence.

On grounds of these observations, the EFSA (2012) is recommended at daily intake of 0.75 mg zeaxanthin per kg body weight.[53]

The xanthophylls, zeaxanthin, and lutein have been demonstrated to protect against high energy blue light, which are key to vision and high visual acuity. Their antioxidant nature enables repair of photo oxidative damage. Recent studies conclude that both of these xanthophylls have potential to enhance eye health if consumed in daily diet. Another report states that diet added with zeaxanthin and lutein moderately reduced the occurrence of cataract in women (aged 50–79 years) in contrast to the control group placed on diet devoid of these carotenoids.[39]

The pigment composition of *T. arborum* cells under different cultivation conditions was found different. The orange biomass contained 13.19 mg/g total carotenoids with high β-carotene content. It is employed as a colorant in food items, pigmentation of fish, poultry and eggs, in cosmetics, and pharmaceuticals. Microalga *T. arborum* could generate 3.88 mg/g dry biomass of zeaxanthin, which is comparable to that reported in *Dunaliella salina zea1*.[10]

Phytochemicals such as xanthophyll, astaxanthin, cryoptoxanthin, and zeaxanthin metabolites have been used for the treatment of colon cancer.[28] There are several studies that have demonstrated the antiproliferative activity of carotenoids, such as zeaxanthin isolated from microalgae against tumor cell lines.[8]

10.2.2.4 LUTEIN

It is used as feed additive and natural colorant besides its distinguishable role in combating degenerative human diseases. Commercial lutein is obtained from marigold, but demands are not met because of low lutein content (0.03%) and high operational costs. On the other hand, several microalgal species like *Muriellopsis* sp., *Chlorella zofingiensis*, *Chlorella sorokiniana*, *Scenedesmus obliquus*, and *Chlamydomonas* sp. can produce higher amounts of lutein, and hence are commercially applicable substitutes.[61] Although the lutein content is almost equivalent in many of the above-mentioned species, yet only *Muriellopsis* sp. and *Scenedesmus almeriensis* have been tested in large scale systems for mass production.[33]

Picochlorum maculatum (marine pico alga) is a rich source of lutein and zeaxanthin, with feed and nutraceutical applications.[4] Microalga *Botryococcus braunii* produce lutein and finds application in nutraceutical, food and feed applications. *B. braunii*, the chlorophycean colonial micro-alga has three races A, B, and L out of which races B and L produce many carotenoids, lutein being the major one (22–29%). While comparing the bioavailability and antioxidant activity of *S. platensis, H. pluvialis,* and *B. braunii* biomass in the plasma and liver tissues in animals, an interesting observation was made that the biomass prevented lipid peroxidation through free radicals scavenging. Therefore, these microalgae were suggested for food, nutraceutical, and pharmaceutical applications.[44]

Lutein has been clinical proven to prevent cataract and macular degeneration.[17] *Coccomyxa onubensis*, a microalga, accumulates high concentrations of lutein, an average lutein content of 6 mg/g dry weight, making it one of the most promising lutein producing microalgal species (more than 75% of total carotenoid pool).[58,59]

10.2.2.5   LYCOPENE

Lycopene is common precursor from which astaxanthin and lutein are synthesized through two divergent pathways with β-carotene and α-carotene as intermediates, respectively.[1] Though lycopene was commercially sold as an antioxidant, yet there is no sufficient scientific evidence available for treating cardiovascular diseases and prostate cancer.[17]

10.2.2.6   CANTHAXANTHIN

Canthaxanthin is used to treat some blood disorders. However, reports suggest that it is not safe to consume it in large doses daily for purposes like skin tanning, as it may cause blindness or aplastic anemia.[17] Violaxanthin and canthaxanthin are also xanthophylls. Canthaxanthin is approved as food additive E161g in USA, but not in New Zealand and Australia, and only to a limited extent in the EU. Typical areas of application are poultry feeding for coloration of egg yolk and skin, aquaculture of salmonids, and sausage coloration. In the cosmetic sector, canthaxanthin found application in tanning pills especially in Canada. Typical production strains of canthaxanthin are Euglenophyta species, for example, *Nannochloropsis* sp. like *Nannochloropsis salina, Nannochloropsis oculata,* or *Nannochloropsis gaditana*.[27]

10.2.3   PHYCOBILINS

These are pigments bound to some specific proteins called phycobiliproteins that are soluble in water. Found in cyanobacteria, red algae, and cryptomonads, they are generally classified into phycocyanobilin (blue), phycoerythrobilin (red), phycouroblin (yellow), and phycobiviolin (purple). These find application as natural colorants, fluorescent agents, and in pharmaceuticals. The applications in pharmacological field include: anti-inflammatory, antioxidant, hepatoprotective, and neuroprotective agents.

The main species used for phycocyanin (extensively used as colorant) production is *Arthrospira platensis*, and *Porphyridium* for phycoerythrin (fluorescent agent) production.[33] Phycobiliproteins are extensively used as natural colorant in foods, cosmetics, pharmacology, and as fluorescent biomarkers in medicine, biotechnology, and diagnostic reagents. They have highly valuable biological and pharmaceutical properties.[32,47] Phycoerythrin is used as a red fluorescent agent, while phycocyanin obtained from *Spirulina platensis* as a blue fluorescent agent.[37]

Purified phycobiliproteins exhibit high fluorescence due to absence of acceptors to receive the energy harvested by them. The blue phycobiliproteins, allophycocyanins is produced by *Spirulina* sp.; and red B-phycoerethrin and R-phycoerethrin are produced by red microalgae, such as *Porphyridium* sp., *Rhodella* sp., and *Bangia* sp.[27,62]

The C-phycocyanin from *Spirulina plantensis* has showed in vivo inhibition of CCl_4-induced lipid peroxidation in rodent liver. C-phycocyanin of more than 95% purity was tested on the growth of K562 cell line, and it was found that a concentration of 50 μM treated up to 48 h could significantly diminish cell proliferation by 49%. In another study, C-phycocyanin from *S. platensis* was used to treat HepG2 human hepatocarcinoma cell, which downregulated MDR-1. Thus C-phycocyanin and B-phycoerythrin exerted in vitro antiproliferation and anticancer activities on various tumor cell lines. Altogether, these findings indicate that phycobiliproteins obtained from marine microalgae can be exploited for developing pharmaceuticals to combat cancers. Interestingly, these pigments are used as natural colorants in different food products viz., dairy products, chewing gums, sherbaths, and gellies. Also diverse health products are available in the market these days.

B-Phycoerythrin exhibit tremendous bioactivities like antioxidant, antitumor, neuroprotective, hepatoprotective, and antiviral activities.

B-phycoerythrin is more heat stable and pH tolerant, which has paved way for its application as natural pink and purple colorants for various cosmetic products.[48] The C-phycocyanin from *Arthrospira maxima* was found to prevent increase of oxidative markers, and protect renal cells against mercury-caused ($HgCl_2$) oxidative stress and cellular damage. The studies on culturing *Nostoc muscorum* and *Oscillatoria* sp. in an increased nitrate concentration (nitrogen stress) exhibited an increase in phycobilins, and in turn an enhancement in antioxidant and anticancer activities against EACC and HepG2 cell lines, in both cyanobacterial species.[51]

The use of phycocyanin extracted by Super Critical Fluid Extraction method against A549 lung cancer cell line showed cytotoxicity, and hence the possibility of including it in daily diet for the prevention of highly lethal lung cancer.[13]

It was recently reported that feeding of *Spirulina* phycocyanin extract leads to correcting effect within 4 weeks when Wistar rats were exposed to X-rays. HeLa cells treated with highly purified C-phycocyanin showed a significantly decreased survival rate. Highly purified C- phycocyanin induces cell death in K-562 cells.[20]

The C-phycocyanin induces secretion of inflammatory cytokines and increase COX-2 protein expression dose dependently. C-phycocyanin also stimulates phosphorylation of signaling molecules, related to inflammation. These findings concluded that C-phycocyanin exhibit an elaborate molecular mechanism of bioactivity by boosting immunomodulation.[9]

The AFA (*Aphanizomenon flos-aquae*) extract Klamin® is reported to possess neuromodulating capability due to presence of a general neuromodulator named phenylethylamine. It was evident in some studies that this phenylethylamine would be rapidly degraded by monoamine oxidase B enzymes without the inhibition of AFA-PC (phycocyanin) and AFA-MAAs (AFA, *Aphanizomenon-flos-aquae*; MAAs, mycosporine like aminoacids). This investigation thus proposes that this extract can potentially be utilized in fighting mood disorders and neurodegenerative diseases.[49]

C-phycocyanin possesses anti-inflammatory, radical scavenging, lipid-lowering, and antioxidation effects, thus suggesting its role in combating cardiovascular diseases and atherosclerosis.[52] *Spirulina fusiformis Voronikhin*, used as functional food and therapeutic, has bioactives that possess anti-hyperglycemic activity or anti-diabetes activity, and when fed with phycocyanin in mice was showed to decrease the blood glucose level.[50]

The unique characteristics of phycobiliproteins make them most promising candidates for variety of applications. The properties like hepatoprotective, antioxidant, and anti-inflammatory activity suggest them as the prospective molecules for therapeutic, pharmacological, and diagnostic purposes. Research has demonstrated that the antioxidant property of phycoerythrin from *Lyngbya* sp. A09DM could diminish the aging phenomenon in *Caenorhabditis elegans*. Extension of lifespan and enhancement in pharyngeal pumping of *C. elegans* was noticed upon pretreating them with phycoerythrin. This supports the "free-radical theory of aging."[54]

10.3 SUMMARY

The primary and secondary metabolism of diverse microalgal species produces several commercially important chemicals including pigments. The various pigments so obtained from microalgae find application in diverse fields like, food, feed, cosmetics, pharmaceuticals, nutraceuticals, diagnostics, nanotechnology products, and so on. This review focuses on recent advances in the application of microalgal-derived pigments like carotenoids, astaxanthin, fucoxanthin, zeaxanthin, and phycobilins (phycocyanin and phycoerythrin). With proper research and facilitation, these pigments may turn into potential candidates for many health benefits. If wisely handled, these microalgae can be the future biorefineries for the production of useful pigments.

KEYWORDS

- antioxidant
- astaxanthin
- aanthaxanthin
- carotenoid
- chlorophylls
- fucoxanthin
- lutein

REFERENCES

1. Aburai, N.; Ohkubo, S.; Miyashita, H.; Abe, K.. Composition of Carotenoids and Identification of Aerial Microalgae Isolated from the Surface of Rocks in Mountainous Districts of Japan. *Algal Res.* **2013,** *2* (3), 237–243.
2. Ahmed, F.; Fanning, K.; Netzel, M.; Turner, W.; Li, Y.; Schenk, P. M. Profiling of Carotenoids and Antioxidant Capacity of Microalgae From Subtropical Coastal and Brackish Waters. *Food Chem.* **2014,** *165,* 300–306.
3. Arathi, P.; Sowmya, P.; Lakshminarayana, R.. Metabolomics of Carotenoids: The Challenges and Prospects—A Review. *Trends Food Sci. Technol.* **2015,** *45* (1), 105–117.
4. Augustine, A.; Kumaran, J.; Puthumana, J. Multifactorial Interactions and Optimization in Biomass Harvesting of Marine Picoalga Picochlorum Maculatum MACC3 with Different Flocculants. *Aquaculture* **2017,** *474,* 18–25.
5. Barrios, V.; Escobar, C. A Nutraceutical Approach (Armolipid Plus) to Reduce Total and LDL Cholesterol in Individuals with Mild to Moderate Dyslipidemia: Review of the Clinical Evidence. *Atherosclerosis Suppl.* **2016,** *24,* 1–15.
6. Barros, M. P.; Marin, D. P. Combined Astaxanthin and Fish Oil Supplementation Improves Glutathione-based Redox Balance in Rat Plasma and Neutrophils. *Chem. Biol. Interact.* **2012,** *197* (1), 58–67.
7. Capelli, B. Jenkins, U.; Cysewski, G. R. Role of Astaxanthin in Sports Nutrition. *Nutr. Enhanced Sports Perform.* **2014,** *2,* 465–471.
8. Castro-Puyana M.; Pérez-Sánchez A.; Valdés, A. Pressurized Liquid Extraction of Neochloris Oleoabundans for the Recovery of Bioactive Carotenoids with anti-proliferative Activity Against Human Colon Cancer Cells. *Food Res. Int.* **2017,** *99,* 1048–1055.
9. Chen, H.-We.; Tsung-Shi, Y. Purification and Immunomodulating Activity of C-Phycocyanin from Spirulina Platensis Cultured Using Power Plant Flue Gas. *Proc. Biochem.* **2014,** *49* (8), 1337–1344.
10. Chen, L.; Zhang, L. Concurrent Production of Carotenoids and Lipid by a Filamentous Microalga Trentepohlia Arborum. *Biores. Technol.* **2016,** *214,* 567–573.
11. da Silva, F. O.; Tramonte, V. L.C.G. Litopenaeus Vannamei Muscle Carotenoids Versus Astaxanthin: A Comparison of Antioxidant Activity and In Vitro Protective Effects Against Lipid Peroxidation. *Food Biosci.* **2015,** *9,* 12–19.
12. Delgado, A. M. A.; Khandual, S. Chemical Stability of Astaxanthin Integrated Into a Food Matrix: Effects of Food Processing and Methods for Preservation. *Food Chem.* **2017,** *225,* 23–30.
13. Deniz, I.; Ozen, M. O.; Yesil-Celiktas, O. Supercritical Fluid Extraction of Phycocyanin and Investigation of Cytotoxicity on Human Lung Cancer Cells. *J. Supercrit. Fluids* **2016,** *108,* 13–18.
14. Eaton, L.; Clezy, K. The Behavioural Effects of Supplementing Diets with Synthetic and Naturally Sourced Astaxanthin in an Ornamental Fish (Puntius titteya). *Appl. Animal Behav. Sci.* **2016,** *182,* 94–100.
15. Foo, S. C.; Yusoff, F. Md. Production of Fucoxanthin-rich Fraction (FxRF) from a Diatom, Chaetoceros Calcitrans (Paulsen) Takano 1968. *Algal Res.* **2015,** *12,* 26–32.

16. Goiris, K.; Muylaert, K.; De Cooman, L. Microalgae as a Novel Source of Antioxidants for Nutritional Applications. In: *Handbook of Marine Microalgae*; 1st ed; Academic Press: New York, 2015, pp. 269–280.

17. Gong, M.; Bassi, A. Carotenoids from Microalgae: A Review of Recent Developments. *Biotechnol. Adv.* **2016,** *34* (8), 1396–1412.

18. Guerra, B. A.; Bolin, A. P.; Otton, R. Carbonyl Stress and a Combination of Astaxanthin/Vitamin C Induce Biochemical Changes in Human Neutrophils. *Toxicology* In Vitro **2012,** *26* (7), 1181–1190.

19. Hashtroudi, M. S.; Shariatmadari, Z. Analysis of Anabaena Vaginicola and Nostoc Calcicola from Northern Iran, as rich Sources of Major Carotenoids. *Food Chem.* **2013,** *136* (3), 1148–1153.

20. Hernandez, F. Y. F.; Khandual, S. Cytotoxic Effect of Spirulina Platensis Extracts on Human Acute Leukemia Kasumi-1 and Chronic Myelogenous Leukemia K-562 Cell Lines. *Asian Pacific J. Trop. Biomed.* **2017,** *7* (1), 14–19.

21. Hu, X.; Tao, N.; Wang, X. Marine-derived Bioactive Compounds with Anti-obesity Effect: A Review. *J. Funct. Foods* **2016,** *21*, 372–387.

22. Jena, J.; Pradhan, N. Pigment Mediated Biogenic Synthesis of Silver Nanoparticles Using Diatom Amphora sp. and its Antimicrobial Activity. *J. Saudi Chem. Soc.* **2015,** *19* (6), 661–666.

23. Kaur, A.; Gupta, V. Nutraceuticals in Prevention of Cataract–An Evidence Based Approach. *Saudi J. Ophthalmol.* **2016,** *31* (1), 30–37.

24. Kim, D.; Vijayan, D. Cell-wall Disruption and Lipid/Astaxanthin Extraction from Microalgae: Chlorella and Haematococcus. *Biores. Technol.* **2016,** *199*, 300–310.

25. Kiperstok, A.; Sebestyén, P. Biofilm Cultivation of Haematococcus Pluvialis Enables a Highly Productive One-Phase Process for Astaxanthin Production Using High Light Intensities. *Algal Res.* **2017,** *21*, 213–222.

26. Ko, J.-C.; Chen, J. Astaxanthin Down-regulates Rad51 Expression via Inactivation of AKT Kinase to Enhance Mitomycin C-induced Cytotoxicity In Human Non-Small Cell Lung Cancer Cells. *Biochemical Pharmacology,* **2016,** *105*, 91-100.

27. Koller, M.; Muhr, A.; Braunegg, G. Microalgae as Versatile Cellular Factories for Valued Products. *Algal Res.* **2014,** *6*, 52–63.

28. Kuppusamy, P.; Yusoff, M. Nutraceuticals as Potential Therapeutic Agents for Colon Cancer: A Review. *Acta Pharm. Sinica B* **2014,** *4* (3), 173–181.

29. Kyriakopoulou, K.; Papadaki, S. Life Cycle Analysis of β-Carotene Extraction Techniques. *J. Food Eng.* **2015,** *167*, 51–58.

30. Li, J.; Liu, Y.; Cheng, J. Biological Potential of Microalgae in China for Biorefinery-based Production of Biofuels and High Value Compounds. *New Biotechnol.* **2015,** *32* (6), 588–596.

31. Machado, R. S.; Trevisol, Thalles C. Technological Process for Cell Disruption, Extraction and Encapsulation of Astaxanthin from *Haematococcus pluvialis*. *J. Biotechnol.* **2016,** *218*, 108–114.

32. Manirafasha, E.; Ndikubwimana, T. Phycobiliprotein: Potential Microalgae Derived Pharmaceutical and Biological Reagent. *Biochem. Eng. J.* **2016,** *109*, 282–296.

33. Markou, G.; Nerantzis, E. Microalgae for High-value Compounds and Biofuels Production: A Review with Focus on Cultivation Under Stress Conditions. *Biotechnol. Adv.* **2013,** *31* (8), 1532–1542.

34. Millao, S.; Uquiche, E. Antioxidant Activity of Supercritical Extracts From Nanno-chloropsis Gaditana: Correlation with iIts Content of Carotenoids and Tocopherols. *J. Supercrit. Fluids* **2016**, *111*, 143–150.

35. Minhas, A.; Hodgson, P. The Isolation and Identification of New Microalgal Strains Producing Oil and Carotenoid Simultaneously with Biofuel Potential. *Biores. Technol* **2016**, *211*, 556–565.

36. Mohammadpour, R.; Janfaza, S. Light Harvesting and Photocurrent Generation by Nanostructured Photoelectrodes Sensitized with a Photosynthetic Pigment: A New Application for Microalgae. *Bioresource Technol.* **2014**, *163*, 1–5.

37. Munier, M.; Jubeau, S.; Wijaya, A. Physicochemical Factors Affecting the Stability of Two Pigments: R-phycoerythrin of Grateloupia turuturu and B-phycoerythrin of Porphyridium cruentum. *Food Chem.* **2014**, *150*, 400–407.

38. Muradian, K.; Vaiserman, A.; Min, K. J; Fraifeld, V. E. Fucoxanthin and Lipid Metabolism: A Minireview. *Nutr. Metab. Cardiovasc. Dis.* **2015**, *25* (10), 891–897.

39. Nwachukwu, I. D.; Udenigwe, C. C.; Aluko, Rotimi E. Lutein and Zeaxanthin: Production Technology, Bioavailability, Mechanisms of Action, Visual Function, and Health Claim Status. *Trends Food Sci. Technol.* **2016**, *49*, 74–84.

40. Pádua, D.; Rocha, E.; Gargiulo, D.; Ramos, A. A. Bioactive Compounds from Brown Seaweeds: Phloroglucinol, Fucoxanthin and Fucoidan as Promising Therapeutic Agents Against Breast Cancer. *Phytochem. Lett.* **2015**, *14*, 91–98.

41. Paliwal, C.; Ghosh, T.; George, B. Microalgal Carotenoids: Potential Nutraceutical Compounds with Chemotaxonomic Importance. *Algal Res.* **2016**, *15*, 24–31.

42. Panis, G.; Carreon, J. Commercial Astaxanthin Production Derived by Green Alga Haematococcus Pluvialis: A Microalgae Process Model and a Techno-economic Assessment All Through Production Line. *Algal Res.* **2016**, *18*, 175–190.

43. Pham, M.; Byun, H. Effects of Dietary Carotenoid Source and Level on Growth, Skin Pigmentation, Antioxidant Activity and Chemical Composition of Juvenile Olive Flounder Paralichthys Olivaceus. *Aquaculture* **2014**, *431*, 65–72.

44. Rao, A.; Baskaran, V.; Sarada, R.; Ravishankar, G. A. In Vivo Bioavailability And Antioxidant Activity of Carotenoids from Microalgal Biomass—A Repeated Dose Study. *Food Res. Int.* **2013**, *54*(1), 711-717.

45. Rodrigues, D. B.; Menezes, C. R. Bioactive Pigments from Microalgae Phormidium Autumnale. *Food Res. Int.* **2015**, *77*, 273–279.

46. Rodríguez-Sánchez, R.; Ortiz-Butrón, R.; Blas-Valdivia, V. Phycobiliproteins or C-phycocyanin of Arthrospira (Spirulina) Maxima Protect Against $HgCl_2$-caused Oxidative Stress and Renal Damage. *Food Chem.* **2012**, *135* (4), 2359–2365.

47. Ruiz-Ruiz, F.; Benavides, J.; Rito-Palomares, M. Scaling-up of a B-phycoerythrin Production and Purification Bioprocess Involving Aqueous Two-phase Systems: Practical Experiences. *Proc. Biochem.* **2013**, *48* (4), 738–745.

48. Samarakoon, K.; Jeon, Y. Bio-functionalities of Proteins Derived from Marine Algae–A Review. *Food Res. Int.* **2012**, *48* (2), 948–960.

49. Scoglio, S.; Benedetti, Y.; Benvenuti, F. Selective Monoamine Oxidase B Inhibition by an Aphanizomenon Flos-qquae Extract and by its Constitutive Active Principles Phycocyanin and Mycosporine-like Amino Acids. *Phytomedicine* **2014**, *21* (7), 992–997.

50. Setyaningsih, I.; Bintang, M.; Madina, N. Potentially Antihyperglycemic from Biomass and Phycocyanin of Spirulina Fusiformis Voronikhin by In Vivo Test. *Procedia Chem.* **2015**, *14*, 211–215.

51. Shanab, M. M.; Mostafa, Soha S. M. Aqueous Extracts of Microalgae Exhibit Antioxidant and Anticancer Activities. *Asian Pacific J. Trop. Biomed.* **2012**, *2* (8), 608–615.

52. Sheu, M.; Hsieh, Y.; Lai, C. Antihyperlipidemic and Antioxidant Effects of C-Phycocyanin in Golden Syrian Hamsters Fed with a Hypercholesterolemic Diet. *J. Tradit. Complement. Med.* **2013**, *3* (1), 41.

53. Singh, D.; Puri, M. Characterization of a New Zeaxanthin Producing Strain of Chlorella Saccharophila Isolated from New Zealand Marine Waters. *Biores. Technol.* **2013**, *143*, 308–314.

54. Sonani, R.; Singh, N. Concurrent Purification and Antioxidant Activity of Phycobiliproteins from Lyngbya sp. A09DM: An Antioxidant and Anti-aging Potential of Phycoerythrin in Caenorhabditis Elegans. *Proc. Biochem.* **2014**, *49* (10), 1757–1766.

55. Turemis, M.; Rodio, G.; Pezzotti, G. A Novel Optical/Electrochemical Biosensor for Real Time Measurement of Physiological Effect of Astaxanthin on Algal Photoprotection. *Sensors Actuators B: Chem.* **2017**, *241*, 993–1001.

56. Urreta, I.; Ikaran, Z. Revalorization of Neochloris Oleoabundans Biomass As Source of Biodiesel By Concurrent Production of Lipids and Carotenoids. *Algal Res.* **2014**, *5*, 16–22.

57. Vanthoor-Koopmans, M.; Wijffels, R. Biorefinery of Microalgae For Food And Fuel. *Biores. Technol.* **2013**, *135*, 142–149.

58. Vaquero, I.; Mogedas, B. Light-mediated Lutein Enrichment of an Acid Environment Microalga. *Algal Res.* **2014**, *6*, 70–77.

59. Vaquero, I.; Ruiz-Domínguez, M. Cu-mediated Biomass Productivity Enhancement And Lutein Enrichment of the Novel Microalga Coccomyxa Onubensis. *Proc. Biochem.* **2012**, *47* (5), 694–700.

60. Vigani, M.; Parisi, C.; Rodríguez-Cerezo, E. Food and Feed Products from Microalgae: Market Opportunities and Challenges for the EU. *Trends Food Sci. Technol.* **2015**, *42* (1), 81–92.

61. Xie, Y.; Ho, S.; Chen, C. Phototrophic Cultivation of a Thermo-tolerant Desmodesmus Sp. for Lutein Production: Effects of Nitrate Concentration, Light Intensity and Fed-batch Operation. *Biores. Technol.* **2013**, *144*, 435-444.

62. Yen, H.; Hu, I. Microalgae-based Biorefinery-from Biofuels to Natural Products. *Biores. Technol.* **2013**, *135*, 166–174.

63. Zhou, X.; Zhang, F.; Hu, X.; Chen, J.. Inhibition of Inflammation by Astaxanthin Alleviates Cognition Deficits in Diabetic Mice. *Physiol. Behav.* **2015**, *151*, 412–420.

INDEX